NATO ASI Series

Advanced Science Institutes Series

A series presenting the results of activities sponsored by the NATO Science Committee, which aims at the dissemination of advanced scientific and technological knowledge, with a view to strengthening links between scientific communities.

The Series is published by an international board of publishers in conjunction with the NATO Scientific Affairs Division.

A Life Sciences B Physics	Plenum Publishing Corporation London and New York
C Mathematical and Physical Sciences D Behavioural and Social Sciences E Applied Sciences	Kluwer Academic Publishers Dordrecht, Boston and London
F Computer and Systems Sciences G Ecological Sciences H Cell Biology I Global Environmental Change	Springer-Verlag Berlin Heidelberg New York Barcelona Budapest Hong Kong London Milan Paris Santa Clara Singapore Tokyo

Partnership Sub-Series

1. Disarmament Technologies	Kluwer Academic Publishers
2. Environment	Springer-Verlag / Kluwer Academic Publishers
3. High Technology	Kluwer Academic Publishers
4. Science and Technology Policy	Kluwer Academic Publishers
5. Computer Networking	Kluwer Academic Publishers

The Partnership Sub-Series incorporates activities undertaken in collaboration with NATO's Cooperation Partners, the countries of the CIS and Central and Eastern Europe, in Priority Areas of concern to those countries.

NATO-PCO Database

The electronic index to the NATO ASI Series provides full bibliographical references (with keywords and/or abstracts) to about 50 000 contributions from international scientists published in all sections of the NATO ASI Series. Access to the NATO-PCO Database is possible via the CD-ROM "NATO Science & Technology Disk" with user-friendly retrieval software in English, French and German (© WTV GmbH and DATAWARE Technologies Inc. 1992).

The CD-ROM can be ordered through any member of the Board of Publishers or through NATO-PCO, B-3090 Overijse, Belgium.

Series F: Computer and Systems Sciences, Vol. 159

Springer

Berlin
Heidelberg
New York
Barcelona
Budapest
Hong Kong
London
Milan
Paris
Singapore
Tokyo

Fractal Image Encoding and Analysis

Edited by

Yuval Fisher

Institute for Nonlinear Science
University of California, San Diego
La Jolla, CA 92093-0402, USA

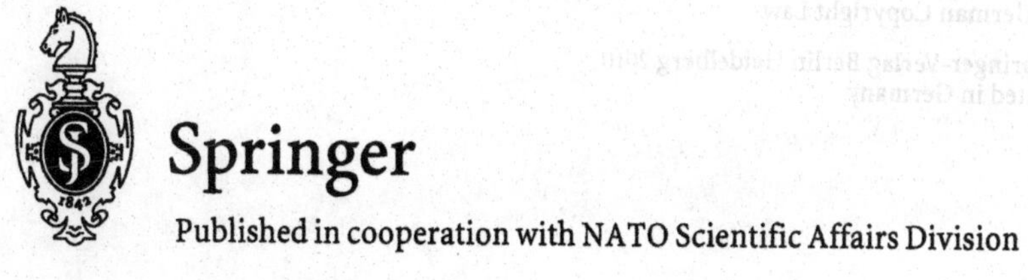

Springer

Published in cooperation with NATO Scientific Affairs Division

Proceedings of the NATO Advanced Study Institute on Fractal Image Encoding and Analysis, held in Trondheim, Norway, July 8–17, 1995

Library of Congress Cataloging-in-Publication Data

NATO Advanced Study Institute on Fractal Image Encoding and Analysis (1995: Trondheim, Norway)
Fractal image encoding and analysis/edited by Yaval Fisher.
p. cm. -- (NATO ASI series. Series F, Computer and systems sciences; no. 159)
Includes bibliographical references and index.

1. Image processing--Digital techniques--Mathematics. 2. Fractals. 3. Image compression. 4. Data encryption (Computer science) I. Fisher, Yuval. II. Title. III. Series.
TA1637.N38 1998 621.36´7´02855746--DC21 98-22446 CIP

ACM Subject Classification (1998): I.3.7, I.4, I.5.4, E.4, I.2.10

ISBN 978-3-642-08324-2

Printed in Germany

Preface

From a handful of papers in the late 1980s to more papers than normal people ought to read, the field of fractal encoding and analysis has blossomed. This book contains the fruit of that blossom. It contains, roughly, a portion of the proceedings of the NATO Advanced Study Institute on Fractal Image Encoding and Analysis, held in Trondheim, Norway on July 8–17, 1995. Many excellent talks were given at the conference, and this book contains a selection of the invited speakers' presentations. A large group of participant speakers could not be accommodated in this book, and their papers are published separately in a special April 1997 issue of *Fractals*, published by World Scientific Publishers.

This book is also a sort of "enhanced" sequel to *Fractal Image Compression: Theory and Application,* also published by Springer-Verlag. It's enhanced because it contains a good amount of material on fractal image analysis, a topic that the previous book didn't cover at all. It's also enhanced because it contains a wealth of new and interesting results in fractal image encoding.

I would like to express my thanks to NATO, Iterated Systems, Inc., and the institute for Nonlinear Science at the University of California, San Diego, who sponsored the conference. My thanks to the conference co-directors, Jacques Lévy-Véhel, Claude Tricot, and especially Geir Øien, who helped with the details of arranging a conference. I would also like to express my gratitude to the conference participants, and to the book's authors, whose contributions made this book possible. I'm indebted also to Beryl Nasworthy for her invaluable editorial and administrative help, and to Doug Ridgway and Frank Dudbridge, for helping me with the manuscript.

La Jolla, California, January 1998 Yuval Fisher

Contents

I Fractal Image Encoding

Part I

Fractal Image Encoding

Chapter 1

Why Fractal Block Coders Work

Geoffrey Davis

1.1 Introduction

Most commonly-used lossy image compression schemes such as JPEG fit into a standard, well-established paradigm. First an invertible linear transform is performed on an image; then the transform coefficients are scalar quantized, entropy-coded, and stored. The image is decoded by reversing each of the three steps. The principles underlying the design and implementation of these transform coders are well-understood and have been explored thoroughly since the introduction of such coders in 1956 [144]. We discuss the implicit image models underlying such coders in the next section.

Fractal image compression, introduced much more recently [18], operates on a very different set of principles which are motivated by the theory of iterated function systems (IFS). Fractal coders store images as fixed points of maps on the plane rather than as a set of quantized transform coefficients. Fractal compression is related to vector quantization, but fractal coders use a self-referential vector codebook, drawn from the image itself, rather than a fixed codebook.

IFS theory motivates a broad class fractal compression schemes but does not give much insight as to *why* particular fractal schemes work well. IFS theory also gives little guidance as to how efficient schemes should be implemented. Basic components of fractal schemes, such as decoder convergence properties, methods of estimating quantization error, and bit allocation methods are poorly understood.

In this chapter we introduce a wavelet-based framework for analyzing fractal block coders. The wavelet framework simplifies the analysis of these coders considerably and, more importantly, gives a clear picture of why they have proven effective. We show that fractal block coding schemes function essentially by extrapolating Haar wavelet coefficients across scales.

Our analysis gives insight into a number of important implementation issues which are not well-understood, including bit allocation methods, error estimation, quantization vector search strategies, and super-resolution of images. Using the insights from our analysis we derive a wavelet-based analog of fractal compression which yields a large reduction in visual artifacts and peak signal-to-noise ratios of 0.75 dB to 1 dB higher than the best fractal compressors currently in the literature at the same bit rate.

The balance of the chapter is organized as follows. In the next section we give an overview of linear transform coders and the image models which motivate their use. In section 1.3 we give a brief description of fractal block coding schemes and the corresponding assumptions made about images. In section 1.4 we introduce a wavelet framework for analyzing fractal block coders and we discuss the implications of our analysis. Finally, in section 1.5 we introduce a generalization of fractal block coders which yields compressed images of substantially higher quality than block coders.

1.2 Linear Transform Coders

Image compression schemes operate on the assumption that there is underlying structure to images that can be exploited to obtain a more efficient image representation. For the discussion below, we assume that we are attempting to encode an N pixel by N pixel greyscale image for which each pixel brightness is specified by a b bit value. Linear transform coders treat these digital images as realizations of an N^2-dimensional vector-valued random variable $\mathbf{X}$ which has a density function with a particular structure.

Transform coders proceed in three stages. First a linear transform is performed on the image vector to decorrelate the pixels and to pack the image energy into a small number of coefficients. Next the decorrelated transform coefficients are quantized to a finite set of symbols. In the final step the symbols are entropy coded and stored. We give a brief overview of these steps to illustrate the underlying image model used by transform coders. We will contrast this with model used for fractal coders in section 1.3. Further details on transform coders can be found in [97] and [248].

1.2.1 Entropy Coding

A very simple image model for coding is that image pixels are realizations of independent, identically distributed (i.i.d.) random variables. The probability density function for these random variables is determined empirically by counting frequencies of occurrence of various pixel values. Entropy coders provide efficient bit representations of pixel values by assigning short bit strings to frequently occurring values and longer strings to less frequently occurring values. The Shannon source coding theorem gives a lower bound on the average number of bits required to encode realizations of i.i.d. random variables, and this lower bound can be achieved asymptotically via arithmetic coding [29].

Treating pixel values as i.i.d. random variables and entropy coding them in practice yields a modest decrease in storage requirements, and it does so losslessly. However, the assumption that pixel values are independent precludes making use of the considerable spatial redundancy present within images. To obtain higher compression rates we need to use a better model.

1.2.2 The Karhunen-Loève Transform

An observation that motivates transform coders is that our visual surroundings can be closely approximated by smooth surfaces separated by discontinuities that lie along continuous curves. Because of the physical structure underlying real-world images, pixel values are far from independent, especially pixels which are close together. Pixels in smooth regions can be predicted with considerable accuracy given the values of their neighbors, so it is unnecessary to store all pixels independently. By exploiting this spatial redundancy we can obtain a considerable improvement in performance over entropy coding alone. Evidence that the human visual system makes use of these types of local smoothness assumptions in low-level visual processing [180] further motivates this approach.

The reason for transforming the image data is to remove or at least to reduce the spatial redundancy within images. Ideally we would like to transform the image vector $\mathbf{x}$ to a new vector $\mathbf{y} = T\mathbf{x}$ for which the components are independent and identically distributed. Except in special cases, however, the best we can do with a linear transform is to produce a transformed image for which pixels are uncorrelated. We can estimate pixel correlations by assuming that the image is a realization of a wide-sense stationary process. This essentially amounts to assuming that the marginal densities for pixel values are identical and that the joint density for any pair of pixels is a function only of their relative positions. In practice we find nearby pixels are strongly correlated and that this correlation decreases with the distance between pixels.

By applying the Karhunen-Loève (K-L) transform to the image, we obtain a set of transform coefficients which are decorrelated, i.e. we remove the first-order linear redundancy in the pixels [97]. When the image is the realization of a mean-zero Gaussian process and high resolution quantization is used, the quantized linear transform coefficients have the smallest expected mean squared distortion when the transform is the K-L transform. Good results are obtained in practice under less restrictive conditions, i.e. for real-world images with coarser quantization. The geometrical intuition behind this process is illustrated in Figure 1.1. Suppose that adjacent pairs of pixels in our image $\mathbf{x}$ are strongly correlated . If we plot the pairs (x_{2k}, x_{2k+1}) we see that these points are clustered in an elliptical cloud having a major axis with positive slope. The number of bits required to store each pixel value separately is given by the entropy of the marginal densities of x_{2k} and x_{2k+1} (which are assumed to be identical). The K-L transform rotates this ellipse so that its major axis is horizontal, transforming the pairs (x_{2k}, x_{2k+1}) to the pairs (y_{2k}, y_{2k+1}). The energy of the transformed pairs is largely confined to the first coordinate. The entropy of the transformed coordinate y_{2k+1} is considerably less than that of x_{2k+1}, giving a substantial savings in coding cost. The entropy of y_{2k} increases slightly over x_{2k}, but not enough to offset the savings from the second coordinate.

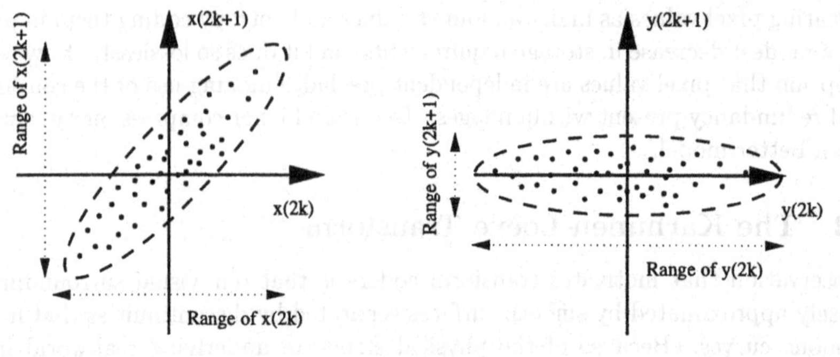

Figure 1.1: Correlated pixels plotted on the left are decorrelated by the K-L transform. After the transformation, the range of data to be quantized is reduced considerably for the second coordinate.

The K-L transform is not used in practice because the transform is image dependent and slow. Instead various approximations are used. The discrete cosine transform (DCT) is used in JPEG, and a variety of compactly supported wavelet bases are used in wavelet-based coders. Some implicit assumptions about the structure of the interpixel correlations underly the choice of these particular bases to approximate the K-L basis. The DCT, for example, approximates the K-L basis for a first-order Gauss-Markov process with positive correlation close to 1 [2].

1.2.3 Image Model for Transform Coders

In summary, transform coders exploit very simple image structures, namely first-order linear correlation between pixels. The use of the K-L transform is motivated by the assumption that the degree of interrelationship between pixels depends mainly on their relative positions. Approximate K-L transforms assume some particular form of interrelationship; in the case of the DCT it is assumed that pixel correlations fall off quickly as the distance between them increases. These assumptions are quite natural and readily verifiable. In general this simple model works very well as is evidenced by the success of JPEG. For images for which these assumptions do not work well, e.g. in the case of images with substantial nonlinear interrelationships between pixels, transform coders do not perform as well as otherwise, but neither do they fail completely. In practice the main shortcoming of the transform coders is that they assume too little about the structure of images and fail to exploit higher order redundancies.

1.3 Fractal Block Coders

Fractal image compression techniques, introduced by Barnsley and Jacquin [18], operate on very different principles than transform coders. In this section we describe a basic fractal block coder and discuss the image model which motivates it. A more thorough discussion of fractal coding techniques can be found in [87][126].

1.3.1 Image Model for Fractal Coders

The motivation for fractal image compression is that images contain not only the spatial redundancy incorporated into the transform coder image model, but also redundancy in *scale*. For example, we observe that straight edges in images are invariant under rescaling, as are constant regions. An analysis of natural scenes (i.e. images containing no man-made objects) in [79] suggests that such images have approximately scale-invariant power spectra and phase spectra, providing further evidence of redundancy in scale.

We distinguish this redundancy in scale from the spatial correlation discussed with transform coders because these features which are redundant across scales can involve nonlinear relationships between the coefficients of our approximate K-L transform. For example, although sharp discontinuities in an image have energy which is spread across many coefficients in both the wavelet and discrete cosine bases, these coefficients in general have very low correlation. Because the pixel interrelationships induced by edges are poorly captured by first-order linear correlations, linear transform coders are unable to make use of the redundancy from edges.

Fractal compression takes advantage of this redundancy in scale by using coarse-scale image features to quantize fine-scale features. Fractal block coders perform a vector quantization of image blocks. The codebook consists of larger blocks from the image which are locally averaged and subsampled. This codebook is effective for coding constant regions and straight edges due to the scale invariance of these features. An important advantage over standard vector quantization coders is that fractal coders do not require separate storage of a fixed vector codebook. Fractal encoding algorithms entail the construction of a map from the plane to itself of which the unique fixed point is an approximation to the image to be coded. Compressed images are stored by saving the parameters of this map and recovered by iteratively applying the map to find its fixed point.

The image model underlying fractal block compression routines is that image block features at fine scales are present at coarser scales up to a rigid motion and an affine transform of intensities. This is trivially true for the finest scale, since all single pixels (the finest-scale features) are given by affine transforms of averaged and subsampled 2×2 blocks. For coarser scales, however, this assumption of local block self-similarity is a stronger assumption than is made for transform coders. Because considerable structure is assumed present in images, when it is actually there (as in the case of the Serpinski gasket[87]), fractal schemes yield extremely high compression ratios with little or no distortion. On the other hand, when this structure is absent, such as in the case of an image consisting only of fine-scale Haar wavelets, fractal coders break down. In practice, neither of these extremes is encountered.

The heuristic explanation that fractal compression takes advantage of local self-similarity does a good job of explaining the performance of these coders for edges and constant regions. Explaining why fractal compression continues to function for complex, textured regions is more difficult. Such regions often possess no obvious self-similarities. Our analysis using the wavelet framework developed below sheds new light on this question. We suggest that fractal compressors' performance on such regions results from energy packing properties of the Haar transform.

1.3.2 Mechanics of Fractal Block Coders

We now describe a simple fractal block coding scheme based on those in [125] and [84]. Let $\mathcal{I}$ be a $2^N \times 2^N$ pixel image, and let $\mathbf{B}^J_{K,L}\mathcal{I}$ be the $2^J \times 2^J$ subblock of $\mathcal{I}$ with lower left hand corner at $(2^J K, 2^J L)$. $\mathbf{B}^J_{K,L}\mathcal{I}$ is the result of applying the linear "get-block" operator $\mathbf{B}^J_{K,L} : \mathbb{R}^{2^{2N}} \to \mathbb{R}^{2^{2J}}$ to the image $\mathcal{I}$. The adjoint of the get-block operator, $(\mathbf{B}^J_{K,L})^*$, is a "put-block" operator which maps a $2^J \times 2^J$ subblock to a $2^N \times 2^N$ image containing the block with its lower left hand corner at the point (K, L). We will use capital letters to denote block coordinates and lower case to denote individual pixel coordinates. To simplify our notation we will use a capital Greek multi-index, usually Γ, to abbreviate the block coordinates K, L and a lower-case Greek multi-index to abbreviate pixel coordinates.

We partition $\mathcal{I}$ into a set of non-overlapping $2^R \times 2^R$ *range blocks*. The goal of the compression scheme is to approximate each range block with a block from a codebook constructed from a set of $2^D \times 2^D$ *domain blocks*, where $D > R$ and $D, R \in \mathbf{Z}$. This approximation induces a map from the image to itself (i.e. from the domain blocks to the range blocks) of which the image is an approximate fixed point. Under certain conditions we can show this map to be a contraction map. We store the image by storing the parameters of this map, and we recover the image by iterating the map to its fixed point.

Iterated function system theory motivates this general method of storing images, but gives little specific guidance as to what types of map should be used. The basic form of the block coder described below is the result of considerable empirical work. We show below that this particular coder in fact arises quite naturally in a wavelet framework, and we use our framework to derive some simple extensions which give greatly improved performance.

The range block partition consists of the set $\mathbf{B}^R_{2^R K, 2^R L}\mathcal{I}$, with $(K, L) \in \mathcal{R}$ where $\mathcal{R} = \{(K, L) : K, L \in \mathbf{Z} \text{ and } 0 \le K, L < 2^{N-R}\}$. By construction, the set of integer-valued pairs (K, L) corresponds to a disjoint partition of the image. The domain blocks used for approximating range blocks are drawn from the *domain pool*, the set $\mathbf{B}^D_{2^D K, 2^D L}\mathcal{I}$, with $(K, L) \in \mathcal{D}$ and $D > R$. We discuss several different domain pools below. One commonly used domain pool [125] is the set of all unit translates of $2^D \times 2^D$ blocks $\mathcal{D} = \{(K, L) : 2^D K, 2^D L \in \mathbf{Z} \text{ and } 0 \le K, L < 2^{N-D}\}$.

We now define several operators which are used to construct the codebook from the domain blocks. The "average-and-subsample" operator $\mathbf{A}$ maps $2^J \times 2^J$ image blocks to $2^{J-1} \times 2^{J-1}$ blocks by averaging each pixel in $\mathbf{B}^J_\Gamma$ with its neighbors and then subsampling. We define $(\mathbf{A}\mathbf{B}^J_\Gamma\mathcal{I})(k, l) = \frac{1}{4}[(\mathbf{B}^J_\Gamma\mathcal{I})(2k, 2l) + (\mathbf{B}^J_\Gamma\mathcal{I})(2k+1, 2l) + (\mathbf{B}^J_\Gamma\mathcal{I})(2k, 2l+1) + (\mathbf{B}^J_\Gamma\mathcal{I})(2k+1, 2l+1)]$ where $\mathbf{B}^J_\Gamma\mathcal{I}(k, l)$ is the pixel at coordinates (k, l) within the subblock $\mathbf{B}^J_\Gamma\mathcal{I}$. The operator $\mathbf{L}_i$, $1 \le i \le 8$ maps a square block to one of the 8 isometries obtained from compositions of reflections and 90 degree rotations.

The elements of the codebook used for quantizing range blocks are constructed from linear combinations of a codebook element and a subblock of the matrix $\mathbf{1}$, the $2^N \times 2^N$ matrix of ones. This subblock of the matrix of 1's permits control of the DC component of the codebook elements. The ultimate goal is to find an encoding of range blocks so that the l^2 distortion of the reconstructed image is minimized. Due to the iterative nature of the reconstruction process, minimizing the l^2 block approximation error does not necessarily lead to minimal reconstruction errors. In practice, however, the l^2 block approximation error is minimized because it has been found empirically to give good results, and because determining the error in the reconstructed image is quite difficult. Our analysis using the wavelet framework described below gives rise to an improved codebook element selection criterion [54] as well as more perceptually appropriate norms.

Each range block is approximated by

$$\mathbf{B}_\Gamma^R \mathcal{I} \approx g_\Gamma \mathbf{L}_{P(\Gamma)} \mathbf{A}^{D-R} \mathbf{B}_{\Pi(\Gamma)}^D \mathcal{I} + h_\Gamma \mathbf{B}_\Gamma^R \mathbf{1}, \tag{1.1}$$

where $\Pi(\Gamma)$ assigns an element from the domain pool to each range element, $P(\Gamma)$ assigns each range element a symmetry operator index, and $\mathbf{A}^{D-R}$ denotes the operator $\mathbf{A}$ applied $D-R$ times. The scalars g_Γ and h_Γ, the domain block index $\Pi(\Gamma)$, and the symmetry operator index $P(\Gamma)$ are chosen to minimize the l^2 approximation error $\|\mathbf{B}_\Gamma^R \mathcal{I} - (g_\Gamma \mathbf{L}_{P(\Gamma)} \mathbf{A}^{D-R} \mathbf{B}_{\Pi(\Gamma)}^D \mathcal{I} + h_\Gamma \mathbf{B}_\Gamma^R \mathbf{1})\|$.

The image $\mathcal{I}$ can be written as a sum of its range blocks, $\mathcal{I} = \sum_{\Gamma \in \mathcal{R}} (\mathbf{B}_\Gamma^R)^* \mathbf{B}_\Gamma^R \mathcal{I}$, so we have

$$\begin{aligned} \mathcal{I} &\approx \sum_{\Gamma \in \mathcal{R}} g_\Gamma (\mathbf{B}_\Gamma^R)^* \mathbf{L}_{P(\Gamma)} \mathbf{A}^{D-R} \mathbf{B}_{\Pi(\Gamma)}^D \mathcal{I} + \sum_{\Gamma \in \mathcal{R}} h_\Gamma (\mathbf{B}_\Gamma^R)^* \mathbf{B}_\Gamma^R \mathbf{1} \\ &= \mathbf{G}\mathcal{I} + \mathcal{H}. \end{aligned}$$

To store the image we store for each $\Gamma \in \mathcal{R}$ the scalars g_Γ and h_Γ, the symmetry operator index $P(\Gamma)$, and the domain block $\Pi(\Gamma)$ used for quantization.

Images are recovered from this stored information by an iterative procedure. We start with an arbitrary image $\mathcal{I}_0$, and we compute $\mathcal{I}_n = \mathbf{G}\mathcal{I}_{n-1} + \mathcal{H}$. It can be shown that the images $\mathcal{I}_n$ converge pointwise when the gains $|g_\Gamma| < 1$ [87]. Numerical experiments show that this upper bound is a sufficient but not a necessary condition for convergence, and that the use of larger bounds on the scaling factors can yield improved compression results [122].

1.4 A Wavelet Framework

1.4.1 The Discrete Wavelet Transform

The wavelet transform is a natural tool for analyzing fractal block coders since wavelet bases possess the same type of dyadic self-similarity that fractal coders try to exploit. In particular, the Haar wavelet basis possesses a regular block structure which is equivalent to the structure imposed on the image by the partition into range blocks. For simplicity we will consider orthogonal wavelets only here; it is straightforward to generalize the results to the biorthogonal case.

Orthogonal wavelet bases of $L^2(\mathbb{R})$ have a simple regular structure which is important for our analysis below. In 1-D, the basis consists of translates of a function $\phi(x)$, called a *scaling function*, together with translations and dyadic scalings of a function $\psi(x)$, called a *wavelet*. We can construct a similar basis in 2-D using the scaling function $\phi(x,y) = \phi(x)\phi(y)$ together with dyadic scalings and integer translations of the three wavelets $\psi_H(x,y) = \phi(x)\psi(y)$, $\psi_V(x,y) = \psi(x)\phi(y)$, and $\psi_D(x,y) = \psi(x)\psi(y)$ (the subscripts come from the fact that these wavelets are best-suited to representing edges which are oriented horizontally, vertically, and diagonally, respectively). We will use the subscript ω to represent one of the three orientations in $\Omega = \{H, V, D\}$. See [136] and [165] for a detailed overview of wavelets.

The discrete wavelet transform of a $2^N \times 2^N$ image $\mathcal{I}$ expands the image into a linear combination of the basis functions in the set $\mathcal{W}_J$, the functions $\phi^J_{k,l} = 2^J \phi(2^J x - k, 2^J y - l)$ and $\psi^j_{\omega,k,l} = 2^j \psi_\omega(2^j x - k, 2^j y - l)$, for $J \le j < N$. We will use a single lower-case Greek multi-index, usually γ, to abbreviate the orientation and translation subscripts of ϕ and ψ.

An important observation about these wavelet bases, and in particular the Haar basis, is that they preserve the spatial localization of image features. For example, the coefficient of the Haar scaling function $\phi^J_{k,l}$ is proportional to the average value of an image in the $2^J \times 2^J$ block of pixels with lower left corner at $2^J k, 2^J l$. The wavelet coefficients associated with this region are organized into three quadtrees. We call this set of trees a *wavelet subtree*.

At the root of a wavelet subtree are the coefficients of the wavelets $\psi^J_{\omega,k,l}$, where $\omega \in \{H, V, D\}$. These coefficients correspond to the block's coarse-scale information. Each wavelet $\psi^j_{\omega,k,l}$ has four children which correspond to the same spatial location and the same orientation, the wavelets of the next finer scale, $\psi^{j+1}_{\omega,2k,2l}$, $\psi^{j+1}_{\omega,2k+1,2l}$, $\psi^{j+1}_{\omega,2k,2l+1}$, and $\psi^{j+1}_{\omega,2k+1,2l+1}$. A wavelet subtree consists of the coefficients of the roots, together with all of their descendents. Two wavelet subtrees are shown as shaded regions in Figure 1.2.

1.4.2 A Wavelet Analog of Fractal Block Coding

We now describe a wavelet-based analog of fractal block coding introduced in [56]. In fractal block coders we approximate a set of $2^R \times 2^R$ range blocks using a set of $2^D \times 2^D$ domain blocks. The wavelet analog of an image block, a set of pixels associated with a small region in space, is a wavelet subtree together with its associated scaling function coefficient.

We define a linear "get-subtree" operator $\mathbf{S}^J_{K,L} : \mathbb{R}^{2^{2N}} \to \mathbb{R}^{2^{2(N-J)}-1}$ which extracts a subtree with root wavelets $\psi^J_{\omega,K,L}$ from the wavelet transform of an image. The adjoint of $\mathbf{S}^J_{K,L}$ is a "put-subtree" operator which inserts a subtree into an all-zero wavelet transform at scale J, offset (K, L). For the Haar basis, subblocks and their corresponding subtrees and associated scaling function coefficients contain identical information, i.e. the transform of a range block $\mathbf{B}^R_\Gamma \mathcal{I}$ yields the coefficients of subtree $\mathbf{S}^{N-R}_\Gamma \mathcal{I}$ and the scaling function coefficient $\langle \mathcal{I}, \phi^{N-R}_\Gamma \rangle$. For the remainder of this section we will take our wavelet basis to be the Haar basis. The actions of the get-subtree and put-subtree operators are illustrated in Figure 1.2.

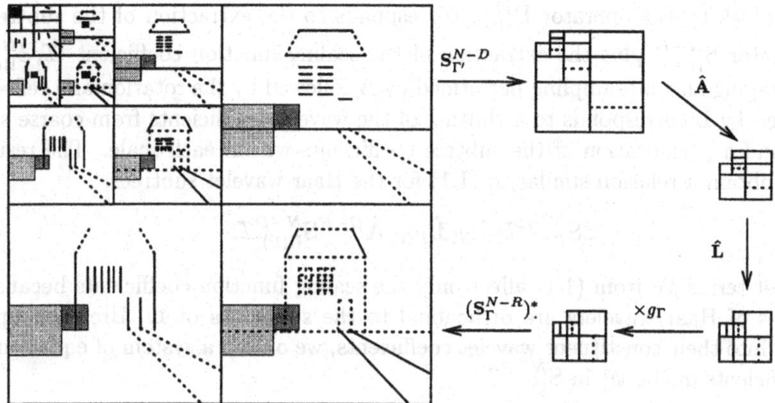

Figure 1.2: We approximate the darkly shaded range subtree $\mathbf{S}_{\Gamma}^{N-R}\mathcal{I}$ using the codebook element $g_\Gamma \hat{\mathbf{L}}\hat{\mathbf{A}}\mathbf{S}_{\Gamma'}^{N-D}\mathcal{I}$ which is derived from the lightly shaded domain subtree $\mathbf{S}_{\Gamma'}^{N-D}\mathcal{I}$. $\hat{\mathbf{A}}$ truncates the finest scale coefficients of the domain subtree and multiplies the coefficients by $\frac{1}{2}$, and $\hat{\mathbf{L}}$ rotates it. When storing this image we save the coarse-scale wavelet coefficients in bands 2 and below, and we save the encodings of all subtrees with roots in band 3.

We now examine the wavelet-domain behavior of the linear operators used in fractal compression. We first consider the wavelet analog $\hat{\mathbf{A}}$ of the average-and-subsample operator $\mathbf{A}$. Averaging and subsampling of the finest-scale Haar wavelets sets them to 0. For scales other than the finest, the local averaging has no effect, and subsampling ψ_γ^j yields the Haar wavelet at the next finer scale, ψ_γ^{j+1}, multiplied by $\frac{1}{2}$. Similarly, averaging and subsampling the scaling function ϕ_γ^j yields the scaling function at the next finer resolution, ϕ_γ^{j+1} except for the finest-scale scaling functions, which are set to 0. The action of the averaging and subsampling operator is thus seen to be a shift of coefficients from coarse scales to fine, a multiplication by $\frac{1}{2}$, and a truncation of the finest-scale coefficients. We see, then, that the effect of $\hat{\mathbf{A}}$ is to prune the leaves of a subtree and shift all coefficients to the next finer scale. The action of $\hat{\mathbf{A}}$ is illustrated in Figure 1.2.

For symmetrical and antisymmetrical wavelets, the only wavelets we will consider here, the horizontal reflection of a block corresponds to a horizontal reflection of wavelet coefficients within each scale of a subtree, and 90 degree block rotations correspond to 90 degree rotations of wavelet coefficients within each scale and a switching of the ψ_H coefficients with ψ_V coefficients. Hence the wavelet analogs $\hat{\mathbf{L}}_i$ of the block symmetry operators $\mathbf{L}_i$ permute wavelet coefficients within each scale. Figure 1.2 illustrates the action of a symmetry operator on a subtree. Note that the only symmetric, compactly supported, orthogonal wavelet basis is the Haar basis [248], which is why we the extension of our analysis to biorthogonal bases is important. See [57] for the details of the biorthogonal case.

The important thing to note is that the operators we use preserve subtrees, and the basic steps in fractal coding have simple wavelet analogs. The extraction of a domain block by the operator $\mathbf{B}^D_{\Pi(\Gamma)}$ corresponds to the extraction of the subtree by the operator $\mathbf{S}^{N-D}_{\Pi(\Gamma)}$ plus the extraction of the scaling function coefficient $\langle \mathcal{I}, \phi^{N-D}_{\Pi(\Gamma)} \rangle$. The averaging and subsampling performed by $\mathbf{A}$ followed by the rotation and reflection performed by $\mathbf{L}$ corresponds to a shifting of the wavelet coefficients from coarse scales to fine and a permutation of the subtree coefficients within each scale. The result is that we obtain a relation similar to (1.1) for the Haar wavelet subtrees,

$$\mathbf{S}^{N-R}_{\Gamma}\mathcal{I} \approx g_\Gamma \hat{\mathbf{L}}_{P(\Gamma)} \hat{\mathbf{A}}^{D-R} \mathbf{S}^{N-D}_{\Pi(\Gamma)} \mathcal{I}. \tag{1.2}$$

The offset terms h_Γ from (1.1) affect only the scaling function coefficients because all translates of Haar wavelets are orthogonal to the subblocks of $\mathbf{1}$. Breaking up the subtrees into their constituent wavelet coefficients, we obtain a system of equations for the coefficients of the ψ^j_γ in $\mathbf{S}^{N-R}_\Gamma \mathcal{I}$,

$$\begin{aligned} \langle \mathcal{I}, \psi^j_\gamma \rangle &\approx \frac{g_\Gamma}{2^{D-R}} \langle \mathcal{I}, \psi^{j-(D-R)}_{\gamma'} \rangle \\ &= \frac{g_\Gamma}{2^{D-R}} \langle \mathcal{I}, \mathbf{T}(\psi^j_\gamma) \rangle. \end{aligned} \tag{1.3}$$

Here $\mathbf{T}$ is the map induced by the domain block selection followed by averaging, subsampling, and rotating which, as we noted above, maps wavelet coefficients to wavelet coefficients. We obtain a similar relation for the ϕ's,

$$\begin{aligned} \langle \mathcal{I}, \phi^{N-R}_\Gamma \rangle &\approx \frac{g_\Gamma}{2^{D-R}} \langle \mathcal{I}, \phi^{N-D}_{\Pi(\Gamma)} \rangle + h_\Gamma \\ &= \frac{g_\Gamma}{2^{D-R}} \langle \mathcal{I}, \mathbf{T}(\phi^{N-R}_\Gamma) \rangle + h_\Gamma \end{aligned} \tag{1.4}$$

The system of equations (1.3) and (1.4) shows that fractal image compression is a map from coarse-scales to fine. The relationships are complicated by the fact that the wavelets and scaling functions on the right hand sides of (1.3) and (1.4) are *not* necessarily members of the basis $\mathcal{W}_{N-D}$, since general domain pools may contain domain blocks that require non-integer translates of wavelets in the subtrees. We discuss this idea of a mapping from coarse to fine scales in greater detail in the next section.

We obtain a wavelet-based analog of fractal compression by replacing the Haar basis used in (1.3) and (1.4) with a symmetric biorthogonal wavelet basis. We discuss this extension, which we call *self-quantization of subtrees* below. When using a biorthogonal basis, the expressions for the wavelet and scaling function coefficients involve the duals $\tilde{\phi}$ and $\tilde{\psi}$ of ϕ and ψ.

There are two effects of this change of basis. First, non-Haar wavelet subtrees are a generalization of image blocks. When the basis elements are smooth, the range subtrees partition the image into sets with smoothly overlapping boundaries. This results in fewer visible artifacts than the partition into blocks. Second, we replace the local averaging and subsampling procedure with more general filtering. With a careful choice of basis, this more generalized filtering results in a greater concentration of energy in the transformed image which makes quantization easier. Numerical experiments [55] show that changing from the Haar basis to a smooth basis results in considerable improvement in compressed image quality due in part to the elimination of the block boundary artifacts (see Figure 1.5.3 below).

1.4.3 Convergence Criterion

Our generalized quantization scheme yields a relation for an image $\mathcal{I}$ of the form

$$(\mathbf{W}\mathcal{I}) = \mathbf{G}(\mathbf{W}\mathcal{I}) + \mathcal{H} \tag{1.5}$$

where $\mathbf{W}\mathcal{I}$ is the discrete wavelet transform of $\mathcal{I}$. Given $\mathbf{G}$ and $\mathcal{H}$ we decode $\mathcal{I}$ by the same iterative process used in fractal schemes. Our first goal is to show that the above encoding and iterative decoding process yields a well-defined image for non-Haar wavelets. Although showing that $(\mathbf{I} - \mathbf{G})$ is non-singular is sufficient to ensure decodability, inverting $(\mathbf{I} - \mathbf{G})$ is prohibitively computationally expensive, requiring $O(2^{6N})$ work for a $2^N \times 2^N$ pixel image. The iterative decoding scheme converges if $\mathbf{G}^n \to 0$ as $n \to \infty$, which will happen if and only if all eigenvalues of $\mathbf{G}$ are strictly less than 1 in magnitude. The matrix $\mathbf{G}$ is not in general diagonalizable, so we are interested in its Jordan canonical form.

By examining (1.3) and (1.4) we find that the entries of the matrix $\mathbf{G}$ in the row corresponding to the basis function $w \in \mathcal{W}_{N-R}$ are the projections of $\mathbf{T}(w)$ onto the basis multiplied by $2^{R-D} g_\Gamma$, i.e. $\mathbf{G}_{k,l} = 2^{R-D} g_{\Gamma_k} \langle w_l, \mathbf{T}(w_k) \rangle$ for $w_k, w_l \in \mathcal{W}$. When $\mathbf{T}(w_k) \in \mathcal{W}$, the matrix $\mathbf{G}$ has only one nonzero entry per row. For general domain pools, the wavelets in the domain subtrees are translated by non-integer amounts, so the rows of $\mathbf{G}$ contain numerous non-zero entries. By the Geršgorin circle limit theorem [116], all eigenvalues of $\mathbf{G}$ will have magnitudes strictly less than 1 if for all g_Γ we have

$$g_\Gamma < \min_{w'} \left(\frac{2^{D-R}}{\sum_{w \in \mathcal{W}_{N-R}} |\langle w, w' \rangle|} \right). \tag{1.6}$$

where the where the minimum is taken over all wavelets contained in the subtrees in the domain pool and all their associated scaling functions.

We have thus shown that by imposing a suitable bound on the gains g_Γ, fractal block coding techniques can be extended to non-Haar wavelet subtrees with *arbitrary* domain pools. By bounding the magnitudes of the eigenvectors of $\mathbf{G}$ below 1, we ensure that the iterative decoding algorithm converges, and we can bound the final l^2 error of the decoded image in terms of the l^2 block approximation errors incurred in (1.2). Our sufficient conditions for convergence is based on the l^1 norm of sampled reproducing kernels for ϕ and ψ. This bound on the g_Γ's is not a necessary condition for convergence. Indeed, the sum in the denominator is in general considerably larger than 2^{D-R}, so for the Haar basis this is a weaker result than can be obtained for subblock methods via pointwise methods [87].

1.4.4 Why Fractal Block Coders Work

We have observed that the block quantization performed in fractal encodings is equivalent to a quantization of Haar wavelet subtrees. As we discussed earlier, the Haar wavelet transform has energy packing properties similar to those of the Karhunen-Loève transform when image correlation structures resemble those of Markov processes [133]. The energy of such processes is concentrated primarily in the coarse-scale Haar transform coefficients.

This energy-packing property of the Haar subtrees is important for explaining *why* fractal compression works as well as it does. Images do contain some self-similar features such as straight edges and constant regions, as well as features which at fine scales are approximately self-similar, such as smooth regions with quickly decaying Taylor series and low-curvature edges. However, these types of features are by no means the only ones present. How do we explain the ability of fractal compressors to provide a reasonable encoding of non-self-similar regions?

The energy-packing in the Haar subtrees presents a another way to explain the performance of fractal compressors. Consider the problem of quantizing a textured region from an image. There is little reason to believe that such a region in general possesses the type of local self-similarity which fractal compressors seek to exploit. The explanation for fractal coders' performance for such regions is somewhat different than for straight edges. Real-world images are representations of physical structures. Physical continuity of these structures produces a local pixel correlation structure much like that of Markov processes. Because of the energy-packing properties of the Haar transform, Haar subtrees tend to have energy concentrated in the coarse-scale coefficients, which means that they lie close to the very low-dimensional subspace consisting of subtrees with all-zero fine-scale coefficients. Matching subtrees which lie in this low-dimensional subspace is a *much* easier problem than matching arbitrary subtrees. A match does not imply any type of structural self-similarity. Instead, the matching results from the fact that typical image blocks occupy a very small subset of the space of all possible blocks. See [57] for additional details.

In the next section we present some modifications to fractal block coding methods which eliminate the constraint on the gains. We see that it is more useful to characterize the type of map we want to construct as one which transfers information from coarse scales to fine rather than as a generic contraction map. With our refinements, the decoding process converges unconditionally and has a fast implementation. Our generalization, which we call self-quantization of subtrees, permits the use of wavelet bases which have better energy packing properties than the Haar basis, resulting in much more efficient encodings.

1.5 Self-Quantization of Subtrees

1.5.1 Extrapolation in Scale

The approximation in (1.4) gives an implicit scheme for obtaining scaling function coefficients, and it requires that we store a constant h_Γ for each scaling function coefficient. We can greatly simplify our scheme by storing the coefficients $\langle \mathcal{I}, \phi_\Gamma^{N-R} \rangle$ directly. This change has other benefits as well. The scaling function coefficients contain considerable redundancy of which fractal schemes fail to take advantage. We can store the scaling function coefficients efficiently by changing from a scaling function basis to a wavelet basis and storing these coarse-scale wavelet coefficients.

We can further simplify our scheme by restricting the domain pool to the disjoint partition $\mathcal{D} = \{(K, L) : K, L \in \mathbf{Z} \text{ and } 0 \le K, L < 2^{N-D}\}$. With this restriction, all wavelets corresponding to coefficients in the domain pool subtrees are members of the basis $\mathcal{W}_{N-D}$. Hence in the system (1.3) each wavelet coefficient of scale $N-R$ and finer depends only on a coefficient of a coarser-scale wavelet.

We say that a map from one set of wavelet coefficients to another is *R-scale-extending* if each wavelet coefficient of scale $j \geq N-R$ in the range is dependent only on wavelet coefficients of scale $< j$. The system given in (1.3) is a particularly simple form of scale-extending map. One can show that (1.3) also yields an R-scale-extending map for the larger domain pool $\mathcal{D} = \{(K,L) : 2^{D-R}K, 2^{D-R}L \in \mathbf{Z} \text{ and } 0 \leq K,L < 2^{N-D}\}$. In the case of the the Haar basis this set corresponds to the set of domain blocks which share boundaries with range blocks. This particular restricted domain pool has been studied for fractal block coders in [200] and [14]. The theorem below generalizes the results of [200] and [14], extends them into the general wavelet framework, and gives new insight into why their results hold.

Theorem 1.1 (Reconstruction Theorem) *Let $\mathcal{I}$ be a $2^N \times 2^N$ image for which the scaling function coefficients $\langle \mathcal{I}, \phi_\gamma^{N-R} \rangle$ are known, and suppose that we know that $\mathcal{I}$ is the fixed point of a linear R-scale-extending map* $\mathbf{M}$. *Then we can find $\mathcal{I}$ using R applications of the map* $\mathbf{M}$.

Proof: By applying the wavelet transform to the image

$$\mathcal{I}_R = \sum_\gamma \langle \mathcal{I}, \phi_\gamma^{N-R} \rangle \phi_\gamma^{N-R}, \tag{1.7}$$

we obtain all the coarse-scale wavelet coefficients $\langle \mathcal{I}, \psi_\gamma^j \rangle$ for $j < N-R$ for $\mathcal{I}$. We can now obtain the wavelet coefficients $\langle \mathcal{I}, \psi_\gamma^{N-R} \rangle$ by applying the map $\mathbf{M}$, since these coefficients depend only on the coefficients we already know. Each time we apply the map $\mathbf{M}$ we obtain the wavelet coefficients at the next finer scale, so by induction the result is proved. A more detailed version of this proof may be found in [54].

□

Note that this theorem does not require that the wavelet basis in question be the Haar basis. Indeed, we can represent images using subtree approximations of the form (1.2) for *any* biorthogonal wavelet basis. We call this process of subtree approximation for general bases *self-quantization of subtrees* (SQS).

When we use the disjoint domain pool described above, the matrix $\mathbf{G}$ from (1.5) has a very simple form. The rows of $\mathbf{G}$ corresponding to the scaling functions are all zero since we have transferred all the scaling function information to the coefficients in $\mathcal{H}$. Because the wavelet permutation map $\mathbf{T}$ maps each basis element to a multiple of a basis element, the rows corresponding to wavelets in subtree $S_\Gamma^{N-R}\mathcal{I}$ contain a single nonzero entry with value $\frac{g_\Gamma}{2^{D-R}}$. We order the vector of coefficients from coarse to fine, so $\mathbf{G}$ will be a strictly lower triangular matrix with all zeros on the diagonal. Hence, all eigenvalues of $\mathbf{G}$ are zero.

Because each fine-scale wavelet coefficient depends on only one other coefficient when we use our restricted disjoint domain pool, the iterative technique of our proof yields a fast decoding algorithm which requires $O(1)$ operations per pixel. This iterative decoding algorithm also yields a fast decoding algorithm for the larger domain pool

$\mathcal{D} = \{(K, L) : 2^{D-R}K, 2^{D-R}L \in \mathbf{Z} \text{ and } 0 \leq K, L < 2^{N-D}\}$ since the matrix $\mathbf{G}$ has a block diagonal structure.

The above reconstruction theorem generalizes to allow adaptive image encoding. Using the disjoint domain pool, we can recover an image using a fast algorithm provided that for each self-quantized subtree we store its associated scaling function coefficient. Equivalently, we can recover an image provided we know all coarse-scale wavelet coefficients not contained in the range subtrees.

1.5.2 Convergence of Fractal Schemes

Standard explanations fractal compression schemes contain descriptions of quantizing "fine-scale" features using "coarse-scale" features. Using the ideas in the above theorem, we can make rigorous the notion of the scale of a feature. The detail spaces of a multiresolution analysis [165] correspond to sets of features at a particular scale. Fractal decoding involves a cascading of information from coarse wavelet coefficients to fine.

Fractal compression has been motivated by the theory of iterated function systems [18], which involves the construction of a strictly contractive map from the image to itself. However, the scale-extending maps described above are not in general contractive in any of the l^p norms. It is the flow from coarse to fine which gives the map its desired properties, and not strict l^p contractivity.

The detail space of resolution 2^j of a multiresolution approximations is invariant under translations of 2^{N-j} pixels, but *not* under unit pixel translations. It is precisely this lack of translation invariance which causes convergence problems when we expand the disjoint domain pools studied above to include finer translates of domain subtrees. When we approximate range subtrees using fine translates of domain subtrees, we introduce dependencies of fine-scale wavelet coefficients on coefficients from the same or finer scales. Information no longer flows strictly from coarse to fine under the map. Dependency loops from fine-scales to fine-scales permit the growth of unstable eigenvectors unless these loops are damped by restricting the gains g_Γ. These dependency loops are the reason gains are limited to be less than 1 in magnitude in conventional fractal coders.

The reproducing kernel for the wavelet basis gives information about the overlap of translations of functions in detail spaces of different scales. The restriction on the gains we derived in (1.6) incorporates this measure of overlap.

We can obtain a translation invariant domain pool with unconditional convergence by switching to a basis of sinc wavelets, since the detail spaces for this basis correspond to translation invariant frequency bands . The sinc wavelet basis possesses a number of disadvantages, however. The basis elements do not have compact support, so locally self-similar features can only be approximately isolated by the subtrees. In addition, computations are slow.

Our analysis sheds light on several other aspects of fractal coding. Our convergence proof shows that images are reconstructed from the stored self-similarity relation by cascading information from coarse to fine scales. This suggests that the decoding process will be more sensitive to errors in coarser scales than fine. In [54] we examine a weighted l^2 error metric which we use to select which domain subtree to use for quantization. The addition of perceptual criteria to our subtree error metric involves a straightforward weighting of errors at different scales [62].

Our convergence proof also gives insight into the mechanism underlying fractal interpolation or super-resolution of images. Each iteration of our decoding process generates a new level of wavelet coefficients. By continuing to decode after convergence, we can generate additional fine-scale coefficients. When we invert the resulting transform, we obtain a larger image than we originally encoded with detail that has been interpolated using our self-similarity map.

1.5.3 Results

Figure 1.3 shows the results of using self-quantization of subtree coding for the Lena image. Subtree quantization and bit allocation was performed using a Lagrangian algorithm described in [55]. The quality of the encoded images is vastly improved over the simple quadtree scheme for two reasons. Unlike fractal block coders, the SQS scheme takes advantage of the spatial redundancy in the coarse-scale wavelet coefficients. The SQS scheme also makes use of superior bit allocation routines. Switching from the Haar basis to the biorthogonal spline basis of [7] yields further improvements. By switching to this smooth basis we are able to reduce the file size by over 40 % while maintaining almost the same image PSNR. We can improve image appearances by optimizing perceptually weighted distortion measure rather than PSNR.

The SQS scheme outperforms the wavelet encoding on edges (roughly 15 % of the image), but wavelets and all-zero subtrees are more effective elsewhere. Additional experiments, to be described in a forthcoming paper [57], have shown we may obtain further improvements by using a wavelet and zerotree scheme and dropping self-quantization of subtrees altogether.

1.6 Conclusion

We have introduced a wavelet framework for analyzing fractal coders which provides considerable insight into how these schemes function. The wavelet framework

- reveals a natural relationship between fractal compression and Haar transform coders,
- gives insight as to how fractal coders function on complex, textured regions,
- reveals important structural details of the map generated by fractal coders, and
- shows why there are convergence problems in decoding fractal images.

Using this framework we generalize block-based fractal schemes to subtree-based schemes. We emphasize that this generalization can be made for *any* domain pool provided that the gains in the affine maps are restricted. We obtain simple, finite-step, unconditional convergence properties when we restrict domain pools to those which generate scale-extending maps.

Figure 1.3: The top left shows the 512 × 512 Lena image compressed at 60.6:1 (PSNR = 24.9 dB) using a disjoint domain pool and the quadtree coder from [87]. The top right image has been compressed at 101.7:1 (PSNR = 26.8 dB) using a rate-constrained Haar wavelet-based fractal scheme with the modifications described in section 1.5.1. This scheme uses exactly the same domain pool as the quadtree scheme, but the insight gained from the wavelet framework enables us to make much more efficient use of bits. The bottom left image has been compressed at 100.2:1 (PSNR = 28.1 dB) using self-quantization of subtrees (SQS). Our algorithm is identical to that for the top right image, but we have switched from the Haar basis to a smooth wavelet basis. Blocking artifacts have been completely eliminated. The bottom right image has been compressed at 160.5:1 (PSNR = 26.3 dB) using the same smooth wavelet basis.

The wavelet framework's reveals similarities in fractal coders to existing wavelet schemes, enabling the incorporation of off the shelf bit-allocation techniques in fractal coding schemes. The use of smooth bases gives greatly improved image fidelity with fewer visible artifacts than the Haar basis. The result is compressed image quality that is greatly improved over that of standard fractal block coder methods.

Acknowledgments

This work was supported in part by DARPA as administered by the AFOSR under contract DOD F4960-93-1-0567.

Chapter 2

On Fractal Compression and Vector Quantization

Skjalg Lepsøy, Paolo Carlini, Geir Egil Øien

Fractal image compression can be seen as a type of vector quantization with a codebook that is extracted from the image being coded. In this chapter we study the properties of codebooks populated in this way and make comparisons with fixed codebooks that are trained for a set of representative images.

There are two major results of the experiments presented here. Firstly, the image being coded is not necessarily the best codebook-generating image; the codebook can be as good or better when it is extracted from another image. Secondly, we have compared a fractal coder to a very similar product code vector quantizer with the origin of the codebooks as the most significant difference. In terms of PSNR of decoded images, the product code VQ has been superior to the fractal coder in all 108 tests, over six test images, nine codebook sizes and two block sizes.

When a coding technique is made adaptive, the adaptivity should improve, not deteriorate the results obtainable by the technique. The fractal coder is an image-adaptive vector quantizer, and should as such offer an advantage over its non-adaptive counterpart. Thus, the main message is not that there exists a certain VQ that is slightly better than the fractal coder tested here, but that fractal coding may be the wrong way of codebook adaptation.

2.1 Two Coders to Compare

2.1.1 Product Code Vector Quantization

Vector quantization (VQ) is a technique for lossy compression of sampled signals using methods of table look-up [102, 164, 97]. A vector quantizer groups the signal samples into vectors that are to be approximated. Present both in the encoder and the decoder is a table of eligible approximating vectors, this table is called the *codebook*. Its vectors are called *codevectors* and their addresses in the codebook are called *codewords*. For a given vector of original samples, the encoder searches the codebook for the codevector that gives the best approximation and it emits the corresponding codeword. The decoder receives the codeword and outputs the codevector pointed to.

The specific vector quantizer to be used in the experiments presented here is a *product code VQ* (PCVQ). A PCVQ uses not one, but several codebooks of *parameter vectors*. The *effective codebook* (i.e. the set of eligible approximating vectors) is implied by a function that takes parameters from each of the parameter codebooks and produces a codevector. Suppose that there are J parameter codebooks $\{\mathcal{C}_j\}_{j=1}^{J}$. Letting f denote the function, the effective codebook is then

$$\mathcal{C} = f\left(\prod_{j=1}^{J} \mathcal{C}_j\right),$$

and the codevector is

$$\hat{y} = f(c_1, \ldots, c_J); c_j \in \mathcal{C}_j.$$

The effective codebook is implied by the cartesian product of all the tables, hence the name product code.

2.1.2 A Mean-Removed Gain-Shape VQ

The PCVQ we shall use was first presented by Murakami *et al.* [190]. It is a mean-removed gain-shape VQ; we shall refer to it as the M-VQ. Its operation is detailed in the following, where we attempt to keep the formalism close to the one used for the fractal coder.

The image produced by the decoder of the M-VQ is expressed as the blockwise sum

$$\hat{x} = \sum_{n=1}^{N} P_n \left(\alpha_n a_{i(n)} + \beta_n b_n\right). \tag{2.1}$$

In this decomposition,

- $a_{i(n)}$ is a block with zero mean taken from a codebook.
- $b_n \in \Re^{B_n}$ is a block of the range block size (B_n pixels) with all pixels equal to 1.
- $\alpha_n, \beta_n \in \Re$ are scalars.
- $P_n : \Re^{B_n} \to \Re^M$ puts a block into the correct range block position and sets the rest of the image to zero (the image has M pixels.)

The M-VQ uses three parameter codebooks. Two of the parameter codebooks contain scalars used to quantize the coefficients α_n and β_n,

$$\alpha_n \in \mathcal{A}, \beta_n \in \mathcal{B}.$$

The third parameter codebook contains candidates for the basis block $a_{k(n)}$,

$$\mathcal{C} = \{a_i : i = 1, \ldots, I\}. \tag{2.2}$$

The codebook $\mathcal{C}$ is called the *pattern codebook* of the M-VQ, as its vectors are responsible for the visible pattern (deviations from the mean value) within the blocks in the decoded image $\hat{x}$.

2.1.3 A Fractal Coder

In a fractal image coder an image is sought approximated as the unique *fixed point* of an image transformation. Compression is obtained by quantizing the parameters of the image transformation.

The image transformation is a sum of blockwise mappings, one for each range block. Each blockwise mapping is affine, and so is the image transformation, as it is a sum of affine mappings. The image transformation is denoted by

$$T : \Re^M \to \Re^M,$$

where

$$Tx = \sum_{n=1}^{N} T_n x = \sum_{n=1}^{N} P_n \left(\alpha_n O_n S_{k(n)} F_{l(n)} x + \beta_n b_n \right). \tag{2.3}$$

In this sum,

- $F_{l(n)} : \Re^M \to \Re^{D_n}$ fetches a domain block (D_n pixels) chosen by the encoder.
- $S_{k(n)} : \Re^{D_n} \to \Re^{B_n}$ deforms the domain block to the size and shape of the range block. This mapping may also include shuffling of the pixels internally in the shrunken block.
- $O_n : \Re^{B_n} \to \Re^{B_n}$ subtracts the mean value of a block from all its pixels. (It orthogonalizes the block with respect to the space of the fixed term which is rather trivial in this case.)
- $b_n \in \Re^{B_n}$ is a block of the range block size, with all pixels equal to 1. This is the 'basis' for the fixed term space.
- $\alpha_n, \beta_n \in \Re$ are scalars.
- $P_n : \Re^{B_n} \to \Re^M$ puts a block into the correct range block position and sets the rest of the image to zero.

Except for the orthogonalization operator, such a formulation of the image transformation encompasses various coding schemes as presented by e.g. Jacquin [123, 125] and Fisher *et al.* [84]. See Chapter 4 for a review of the benefits of orthogonalization.

The fractal coder uses three parameter codebooks, as does the M-VQ. Two of the parameter codebooks contain scalars, used to quantize the coefficients α_n and β_n,

$$\alpha_n \in \mathcal{A}, \beta_n \in \mathcal{B}.$$

The third codebook in the fractal coder is a pattern codebook, as its vectors provide the visible pattern within the blocks. The pattern codebook is implied by the 'fetch', 'shrink and shuffle' and the orthogonalization operators, applied to the image that undergoes the transformation T. With L eligible 'fetch' and K eligible 'shrink and shuffle' operators, the pattern codebook is

$$\begin{aligned} \mathcal{C} &= \{OS_kF_lx : k = 1, \dots, K, l = 1, \dots, L\} \\ &= \{A_ix : i = 1, \dots, I\} \end{aligned} \tag{2.4}$$

where x is the image undergoing the transformation and the symbol A_i is used for convenience to denote the combined operator OS_kF_l. The size of the pattern codebook is $I = KL$.

Inserting $A_{i(n)}$ for $O_nS_{k(n)}F_{l(n)}$ in Equation 2.3 we get the expression

$$Tx = \sum_{n=1}^{N} P_n \left(\alpha_n A_{i(n)}x + \beta_n b_n\right) \tag{2.5}$$

for the image transformation.

2.2 Methods of Codebook Population

Both the M-VQ and the fractal coder are product code vector quantizers, where the effective codebook is implied by all parameter combinations in the product $\mathcal{A} \times \mathcal{B} \times \mathcal{C}$. Parameter codebook $\mathcal{A}$ contains the scalars for quantizing coefficient α_n, $\mathcal{B}$ is for quantizing coefficient β_n, and $\mathcal{C}$ is the pattern codebook. In this work we are only concerned with the effects of different pattern codebooks. Hence for simplicity we do not quantize the scalar coefficients, therefore $\mathcal{A} = \mathcal{B} = \Re$. We remark that for the M-VQ, the pattern codebook and the quantizer for the coefficient α_n could be trained or optimized jointly. Readers interested in joint training of pattern codebook and coefficient quantizer are referred to [185].

In this section we elaborate on how the pattern codebook is populated. For simplicity we refer to it as the codebook.

2.2.1 Population of the Codebook in the Fractal Coder

The fractal coder has a codebook that is extracted from the image that undergoes the image transformation. In the encoder this image is the original image, as the collage is this image transformed once. The codebook in the encoder is thus

$$\mathcal{C} = \{A_ix_O\}_{i=1}^{I}$$

as in Equation 2.4, where x_O is the original image.

The decoder usually iterates the image transformation, producing a sequence of images. For the experiment presented here, the only image of interest is the attractor or the limit of the sequence. Therefore the codebook in the decoder is

$$\mathcal{C} = \{A_ix_T\}_{i=1}^{I}$$

where x_T denotes the attractor.

2.2.2 Population of the Codebook in the M-VQ

The M-VQ operates with a fixed codebook that is present both in the encoder and the decoder. We shall experiment with codebooks populated in two different ways. One way is to extract domain blocks from some selected image in the same manner as does the fractal coder, although possibly another image than the one being coded,

$$\mathcal{C} = \{A_i w\}_{i=1}^{I},$$

where w is some image chosen to serve as a source of domain blocks.

The other way is training of the codebook by a procedure that maximizes the reproduction quality when the codebook is used for a set of training images. We shall use the Generalized Lloyd or LBG algorithm for this purpose [158]. This algorithm produces a sequence of codebooks of non-decreasing quality with respect to the training set, and it converges to a locally optimal codebook. Here we shall develop the function to maximize and, as the cost function is not the common euclidean distance, we shall reexamine one of the steps in the LBG algorithm. This serves to clarify the concept of an optimal codebook and to facilitate a discussion of the outcome. Otherwise, the remainder of this subsection is not necessary for the understanding of the following sections.

For training of the codebook one chooses a set of images with characteristics that are held to be typical for images encountered by the coder in the future. Supposing that there are K blocks in these images, one seeks to find the codebook

$$\mathcal{C} = \{a_i\}_{i=1}^{I}$$

with the best possible average performance over the set of the training blocks

$$\{y_k\}_{k=1}^{K}.$$

Let the approximation of a training block y_k be denoted by $\hat{y}_k$. We are to minimize the average squared error by projection of each block onto a two-dimensional subspace (see below.) By orthogonality this average error is

$$\sum_{k=1}^{K} \|y_k - \hat{y}_k\|^2 = \sum_{k=1}^{K} \|y_k\|^2 - \|\hat{y}_k\|^2$$

and as y_k are already given, minimizing this sum amounts to maximizing

$$\sum_{k=1}^{K} \|\hat{y}_k\|^2. \tag{2.6}$$

From Equation 2.1, each training block is approximated by a linear combination of two blocks

$$\hat{y}_k = \alpha_k a_{i(k)} + \beta_k b$$

where one is taken from the codebook and the other has constant greytone value 1. ($a_{i(k)}$ is the codebook block best suited to approximate training block number k.) These are orthogonal, as the codebook block has zero mean, therefore

$$\|\hat{y}_k\|^2 = \|\alpha_k a_{i(k)}\|^2 + \|\beta_k b\|^2 \tag{2.7}$$

which shows that the codebook can be optimized regardless of the training blocks' components along b, i.e. the 'DC' components. The first term is the norm of a training block projected onto a codebook block, therefore (assuming codebook blocks of unit norm)

$$\alpha_k = \langle y_k, a_{i(k)} \rangle$$

and

$$\|\alpha_k a_{i(k)}\|^2 = \langle y_k, a_{i(k)} \rangle^2. \tag{2.8}$$

As the goal is maximization, the codebook block $a_{i(k)}$ is chosen to make this term as large as possible. Thus we have arrived at one term of the objective function, and we can formulate the **codebook training problem:**

> Given the training set $\{y_k\}$, find the codebook $C = \{a_i\}$ that maximizes the objective function
>
> $$\sum_{k=1}^{K} \max_{a \in C} \langle y_k, a \rangle^2, \tag{2.9}$$
>
> under the constraints of unit norm and zero mean,
>
> $$\|a_i\| = 1, a_i \perp b.$$

The constraint that each codebook block have zero mean is obeyed if the mean values are first subtracted from all training blocks [149]. In the following, we denote subtraction of the mean value by O,

$$Oy = y - \bar{y}.$$

We shall seek to maximize the objective function in Equation 2.9 by the LBG or **Generalized Lloyd Algorithm.** It produces a sequence of codebooks and assumes either an initial codebook or an initial subdivision of the training set into bins. The procedure is as follows:

1. For a given codebook $C = \{a_i\}_{i=1}^{I}$, find the optimal subdivision of the training set

$$\langle Oy_k, a_i \rangle^2 = \max_{a \in C} \langle Oy_k, a \rangle^2 \Rightarrow y_k \text{ belongs to bin } Y_i. \tag{2.10}$$

2. For a given subdivision of the training set, find the optimal codebook, i.e. maximize

$$f(C) = \sum_{i=1}^{I} \sum_{y_k \in Y_i} \langle Oy_k, a_i \rangle^2. \tag{2.11}$$

3. If the objective function has increased relatively by more than a threshold, repeat. If not, stop.

Step 1 in this procedure only involves a search for a maximum and a corresponding redistribution of training blocks into bins. Step 2 is less obvious, and below we demonstrate how the codebook vectors are eigenvectors of certain matrices.

Note that in Equation 2.11 any codebook vector enters in only one sum over one bin, it can therefore be optimized regarding only that partial sum. One such sum is actually the squared l_2-norm of the product of a certain operator H_i and the codebook vector:

$$\sum_{y_k \in Y_i} \langle O y_k, a_i \rangle^2 = \|H_i a_i\|^2$$

where the operator H_i is linear and given by

$$H_i a = \begin{bmatrix} \langle O y_{1_i}, a \rangle \\ \vdots \\ \langle O y_{J_i}, a \rangle \end{bmatrix},$$

and $\{y_{1_i}, \dots, y_{J_i}\}$ are all the training blocks in bin Y_i belonging to codebook vector number i.

Assuming some ordering of the pixels into a vector, the squared norm (omitting the index i) is expandable as

$$\|Ha\|^2 = a^T H^T H a.$$

By Schwartz' inequality

$$a^T H^T H a \leq \|a\| \|H^T H a\| \tag{2.12}$$

where equality holds if the vectors a and $H^T H a$ are parallel. This amounts to stating that

$$H^T H a = \lambda a, \tag{2.13}$$

which is the eigenvalue problem for $H^T H$. Recall that the norm of a is constrained to be equal to 1. Equations 2.12 and 2.13 show that the maximum is attained when the codebook vector a is chosen as the eigenvector that belongs to the largest eigenvalue (for $H^T H$, all are real and non-negative.)

2.3 Cross-Transformability

Fractal coding exploits the similarity between range blocks and domain blocks in an image, using some image transformation as the one described in Section 2.1 where each range block is approximated by a transformed domain block. The underlying assumption is that such similarities are manifest in an image. We can call the presence of such similarities *blockwise self-transformability*.

A fractal encoder finds the best pair {range block, domain block} for every range block in an image, the domain block being found in the same image. If indeed self-transformability is a characteristic property in natural images, one expects not to find

Image name	Image number
Airplane	1
Bank	2
Crowd	3
Loco	4
Peppers	5
Mandrill	6
Couple	7
Face	8
Lena	9
Mill	10
TekRose	11

Table 2.1: The eleven images used for the experiment on cross-similarity.

Image size	512×512 pixels
Range block sizes	4×4 or 8×8
Domain block sizes	2×2 larger than the range blocks
$\{A_i\}$ to build codebook	Evenly spaced domains, shrinking by averaging, no shuffling
Codebook sizes	3000 or 4000

Table 2.2: Data for the experiment on cross-similarity

equally good pairs {range block, domain block} when the domain blocks are taken from another image than the range blocks.

In this section we present an experiment where domains and ranges are from different images. Formally, this means encoding an image x_O with an M-VQ that uses a pattern codebook

$$\mathcal{C} = \{A_i w\}_{i=1}^{I},$$

where w is an image different from x_O. If x_O can be encoded well using the domains from w, we might say that x_O and w are *cross-transformable.* This is to be contrasted to encoding the image x_O using a codebook extracted from itself,

$$\mathcal{C} = \{A_i x_O\}_{i=1}^{I},$$

either for an M-VQ populated like this or for a fractal coder.

2.3.1 Details on the Experiment

We have chosen eleven commonly available images to populate codebooks, see Table 2.1. Each codebook is extracted from one of the images. The same eleven images are then coded with these codebooks. Data for the coders used are found in Table 2.2.

2.3.2 Results and Remarks

The outcome of the experiments is presented graphically as the PSNR versus the domain-supplying images, represented by the numbers found in Table 2.1. Whenever the domains are taken from the same image as the ranges, what is presented is the PSNR quality of the collage (if the coder were an M-VQ, the collage is the decoded image.) The results are shown in Figures 2.1 and 2.2. The deterioration from collage to attractor is shown in Figure 2.3.

If blockwise self-transformability were a characteristic property in natural images (as represented by our image set), we might have expected a peak in the PSNR curves where the image supplying domains coincides with the image containing ranges (i.e. for the collages.) Such a peak is pronounced for three of the eleven images; images 2, 5 and 7 (*Bank, Peppers* and *Couple.*) For the other images, the curves show little variation or have maxima other than the collage.

For any image except *Bank*, replacing the codebook extracted from itself by a codebook extracted from *Peppers* or *Couple* (Fig. 2.4) either improves the quality or leaves it almost unchanged. We are tempted to conclude that blockwise self-transformability is not a common property and that self-transformation is not a superior principle for populating codebooks. Our results are at least a strong indication that this may be the case.

2.4 Trained Codebooks versus Self-Transformation

We have seen that coding by self-transformation (as outlined in Subsection 2.2.1) is an alternative way of populating the codebook in a certain product code VQ structure. The two codebooks in question are of the types

$$\mathcal{C} = \{A_i x_T\}_{i=1}^{I}$$

for the fractal coder (x_T is the attractor), and

$$\mathcal{C} = \{a_i\}_{i=1}^{I}$$

for the trained M-VQ.

We shall compare the codebooks yielded by the two techniques in terms of PSNR for different images, codebook sizes, and block sizes, and we shall visualize the codebooks for direct comparison.

2.4.1 Details on the Experiment

Data for the experiment are found in Table 2.3. Note that there is no shuffling of pixels internally in the shrunken domain blocks. Instead we have used more domain blocks to arrive at a specified codebook size. In informal tests we have found that this does not change the results more than a very small fraction of a *dB*; the codebook size is what counts. The training sets have consisted of blocks taken from the training images shown in Figure 2.5 with maximum even spacing between blocks, given the size of the training set. The mean values were subtracted from the training blocks. The test images are shown in Figure 2.6.

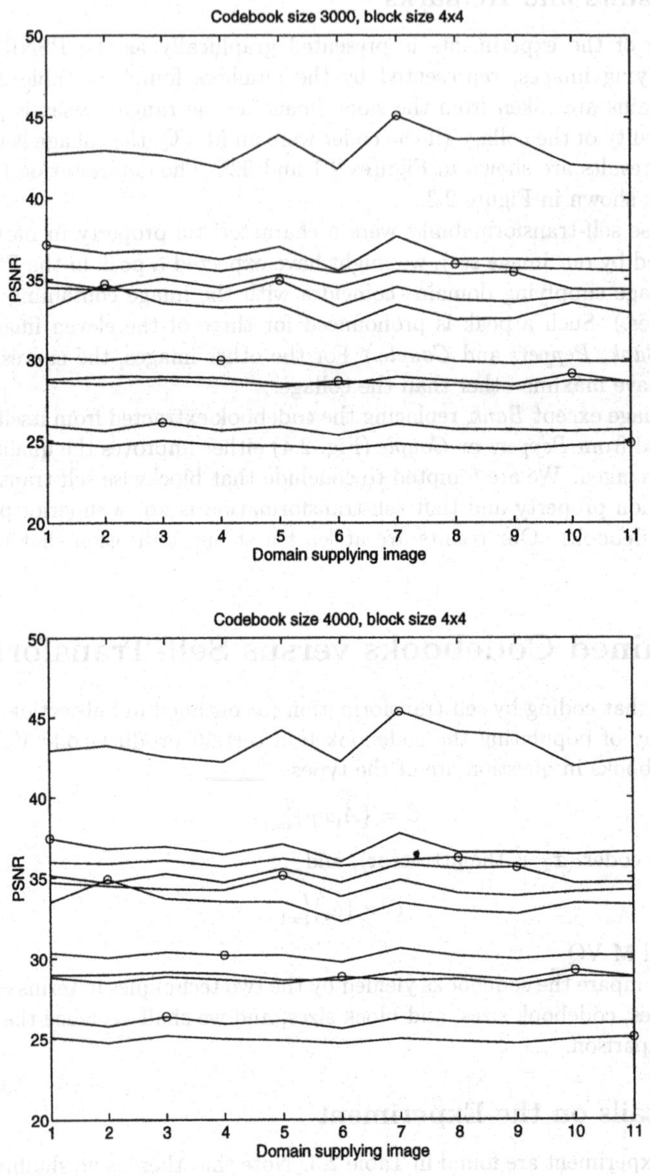

Figure 2.1: Each graph represents the PSNR for one image with different sources for the codebook. The circles indicate the PSNR for the collages (when the image coded is its own source for a codebook). Block size 4×4.

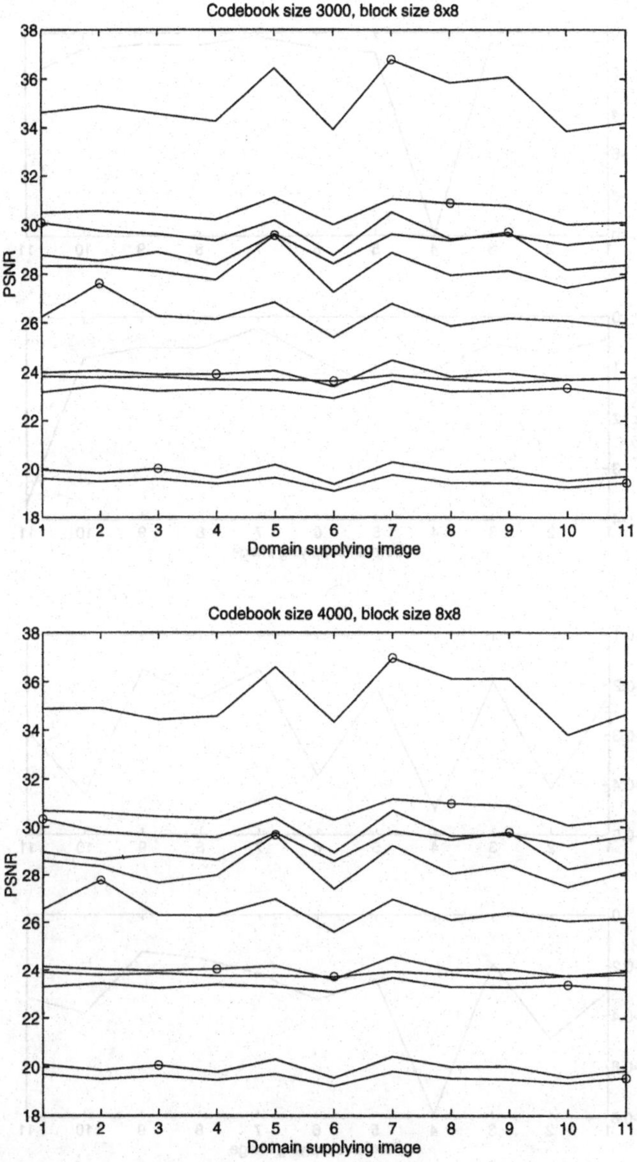

Figure 2.2: Each graph represents the PSNR for one image with different sources for the codebook. The circles indicate the PSNR for the collages (when the image coded is its own source for a codebook). Block size 8×8.

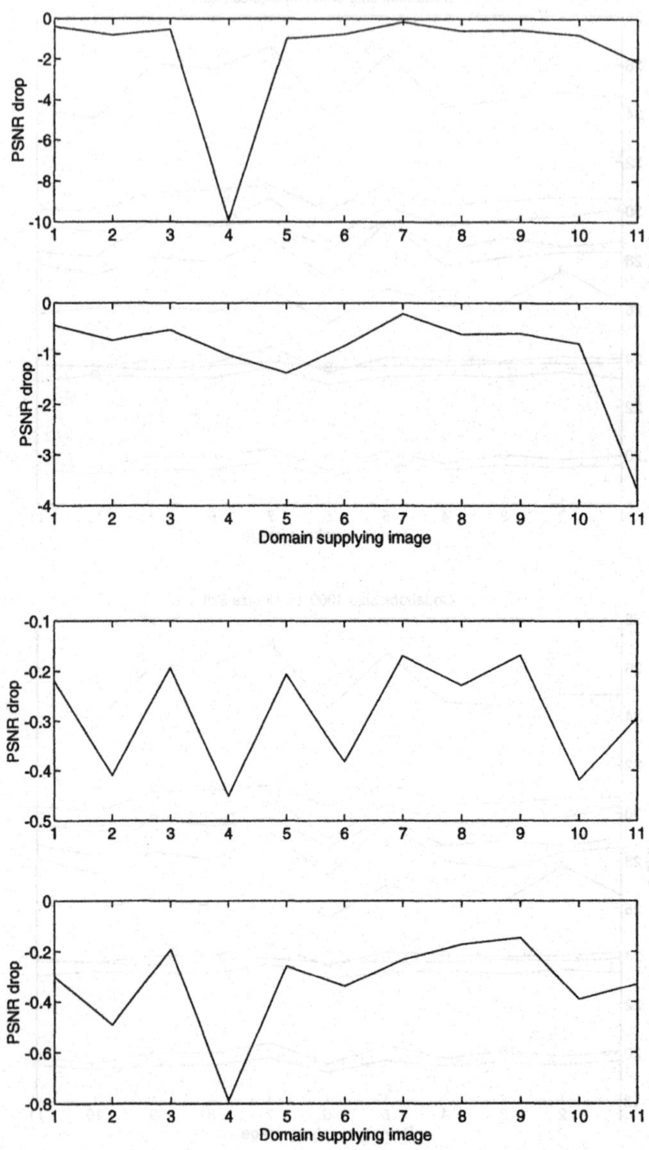

Figure 2.3: Deterioration from collage to attractor as change in PSNR. The two upper graphs hold for block size 4×4 with codebook sizes of 3000 and 4000. The two lower hold for block size 8×8 and codebook size 3000 and 4000. The large drops seen for two images at block size 4×4 are due to non-contractive image transformations.

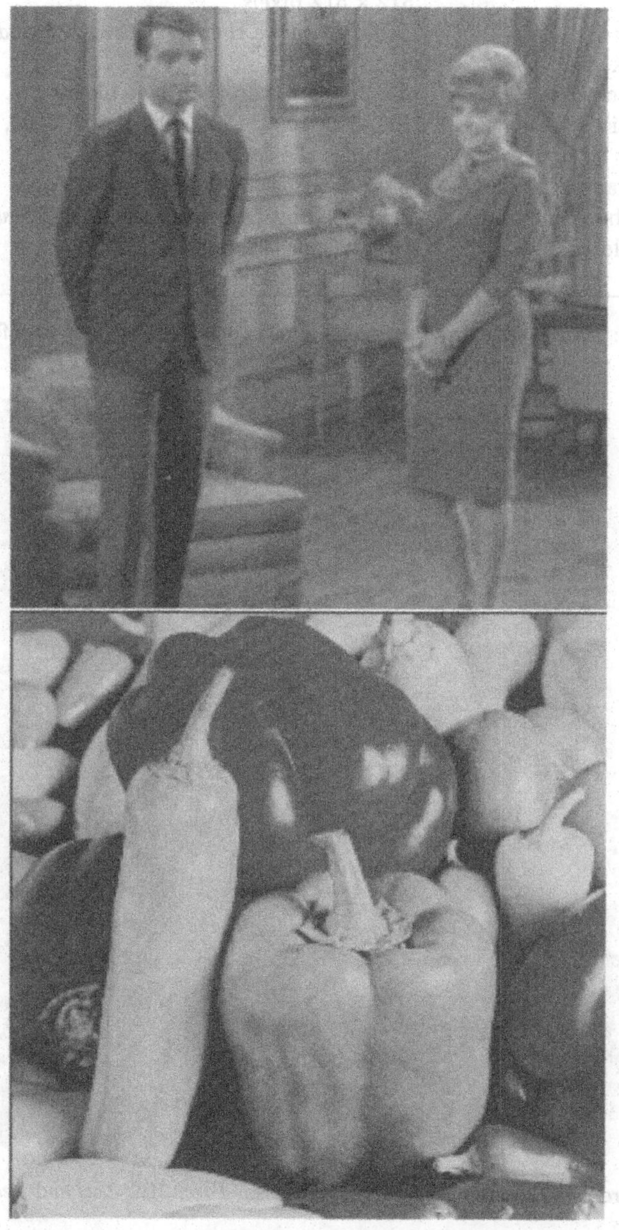

Figure 2.4: Two images well suited as sources for domain blocks. The upper image is *Couple*, the lower image is *Peppers*.

Image size	512×512 pixels
Test images	*Lena, Airplane, Peppers, Crowd, Mandrill, Bank*
Range block sizes	4×4 or 8×8
Domain block sizes	2×2 larger than the range blocks
Codebook sizes	100, 500, 1000, 1500, 2000, 2500, 3000, 3500, 4000
Training set size	50 times the codebook size
Training images	*Loco, TekRose, Mill, Face*
Initial codebook for training	Extracted by $\{A_i\}$ from the image *Couple*
$\{A_i\}$ to build codebook	Evenly spaced domains, shrinking by averaging, no shuffling

Table 2.3: Data for the experiment on trained codebooks vs. self-transformation.

Figure 2.5: The training images. Row-wise: *Face, Mill, Locl* and *TekRose*.

Figure 2.6: The test images. Row-wise: *Lena, Peppers, Mandrill, Airplane, Crowd,* and *Bank.*

2.4.2 Results and Remarks

The outcome of the experiment is presented graphically as the PSNR versus the codebook sizes in the Figures 2.7 – 2.10. An example of what decoded images may look like is given in Figure 2.11.

We see that the fractal coder is inferior to the M-VQ in all of the comparisons. For block sizes of 4×4 pixels, the advantage of M-VQ lies within the range 0.5 – 1.5 *dB*, the smallest found for the image *Bank* at a codebook size of 3500 and the largest found for the images *Mandrill* and *Crowd*, roughly over the whole range of codebook sizes (we disregard the case where the the fractal coder yields a non-contractive transformation — *Peppers* with codebooks of 2000 and 2500 vectors.) For block sizes of 8×8, the advantage of the M-VQ is about the same, with a minimum for *Bank* at 3500 vectors and *Peppers* at 2000 vectors. The maximum advantage for M-VQ, about 1.1 – 1.2 *dB*, is found for *Airplane* at all codebook sizes. These results correspond with those in the previous section, where *Bank* and *Peppers* were found to be two of the three most self-transformable images in the test set.

The results can also be understood in terms of differences in bit rates yielded by the two techniques. To indicate the differences, we shall take the results for the fractal coder with 4000 vectors in the codebook and find the sizes of the M-VQ codebook that produced the same PSNR values. The difference in bit rate would then be

$$\frac{\lceil \log_2 4000 \rceil - \lceil \log_2 m \rceil}{p} \, bpp,$$

	8 × 8 pixels		4 × 4 pixels	
Image	M-VQ size	Δbpp	M-VQ size	Δbpp
Mandrill	300	0.06	500	0.19
Lena	1500	0.02	1000	0.13
Peppers	2100	0.015	500	0.19
Bank	1500	0.02	2000	0.06
Airplane	500	0.05	1000	0.13
Crowd	500	0.05	1000	0.13

Table 2.4: Sizes of M-VQ codebooks equivalent to the fractal codebook with 4000 vectors. Estimated difference in bit rates in favour of the M-VQ.

where m is the M-VQ codebook size equivalent to a 'fractal' codebook of 4000 blocks, and p is the number of pixels in a block, either 16 or 64. The differences are listed in Table 2.4. Effectively, for blocks of 8 × 8 pixels the bit rates for the M-VQ and the fractal coder are very similar. For blocks of 4 × 4 pixels, the difference in bit rate is more pronounced in favour of the M-VQ.

2.4.3 Characteristics of the Two Techniques

The two techniques for codebook population yield codebooks that are qualitatively quite different. Probably the most significant drawback of the fractal coder is that it may produce codebooks with repeated patterns, i.e. with several blocks that are practically similar. This is a waste of space in the codebook. The LBG procedure for training of the M-VQ codebook does not permit repeated patterns. Here, the codebook blocks are computed to represent disjoint subsets of the training set. These subsets are also well separated in the vector space, as the subdivision is based on a nearest-neighbor rule (Equation 2.10.) Each codebook vector is the principal component of its training subset (Equation 2.11), hence it is in a sense the center of its region in the vector space. From this it follows that patterns can not be repeated. Some of the blocks in the M-VQ codebook may resemble each other visually, but only in the parts of the vector space where the density of training vectors is high. These vectors represent patterns that are expected to occur often, and it is advantageous to represent them accurately.

By the same arguments, the M-VQ codebook is general in nature and does not contain types of patterns that are rare in the training set. For example, very detailed, irregular patterns may be visually similar but yet quite different as measured with the squared inner product (see Equation 2.10). Such blocks often go into different training subsets and influence the resulting codebook vectors little. Thus, the M-VQ codebook tends to contain blocks with rather regular patterns such as edges or patterns with a distinct orientation. In contrast, the fractal coder takes blocks indiscriminately from around in the original image and accordingly the codebook contains many irregular or textured blocks if the image itself is rich in textured blocks. Textures are seldom self-similar or self-transformable, and codebook space may be wasted.

The Figures 2.12 and 2.13 shows the blocks in the codebooks for the M-VQ and the fractal coder. The codebooks have been sorted in order to put similarly looking

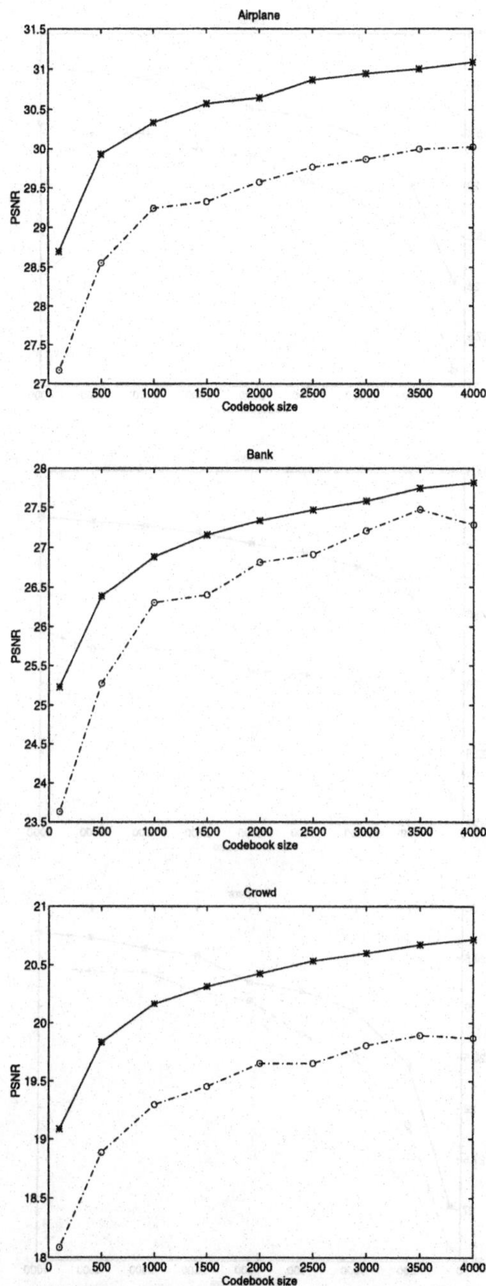

Figure 2.7: PSNR versus codebook size for PCVQ and fractal coding. * indicates PCVQ, o indicates fractal coding. Block size 8×8.

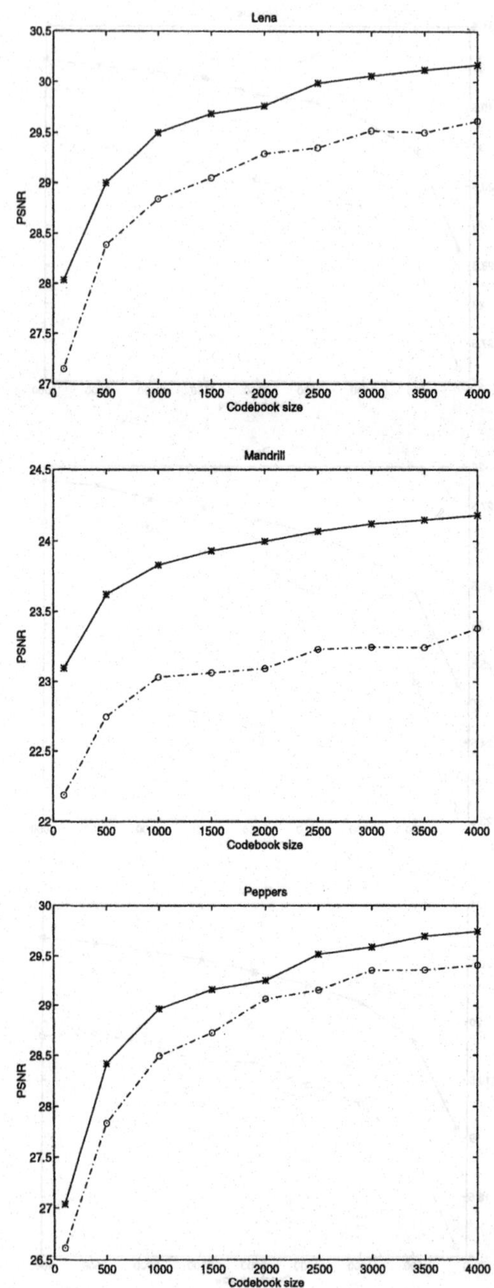

Figure 2.8: PSNR versus codebook size for PCVQ and fractal coding. * indicates PCVQ, o indicates fractal coding. Block size 8×8.

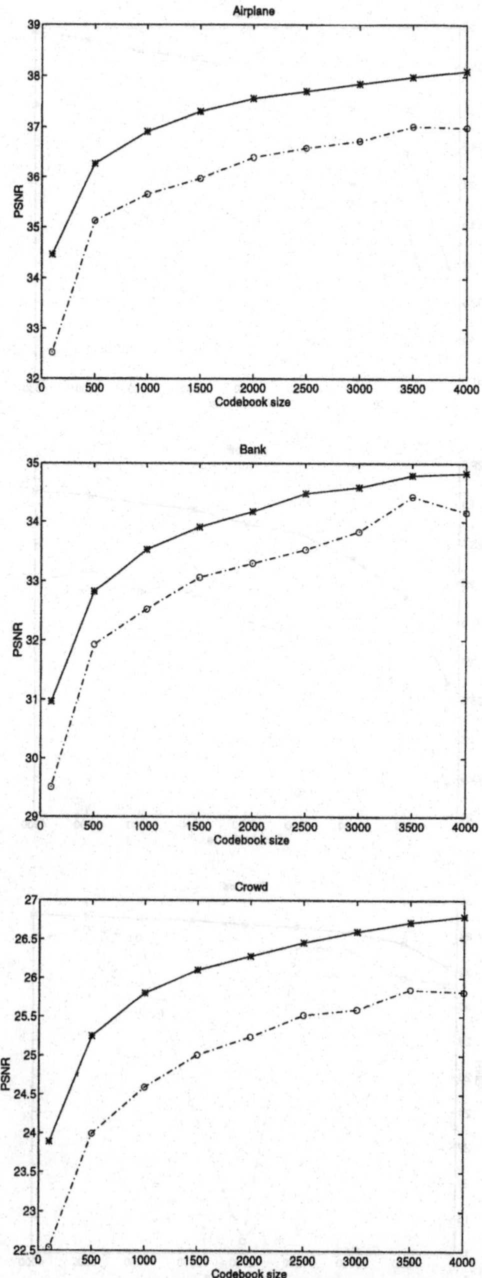

Figure 2.9: PSNR versus codebook size for PCVQ and fractal coding. * indicates PCVQ, o indicates fractal coding. Block size 4×4.

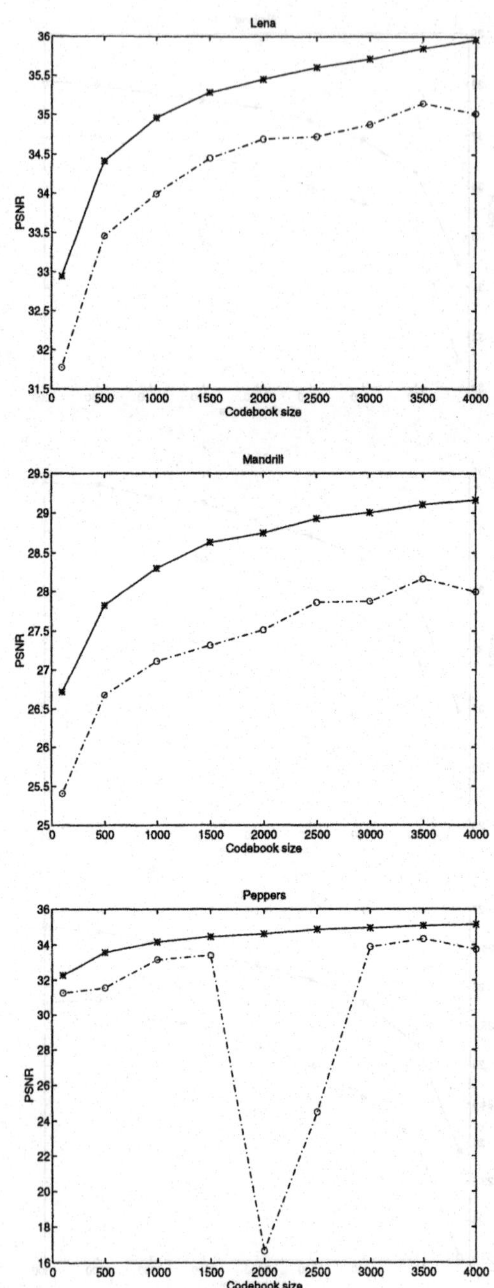

Figure 2.10: PSNR versus codebook size for PCVQ and fractal coding. * indicates PCVQ, o indicates fractal coding. Block size 4×4.

Figure 2.11: The image *Crowd.* The upper image has been encoded with an M-VQ, the lower image has been encoded by a fractal coder. The blocks are 8×8 pixels and the codebooks contain 3000 blocks.

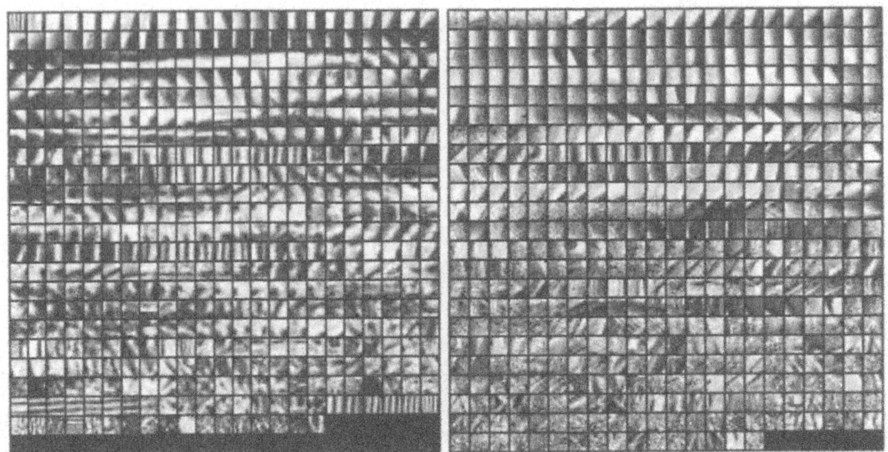

Figure 2.12: Two codebooks. The left is a trained codebook, the right is a fractal codebook extracted from the image *Lena*. The blocks are ordered as to put similarly looking blocks next to each other. Each of these codebooks contain 500 blocks of 8×8 pixels.

blocks next to each other (row-wise). These two figures illustrate the two problems noted above; pattern repetitions and textures. In Figure 2.12 one can see that the M-VQ codebook contains less repetitions of patterns than does the codebook for the fractal coder. The 'fractal codebook' is roughly divided into two parts of equal size; one part with edges and other regular patterns and one part with textures or irregular patterns. In Figure 2.13, the codebook for the fractal coder is taken from the image *Crowd*, an image that does not contain many repetitions of patterns. Here a large part of the codebook is occupied by irregular patterns and textures. The M-VQ codebook is again rich in regular patterns and poor in textures.

It would be interesting to identify visual or subjective classes of patterns for which the fractal coder is clearly better suited than the M-VQ. Figure 2.14 shows the blocks in the images *Lena* and *Peppers* best represented by either method. The 'preference maps' (sets of blocks best encoded by one technique) are not strongly correlated to image content, and the overall score is that roughly 65% of the blocks are handled best by the M-VQ.

As we use no hard definition of classes of patterns, we shall only remark loosely on how textures and edges are reproduced. The M-VQ seems to handle textures better than does the fractal coder. The textured parts of *Lena* are the feather, the hat and the surfaces of the background, the textured parts of *Peppers* are the surfaces on the peppers[1]. None of these show any preference for fractal coding. Some tendency can be seen in the ways edges are reproduced. The two images suggest that sharply defined edges are better reproduced by the M-VQ, whereas blurred edges are better reproduced by the fractal coder.

[1]The surfaces are not entirely smooth. Here we are comparing two pattern codebooks without regarding the 'gain' coefficients, hence all irregular patterns are considered textures, no matter how weak the contrast.

Figure 2.13: Two codebooks. The upper is a trained codebook, the lower is a fractal codebook extracted from the image *Crowd*. The codebooks are ordered as to put similarly looking blocks next to each other. Each of these codebooks contain 1000 blocks of 8×8 pixels.

2.4.4 Conclusion

Coding by self-transformation provides for image-specific codebooks without the necessity for side-information. It is a kind of adaptive product code vector quantization, but in none of the comparisons made here does it offer an improvement over vector quantization with trained and fixed codebooks. The comparisons are made over a set of qualitatively quite different images.

We have thus found strong evidence that even though the fractal coder provides image-specific codebooks, it is not superior to using a fixed codebook in a coder with the same structure.

2.5 Discussion

The fractal coder examined in this chapter does not have all the refinements known from the literature on fractal image coding. The most immediate extension would be to allow for different shapes and sizes of the range blocks. This has been a feature of fractal image coders ever since the first practical method was presented. Another refinement is the use of position-dependent codebooks. This was originally used for reduction of encoder complexity, but more recently it has been applied to improve the compression/quality performance.

Recall that the image output from a fractal decoder is the fixed point x_T of an image mapping T, expressed blockwise as in Equation 2.3. This expression needs not be modified to cater for different block geometries and position-dependent codebooks. The blocks in any subdivision of an image can still be numbered (1 through N), the difference is that the codebook is no longer fixed but depending on the block number n,

$$\begin{aligned} \mathcal{C}_n &= \{OS_{k_n}F_{l_n}x_T : k_n = 1,\ldots,K_n, l_n = 1,\ldots,L_n\} \\ &= \{A_{i_n}x_T : i_n = 1,\ldots,I_n\}\,. \end{aligned}$$

Here, the 'fetch' operators $\{F_{k_n}\}$ that provide domain blocks may change with the range block number n. (For completeness, we also let the 'shrink and shuffle' operators $\{S_{l_n}\}$ vary with the range block.) Such a dependence is also possible with an M-VQ. The output image is still expressed as in Equation 2.1. The block-dependent codebook is

$$\mathcal{C}_n = \{a_{i_n} : i_n = 1,\ldots,I_n\}\,.$$

Below we remark briefly on how the extra degrees of freedom are utilized for fractal coding and how corresponding M-VQ's might be constructed. It would be interesting to find out whether the refinements are equally beneficial to the two coders, i.e. whether the M-VQ continues to be superior.

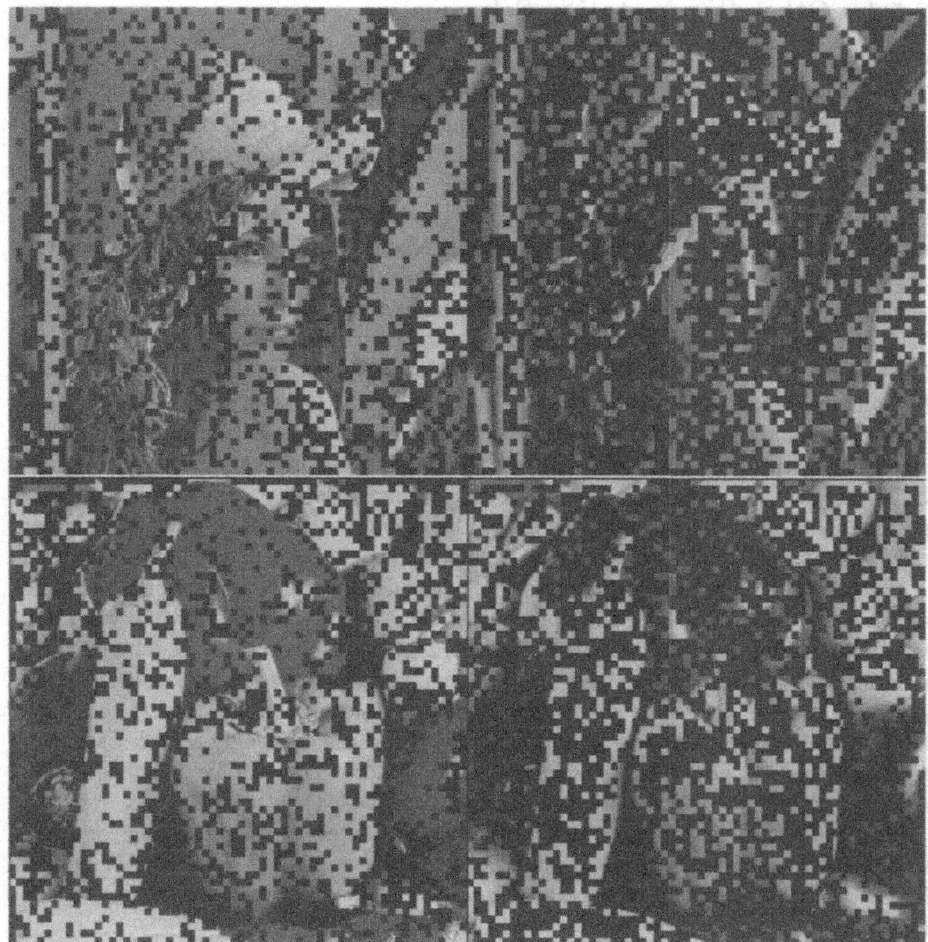

Figure 2.14: Blocks reproduced best by either method. M-VQ to the left, fractal coding to the right. For *Lena*, the M-VQ was superior for 2658 out of 4096 blocks. For *Peppers*, the M-VQ was superior for 2573 out of 4096 blocks. There are 3000 blocks in the codebook, the block sizes are 8×8.

2.5.1 Other Segmentation Schemes

Segmentation is the task of finding the positions, shapes and sizes of the range blocks. The objective is to find a subdivision of an image that allows for a small collage error. One might imagine a very large number of eligible range block geometries, but for two reasons the constraints probably have to be rather severe. Firstly, the number of bits required to encode the block geometries grows with the number of eligible sizes, shapes and positions, and the block transformations would have to be represented by a correspondingly lower number of bits. Secondly, a large number of alternative block geometries implies very complex encoding, as some kind of search seems necessary. Some of the better known schemes are parent–child segmentation [123, 125], quadtrees [84], HV-partitions [84, 85], triangulation [58, 84], and pruning [240]. Most of these are recursive in the sense that the segmentation is gradually refined by splitting blocks into smaller ones. The method of pruning is a sort of region-growing technique that starts with small range blocks and merges adjacent blocks wherever acceptable. This method may result in rather intricate block geometries. It is interesting to note that the paper on pruning reports on compression/quality performance inferior to the results obtained using HV-partitions, i.e. only rectangular range blocks. This indicates that block geometries should perhaps not be too complex.

For an M-VQ with image segmentation, one codebook must in principle be trained and stored for each block size and shape. With some of the schemes known to work well for fractal coding (e.g. HV-partitioning), this may require too many codebooks for a practical vector quantizer. To meet this problem, perhaps some block geometries might share codebooks, with some simple geometrical transformations to convert codebook blocks from one shape or size to another. However, an increasingly complex segmentation scheme will be increasingly difficult to apply for a VQ with fixed codebooks. In a fractal coder, codebooks continue to be easily available.

It is not clear whether the codebook-adaptivity offered by fractal coding continues to be inferior to non-adaptive VQ when segmentation techniques are used. We believe however that the problems of fractal coding with repeated patterns and textures in the codebook do not disappear.

2.5.2 Position-Dependent Codebooks

If the codebook for one range block is different from the codebooks used for other range blocks of the same shape and size, we shall say that it is position-dependent. Position-dependent codebooks were first used to reduce the complexity of a fractal encoder [125]. The image to encode was divided into four subimages, and each range block was sought approximated searching only domain blocks within the subimage of the range block. This was done under the assumption that the best domain blocks would be found close to the range block. Later, Fisher demonstrated that there is no preference for domain blocks close to the range block, i.e. that the difference in position between a range and its best choice for domain is distributed as the distance between two randomly chosen points in a square [82].

Still, it has been maintained that it can be beneficial to construct codebooks letting distances between domains be as small as one pixel and limiting the codebook to domains close to the range. The result by Fisher has been said to be true only if the domains are taken from a grid with some distance between each domain block [24]. The authors have not seen this result published yet, but we take it as an indication that fractal coding may perhaps gain from having position-dependent codebooks.

Vector quantizers may employ changing codebooks by introducing *memory*. This means that the block reproduced by the decoder depends not only on the current received codeword but also on the past history of the codewords received by the decoder [97, Chapter 14]. Examples of VQs with memory are finite-state VQ and predictive VQ. We believe that these techniques provide less arbitrary codebook constructions than do fractal coders with position-dependent codebooks.

2.6 Conclusion

We have studied the codebook population in a fractal image coder. It is shown that by using domain blocks from another image than the one being coded, the PSNR of the collages can be often be improved or left unchanged. Thus one may say that cross-transformability between images is often as strong or stronger than self-transformability. It is also shown that fixed codebooks are superior to codebooks populated by self-transformation (in a series of tests and with the restricted coders considered here.) We take these results as a strong indication that self-transformation is not a good method for codebook adaptation.

Acknowledgements

Skjalg Lepsøy was supported by a postdoctoral grant from the Norwegian Research Council.

The authors wish to thank the participants of the NATO Advanced Study Institute on Fractal Image Encoding and Analysis for stimulating discussions and viewpoints that improved this article.

Chapter 3

On the Use of Subsampling in Fractal Image Compression

Lars Lundheim

3.1 Introduction

A central step in fractal image compression is the decimation done when going from a large domain to a smaller range. Usually this decimation is performed by averaging a group of adjacent pixels. In this paper we consider the alternative approach of subsampling.

In Section 3.2 we briefly recall the fundamental concepts of fractal image compression and present the notation we will use. For a more comprehensive treatment of the fundamentals the excellent introduction by Fisher [162, Chapters 1–3] is recommended.

A preliminary discussion of the decimation alternatives is given in Section 3.3, including an example of a signal which is best modelled using decimation.

The family graph is then introduced. This is a tool to classify the pixels in such a way that their values can be easily computed and eventual contractivity can be tested for. This is demonstrated in Sections 3.5 and 3.9.1 respectively. The complexity of the synthesis algorithm is discussed in Section 3.6.

By an example it is found that the quality degradation from using subsampling instead of decimation is not necessarily large. This is shown in Section 3.9.1.

The paper is not meant to cover all aspects of subsampling in fractal compression. Most of the results presented are based on [163]. The pixel chasing method was found in [162]. Other discussions of subsampling versus decimation have been done by Lepsøy [149], Øien [198], and Baharav et al. [162, Chapter 5].

3.2 Notation and Terminology

Throughout the paper we will use the term *Fractal image compression* to denote methods falling into the very broad class described in [162, Chapter 1], based on "partitioned iterated function systems". In these methods the image is modelled as the fixed point of some transform T on the image space. In this paper we will follow the practice from [162, Chapters 7–9] where $\mathcal{R}^M$ denotes our image space, and all images are represented as one-dimensional column vectors assuming some convenient ordering of the pixels.

Essential to the workings of the transform is the partitioning of the image into a set of *ranges* (or *range blocks*) which are encoded separately from each other. To each range a larger *domain (block)* is assigned. The transform T consists of several sub-transforms — one for each range — and maps the domain to the range by applying in turn a *decimation* giving the domain the same size as the range, an *isometric operation* such as a rotation or a reflection, and finally a non-linearity. In all known cases the non-linear part consists of an affine transform of the form

$$T_n\mathbf{x} = a_n\mathbf{x}_n + \mathbf{t}_n \tag{3.1}$$

where a_n is called the *scaling* and $\mathbf{t}_n$ is called the *offset* or *translation* of sub-transform T_n. Throughout the paper lower case boldface symbols without index ($\mathbf{x}$) denote full images, with an index ($\mathbf{x}_n$) they denote range blocks. Pixels are denoted as lower case symbols with an index. Thus, $\mathbf{x} = [x_1, x_2, \ldots, x_M]^T$. As has been shown e.g. in [162, Chapter 7] the offsets can be collected into a vector $\mathbf{t} \in \mathcal{R}^M$ and the rest of the transforms represented by an $M \times M$ matrix A called the *linear part* of T as follows:

$$T\mathbf{x} = A\mathbf{x} + \mathbf{t}. \tag{3.2}$$

Provided all eigenvalues of A are different from 1 (see [163, Chapter 2]) T will have a unique fixed point $\mathbf{x}^0 = T\mathbf{x}^0$. Fractal *encoding* consists of finding the transform from a predefined class, which has the fixed point yielding the best approximation to a given image $\mathbf{x}$. This involves searching through a collection *(the domain pool)* of possible domains for each range. The values of the scaling and the offset are usually determined by a least squares method. The description of T may then be called a *fractal model* of $\mathbf{x}$. Hopefully this represents less information than $\mathbf{x}$ itself, and data compression is obtained.

When the model is known, the fixed point can be reconstructed by a matrix inversion

$$\mathbf{x}^0 = (I - A)^{-1}\mathbf{t}. \tag{3.3}$$

This reconstruction is called *decoding* and is usually performed by iterating T on some arbitrary image $\mathbf{x}_{\text{start}}$. This requires that T is *eventually contractive* in which case

$$\mathbf{x}^0 = \lim_{n\to\infty} T^n\mathbf{x}_{\text{start}}. \tag{3.4}$$

We will denote iterations of operators by superscript on the operator symbol, e.g. $T^2\mathbf{x} = T(T\mathbf{x})$.

Thus, a rendering as close as desired to $\mathbf{x}^0$ can be found by iterating T a high enough number of times. When $\mathbf{x}_{\text{start}}$ is chosen close to $\mathbf{x}^0$ the iteration will converge faster. This is obvious for strictly contractive operators, but is empirically valid for eventually contractive ones as well.

3.3 Decimation by Subsampling

Several parameters distinguish the various fractal image compression schemes existing, such as the way the image is partitioned into range blocks, the choice and classification of the domain pool, decoding methods etc. In this paper we will pay attention to one particular parameter which can be changed independently of the others. We will take a close look at how the decimation is done when reducing the size of the domains to fit the ranges. Usually this is done by averaging a group of adjacent pixels, as shown in Figure 3.1 b.

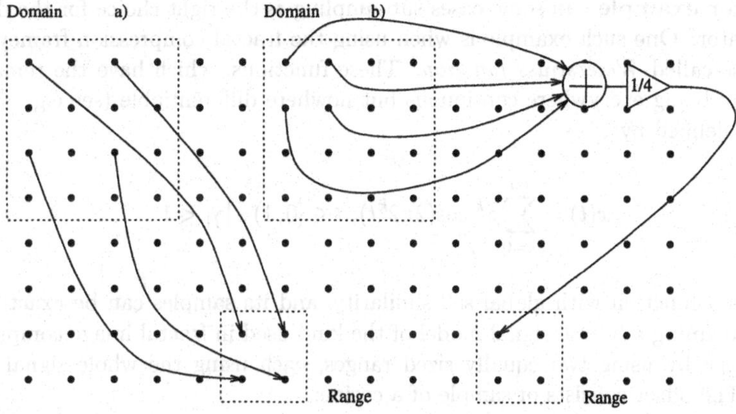

Figure 3.1: Two ways of decimating a domain block: a) Subsampling; b) Averaging. For this example the domains are in both cases four times as big as the ranges.

An alternative approach is to use *subsampling* (or point sampling) where only one point is used from each domain (Figure 3.1 a).

Already now we see that subsampling is in principle a simpler operation that averaging since it involves fewer arithmetic operations. However, this is only a small simplification compared to what can be gained by using the non-iterative method presented later on.

One argument in favour of averaging should be mentioned. Assume that a decimation by four ($D \times D$ blocks are shrinked to $\frac{D}{2} \times \frac{D}{2}$). Let us further assume that the original unsampled image possesses self similarity in the sense that range blocks *can* be peferctly modelled by shrinked and transformed range blocks. (For example a scene with similar-looking objects in different distances from the camera.) If this were the case, the decimation operator should perform a low-pass filtering, ideally removing all spatial frequency components above $0.5f_c$ where f_c is the spatial Nyquist frequency for each direction. If such a filtering is not made, unwanted "aliasing" frequency components, not present in the original will show up corrupting the image. Using averaging in the decimator corresponds, indeed, to a crude low-pass filtering, which will remove most of the aliasing components. One should, however, not overestimate the damage caused by aliasing. Many images have a strong low-pass shape of their frequency content, and in these aliasing artefacts are often not visible when the images are downsampled at modest rates, such as four.

Researchers investigating fractal image compression soon find that the quality of the decoded image using subsampling is somewhat inferior to the results when averaging is used. Furthermore, using the iterative decoding method, slow convergence, or even divergence is often observed. Having a lot of other parameters to tune, they then usually reject subsampling from further study. As we shall see in the rest of this paper, the first shortcoming is not necesserily a problem, and the second one can be overcome altogether.

A Counter-example In some cases subsampling *is* the right choice for the decimation operator. One such example is when using the fractal compression framework to model a so-called *Weierstrass function.* These functions, which have the remarkable property of being everywhere continuous but nowhere differentiable (see e.g. [229, p. 329]), are defined by

$$x(t) = \sum_{k=0}^{\infty} \gamma^k \cos(2\pi 2^k t);\ t \in [0,1)\,,\ |\gamma| < 1. \tag{3.5}$$

This is a function with global self-similarity, and its samples can be exactly computed by defining a fractal signal model of the kind used in fractal image compression. This is done by using two equally sized ranges, each using the whole signal as the domain. The offset consists of sample of a cosine:

$$t_i = \cos\left[2\pi\left(\frac{i-1}{M}\right)\right]. \tag{3.6}$$

It is not difficult to see that this works from the definition of $x(t)$, and a formal proof may be found in [149].

However, the decimation must be done by subsampling! The result is shown in Figure 3.2 a). In b) averaging is used, and we easily see how the function has been smoothed by the inherent lowpass filtering in the decimation.

The chosen example might seem a bit artificial, but it illustrates that the parameters — including the decimation method — of a fractal compression method must be chosen according to the class of signals (images) that are to be coded.

3.4 The Family Graph

We will now introduce a new tool which both gives insight in the mathematical properties of fractal models with subsampled domains and prepares the ground for a fast, non-iterative synthesis method. Let us first look at an extremely simplified example of an image with only $M = 12$ pixels, and with range and domain blocks as shown in Table 3.1.

With scalings as in Table 3.1 and decimation by subsampling the linear part A will

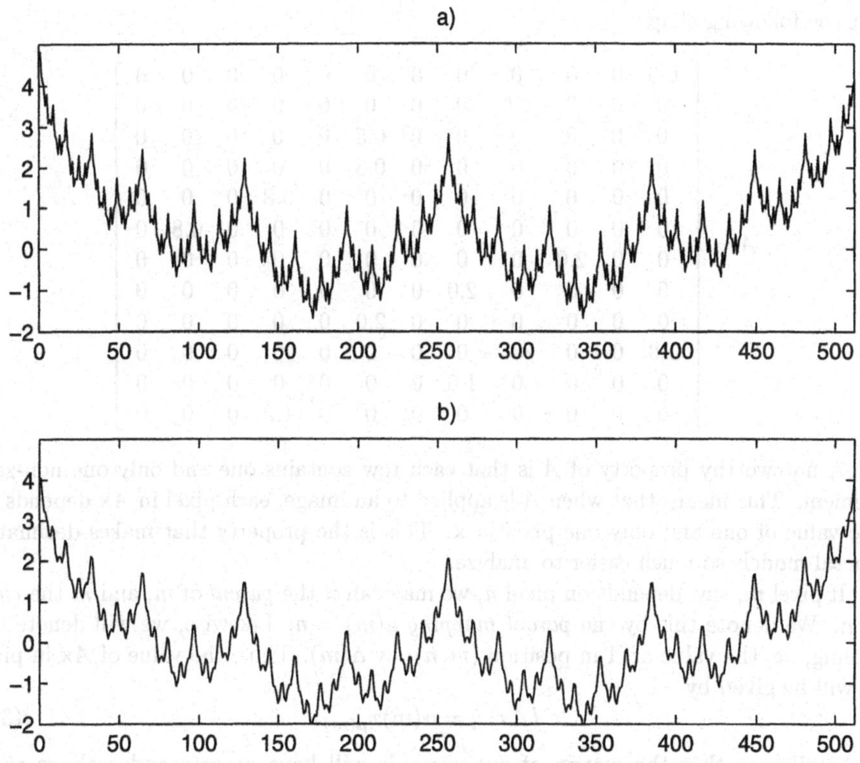

Figure 3.2: Weierstrass function with $\gamma = 0.8$ modelled with a) subsampling, b) averaging.

n	Range block n	Domain block n	Scaling a_n
1	1, 2, 3	$1, 2, \ldots, 9$	0.5
2	4, 5, 6	$7, 8, \ldots, 12$	0.8
3	7, 8, 9	$3, 4, \ldots, 8$	2.0
4	10, 11, 12	$1, 2, \ldots, 12$	1.0

Table 3.1: Range blocks with assigned domain blocks and scalings.

get the following shape

$$A = \begin{bmatrix} 0.5 & 0 & 0 & 0 & 0 & 0 & 0 & 0 & 0 & 0 & 0 & 0 \\ 0 & 0 & 0 & 0.5 & 0 & 0 & 0 & 0 & 0 & 0 & 0 & 0 \\ 0 & 0 & 0 & 0 & 0 & 0 & 0.5 & 0 & 0 & 0 & 0 & 0 \\ 0 & 0 & 0 & 0 & 0 & 0 & 0.8 & 0 & 0 & 0 & 0 & 0 \\ 0 & 0 & 0 & 0 & 0 & 0 & 0 & 0 & 0.8 & 0 & 0 & 0 \\ 0 & 0 & 0 & 0 & 0 & 0 & 0 & 0 & 0 & 0 & 0.8 & 0 \\ 0 & 0 & 2.0 & 0 & 0 & 0 & 0 & 0 & 0 & 0 & 0 & 0 \\ 0 & 0 & 0 & 0 & 2.0 & 0 & 0 & 0 & 0 & 0 & 0 & 0 \\ 0 & 0 & 0 & 0 & 0 & 0 & 2.0 & 0 & 0 & 0 & 0 & 0 \\ 1.0 & 0 & 0 & 0 & 0 & 0 & 0 & 0 & 0 & 0 & 0 & 0 \\ 0 & 0 & 0 & 0 & 1.0 & 0 & 0 & 0 & 0 & 0 & 0 & 0 \\ 0 & 0 & 0 & 0 & 0 & 0 & 0 & 0 & 1.0 & 0 & 0 & 0 \end{bmatrix}$$

A noteworthy property of A is that each row contains one and only one non-zero element. This means that when A is applied to an image, each pixel in $A\mathbf{x}$ depends on the value of one and only one pixel in $\mathbf{x}$. This is the property that makes decimated fractal models so much easier to analyze.

If pixel m, say, depends on pixel n, we may call n the *parent* of m, and m the *child* of n. We denote this by the *parent mapping* $\mu(m) = n$. Likewise, we will denote the scaling, i.e. the value of A in position (m, n), by $\alpha(m)$. Thus, the value of $A\mathbf{x}$ in pixel m will be given by

$$(A\mathbf{x})_m = \alpha(m)x_{\mu(m)}. \tag{3.7}$$

We easily see that the matrix of our example will have parents and scalings as in Table 3.2.

m	1	2	3	4	5	6	7	8	9	10	11	12
$\mu(m)$	1	4	7	7	9	11	3	5	7	1	5	9
$\alpha(m)$	a_1	a_1	a_1	a_2	a_2	a_2	a_3	a_3	a_3	a_4	a_4	a_4

Table 3.2: Parents and scalings for the example operator.

A more revealing way of representing the information of Table 3.2 is to draw a diagram showing how the pixels depend on each other. This gives the graph in Figure 3.3. Each node represents a pixel and the arrow points from parent to child.

Note that the graph consists of two disjoint parts; one containing pixel 1 and 10, and one containing the rest. We say that the graph consists of two distinct *families* of pixels. Observe also that as we trace the family line from an arbitrary pixel m through its chain of ancestors $m, \mu(m), \mu(\mu(m)), \ldots$ we sooner or later end up in a loop where a collection of pixels are their own ancestors. Such a collection we will call the *cycle* of the family. We say that the pixels of the cycle are family members of generation 0, their children of generation 1, and so on. The number of pixels in a cycle will be called the *period* of the cycle.

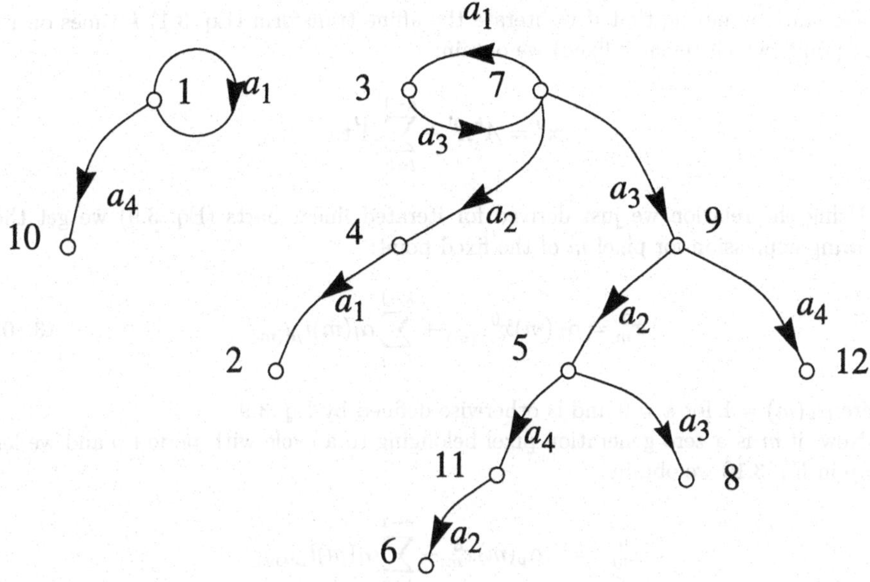

Figure 3.3: The family graph of the linear part A.

It is not difficult to realize that these properties of the family graph all follow logically from the fact that the matrix A has one and only one non-zero element in each row — which again is due to the use of subsampling. For a more formal treatment with all the necessary proofs see [163].

What happens when we iterate the linear part? By applying twice Eq. 3.7 we see that pixel m of $A^2\mathbf{x}$ must be equal to $\alpha(m)\alpha(\mu(m))\mu(\mu(m))$. More generally, A iterated k times is itself a matrix with one and only one non-zero elements in each row, and defines a new parent mapping

$$\mu^k(m) = \overbrace{\mu(\mu(\cdots\mu(m)\cdots))}^{k\text{times}} \tag{3.8}$$

and has scalings

$$\alpha_k(m) = \prod_{l=0}^{k-1} \alpha(\mu^l(m)). \tag{3.9}$$

where the convention $\mu^0(m) = m$ is used to simplify the notation.

3.5 An Efficient Synthesis Algorithm

We shall now see how using the family graph gives a new, non-iterative synthesis method when subsampling is used. In particular, the zero generation pixels will play a crucial role.

We start by noting that if we iterate the affine transform (Eq. 3.1) k times on its fixed point (which remains fixed) we obtain:

$$\mathbf{x}^0 = A^k \mathbf{x}^0 + \sum_{l=0}^{k-1} A^l \mathbf{t}.$$

Using the relation we just derived for iterated linear parts (Eq. 3.9) we get the following expression for pixel m of the fixed point:

$$x_m^0 = \alpha_k(m) x_{\mu^k(m)}^0 + \sum_{l=0}^{k-1} \alpha_l(m) t_{\mu^l(m)} \tag{3.10}$$

where $\alpha_k(m) = 1$ for $k = 0$ and is otherwise defined by Eq. 3.9.

Now, if m is a zero generation pixel belonging to a cycle with period p and we let $k = p$ in Eq. 3.10 we obtain

$$\begin{aligned} x_m^0 &= \alpha_p(m) x_m^0 + \sum_{l=0}^{p-1} \alpha_l(m) t_{\mu^l(m)} \\ &= \frac{1}{1 - \alpha_p(m)} \sum_{l=0}^{p-1} \alpha_l(m) t_{\mu^l(m)}. \end{aligned} \tag{3.11}$$

Thus, we have found an explicit formula for the value of the fixed point in the zero generation pixels. What about the other pixels?

By setting $k = 1$ in Eq. 3.10 we find that

$$x_m^0 = \alpha(m) x_{\mu(m)}^0 + v_m. \tag{3.12}$$

By this equation pixel m can be computed exactly if its parent $\mu(m)$ is known. Actually, we need only to use Eq. 3.10 on *one* pixel of each cycle. Then the rest of the family can be computed by applying Eq. 3.12 repeteadly along each family line originating from the cycle in question.

In Figure 3.4 pseudo code is outlined for an algorithm which computes $\mathbf{x}^0$ in the manner described above. Mind the notation μ^{-1} which is used to denote the child of a pixel. In general this is not unique, but in the algorithm one keeps track of one family line at a time going from an arbitrary pixel through a chain of older and older ancestors till one ends up in a cycle. Whithin this chain of pixels the child mapping μ^{-1} is unique.

3.6 Complexity

For each family the outlined algorithm must perform two kinds of arithmetic computations. First, it must apply Eqs. 3.10 and 3.9 on one zero generation point. If the cycle has period P this takes P mulitplications, P additions, and one division. For the

```
While there are still unknown pixel values do {
    Choose an unknown pixel m_0
    Set m = m_0
    While μ(m) is not marked as visited do {
        Mark μ(m) as visited
        Set m = μ(m)
    }
    if the pixel value x_μ(m) is not known then {
        A cycle is found!
        Compute pixel value x_μ(m) using Eq. 3.10
        Mark pixel μ(m) as known
    }
    Until m = m_0 do {
        Set x_m = α(m)μ(m) + t_m
        Set m = μ^{-1}(m)
        Mark m as known
    }
}
```

Figure 3.4: Pseudo code for fast synthesis algorithm.

rest of the points Eq.3.12 is used, which amounts to one multiplication and addition per pixel. Except for the zero generation pixels this complexity corresponds to one iteration of the mapping T to the whole image. How much extra does computation of the zero generation pixels amount to? That depends on the number of cycles and their periods, and will of course be different for different encodings. In [163, Chapter 6] we have shown that the expected value of P for a cycle grows at most as the square root of the image size. This is also confirmed by statistics taken from a limited ensemble of encodings. Figure 3.5 shows the distribution of period length and number of families in the family graph of 54 different encodings of 512×512 images.

The period length has a mean of 13.5 and standard deviation 31, whereas the number of families has mean 12 and standard deviation 9.1. For a similar set of encodings of 256×256 images the mean number of cycles was 12. Altogether this shows that the extra computations due to the cycles is negligible compared to the computations of the higher generation pixels.

We may conclude that the computational complexity of the proposed method corresponds roughly to one iteration of the mapping T. Then we have only counted the arithmetic operations. In addition comes the overhead due to searching the family graph.

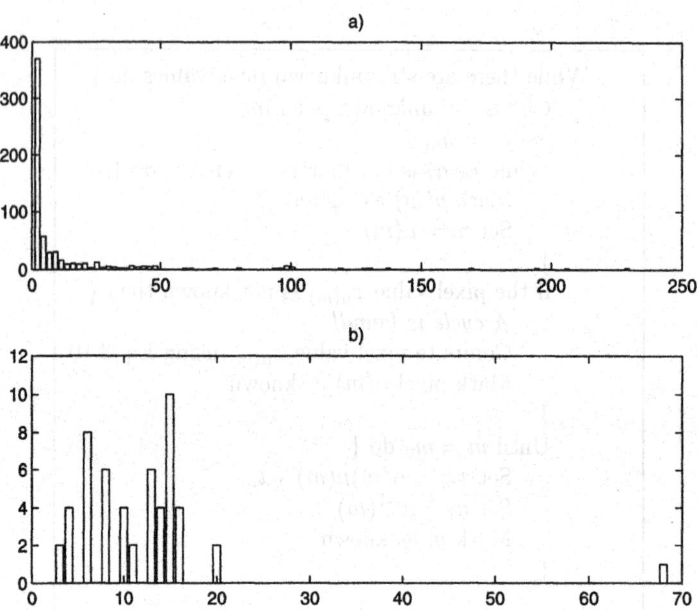

Figure 3.5: Statistics on family graphs from 54 different encodings of 512 × 512 images. a) Distribution of cycle periods. b) Number of families in one graph.

3.7 An Alternative Algorithm

The outlined method above was first described by the author in [163]. Independently a very similar approach, called *pixel chasing* has been suggested by Fisher [162, p. 305].

In this method one starts with an arbitrary image $\mathbf{x}^{(0)}$, corresponding to $\mathbf{x}_{\text{start}}$ of Eq. 3.4). Starting in an arbitrary pixel $x_m^{(0)}$ of this image, one then computes in sequence

$$\begin{aligned} x_m^{(1)} &= \alpha(m)x_{\mu(m)}^{(0)} + t_m \\ x_{\mu(m)}^{(1)} &= \alpha(\mu(m))x_{\mu^2(m)}^{(0)} + t_{\mu(m)} \end{aligned}$$

providing first estimates for the pixel values of the pixels m and $\mu(m)$. Now, better estimates can be found by

$$x_m^{(2)} = \alpha(m)x_{\mu(m)}^{(1)} + t_m$$

and so on, "chasing" the pixels along the family line going backwards from x_m until a cycle is found. When $x_m^{10}, x_{\mu(m)}^{10}, \cdots$ (corresponding to 10 "traditional" iterations of T) are found, a good enough approximation to the fixed point can be assumed to have been reached, and the pixels along the family line are marked as "known". When the procedure is repeated using another starting pixel, say m', there is a certain probability that a "known" pixel is encountered before a cycle is encountered. In this case the rest of the pixels can be computed in the same way as for the non-iterative method presented in this paper. As more and more pixels are marked "known" the probability of finding "old" families increases, and Fisher has found that an average of 3 operations of type 3.12 are necessary to reconstruct the whole image. This is three times more than the non-iterative method. However, it may happen that the overhead is less for pixel chasing. This depends on the hardware used for the implementation. One should also note when comparing the two methods that slow convergence might be a problem for pixel chasing, and that 10 iterations may be insufficient for some pixels.

3.8 Empirical Results

Fast decoding methods have been proposed earlier by others, e.g. [162, Chapter 5] and [162, Chapter 8]. Common to these are that they pose restrictions on the affine operator used. The method of this paper is on the other hand very general. As far as we can see, any fractal coding scheme can take advantage of the "family graph" method by substituting averaging for subsampling when domain blocks are formed. However, it is well known that subsampling leads to somewhat lower PSNR. Exactly how much depends on the setting of the other parameters of the method and the properties of the image. Whether it is worth while to sacrifice quality to decoding speed will of course depend on the application.

It is actually possible to exploit some of the speed-up from subsampling while retaining the superior quality of averaging. Assume an image has been encoded using averaged domains. At the decoding, one first uses the non-iterative method — as if subsampling had been used. This will produce an approximation to the image which can be used as a starting point ($\mathbf{x}_{\text{start}}$ of Eq. 3.4) for the traditional iterative method, resulting in faster convergence than when using an arbitrary starting point. A similar approach has actually been proposed *whithout* the fast non-iterative method, but using subsampling in the first iterations to avoid the extra computation involving averaging [162, Chapter 6].

In this section we will study the effect of using subsampling in one particular approach, the quadtree method by Fisher. The algorithm is explained in detail in [162, Chapter 3]. Basicly it first divides the image into a number of equally sized square range blocks. Then it tries to find an encoding for these. If sufficient quality is not obtained for some of the blocks, they are further subdivided into four new squares. The process is repeated recursively until satisfactory quality is achieved for all blocks, or a minimum allowed block size is reached. The method has several parameters regulating the choice of domain pool, classification of the domains etc. The results below were obtained by using the code in [162]. A single statement was changed in the encoder so that subsampling was used instead of averaging. The parameters of the program were chosen to be the "default values" for 512×512 images recommended in [162]. Postprocessing of decoded images was used. Quality measured by PSNR is shown for various compression rates in Figure 3.6.

In addition to the standard apporach using averaging in both encoder and decoder (marked 'A' in the figure), two variants were studied. First we let the encoder use subsampled domaines and the decoder use the non-iterative method. The curve marked

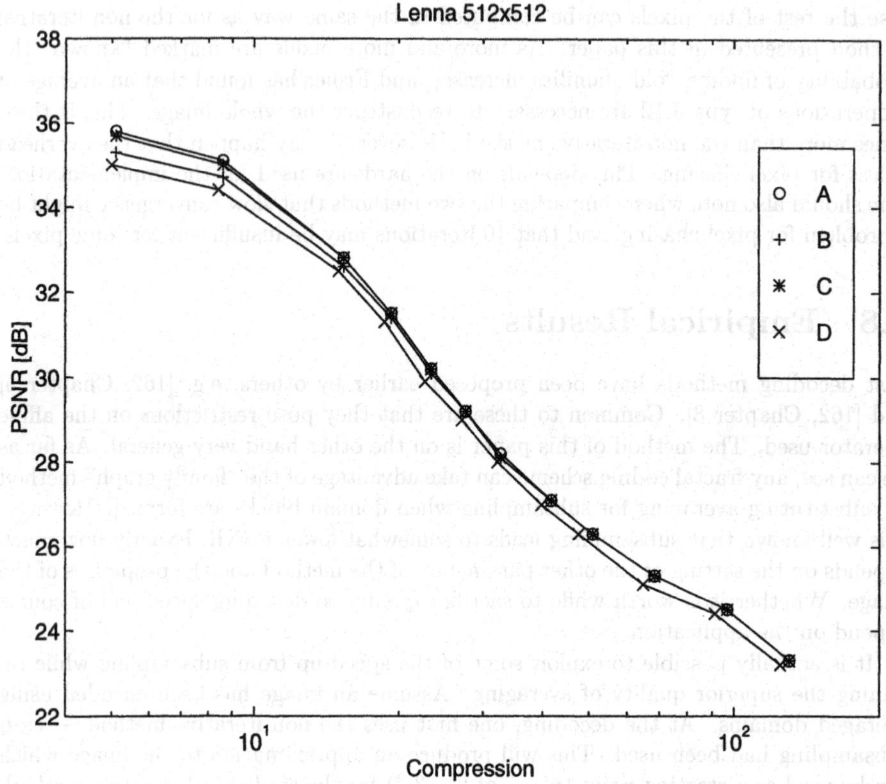

Figure 3.6: Quality vs compression for 512 × 512 Lena. A: Averaging used both in encoding and decoding. B: Averaging used in encoding. Decoding with subsampling, then 1 iteration using averaging. C: As B but using two iterations. D: Subsampling used both in encoding and decoding.

'D' shows the resulting quality. Then we let the decoder operate on encodings using averaging, and used this as a starting point for traditional decoding. The curves marked 'B' and 'C' show the obtained quality after one and two subsequent iterations respectively. With three iterations the PSNR became identical to the one obtained by the traditional method (which used 10 iterations).

When comparing the two 'pure' alternatives 'A' and 'D' we see that for low compression we have a difference of the order of 1 dB (0.8 in this example). At higher compression this difference is smaller.

Subjectively the images resulting from the two methods are very similar. Even when the PSNR difference is at its highest it is not easy to distinguish the two methods. This is seen in Figure 3.8.

3.9 Contractivity

We have noted that the two drawbacks of using decimation is lower PSNR and possibly slower convergence. Some light may be shed on this if we look at the Lipschitz constant of the transform T when the two different types of decimation are used.

Recall that the Lipschitz constsant s is a number that roughly describes how much T reduces ($s < 1$) or expands ($s > 1$) the distance between two images when T is applied to both of them. If $s < 1$ the transform is said to be *(strictly) contractive,* and if the Lipschitz constant for the iterated mapping T^n is strictly less than one for some number n, T is *eventually contractive.* For more details and formalism see [162, Chapter 2].

It was shown in [162, Chapter 7] that when the domain pool consists of non-overlapping blocks of size $D \times D$, and if all ranges have size $R \times R$, the Lipschitz constant for the two cases is expressed by

$$s = \max_{d=1,2,\ldots} \sqrt{\sum_{n \in \mathcal{R}_d} a_n^2} \tag{3.13}$$

when subsampling is used, and

$$s = \frac{R}{D} \max_{d=1,2,\ldots} \sqrt{\sum_{n \in \mathcal{R}_d} a_n^2} \tag{3.14}$$

for averaging. In the equations $\mathcal{R}_d$ denotes the set of all indices n such that range block $\mathbf{x}_n$ uses the d'th domain block. In words Eqs. 3.13–3.14 mean that to find the Lipschitz constant for a transform using subsampling, one must for each domain block find all the ranges using it. Then one computes the root of the sum of the squares of the scalings for these domains. When all domains are checked, the maximum of the found values gives the Lipschitz constant. For averaging the formula is the same, except that it is scaled by the ratio R/D.

From the formulas we may infer that if some domain blocks are particularly popular, this may increase the Lipschitz constant because the sum of squares will have more terms. Furthermore, we may expect a smaller Lipschitz constant when averaging is

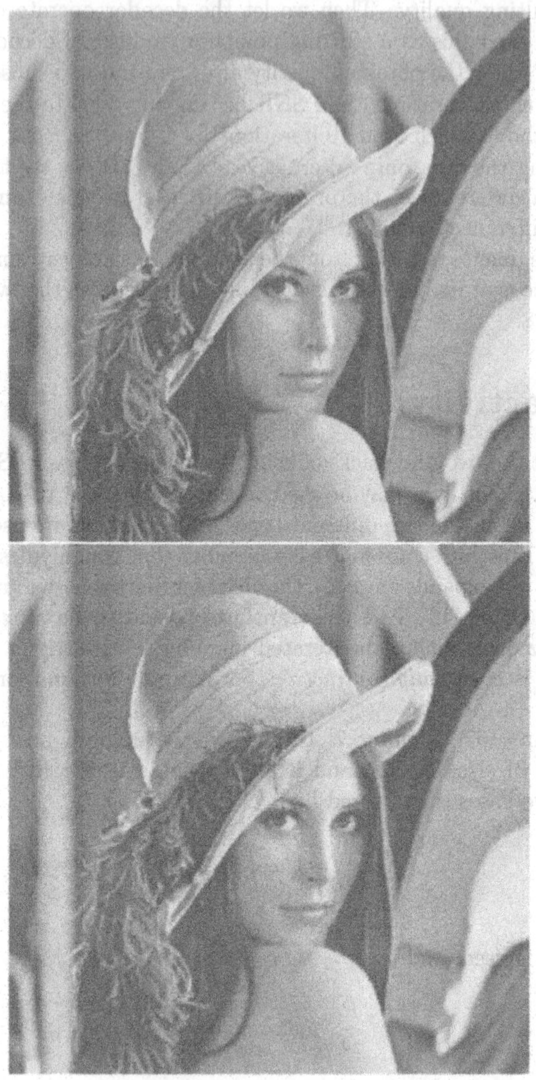

Figure 3.7: The 512 × 512 Lena image coded with averaging, PSNR = 35.8 dB, (top) and subsampling, PSNR = 35.0 dB (bottom). Both images have compression 5.1.

used, particularly if the ranges are much smaller than the domains. This may give one explanation why averaging results in faster convergence than subsampling.

A smaller value of s is also a possible reason for the observed difference in PSNR. In the search for optimal T, fractal encoders normally do not directly optimize the distance between the original signal $\mathbf{x}$ and the fixed point $\mathbf{x}^0$. Instead one looks at the difference between $\mathbf{x}$ and the so-called *collage* $T\mathbf{x}$ (which may be considered a first order approximation to $\mathbf{x}$). That this indeed works is shown by one of the many *collage theorems*. All of these, however, assumes that T is at least eventually contractive. If a small distance between $\mathbf{x}$ and $T\mathbf{x}$ is obtained by the encoder, one can in general say that the result at the decoder will be better the smaller s is.

The above discussion has obviously not universal validity. It is based on the asumption that all range blocks have the same size and that domain blocks are non-overlapping. Moreover since the two expressions in Eqs. 3.13 and 3.14 depend on different choices of domains they cannot be directly compared. However, the two formulas should indicate an average tendency in favour of more contractive operators in the averaging case.

3.9.1 Test of Eventual Contractivity

In the previous section we looked at the contractivity of the operator T, as expressed by its Lipschitz constant. For the subsampling case it is also easy to determine if T is *eventually* contractive, which is the ultimate criterion both for using a collage theorem and for an iterative decoding method.

In [163] it is shown that when subsampling is used, eventual contractivity is equivalent to the linear part A having all its eigenvalue inside the unit circle of the complex plane. It is also shown that fast convergence cannot be expected if any eigenvalue has a magnitude close to 1.

It is interesting to note that the collection of eigenvalues for A is closely related to the family graph defined earlier. It is shown in [163] that to each cycle of the family graph there corresponds a separate set of non-zero eigenvalues. The number of eigenvalues in each set is equal to the period of the corresponding cycle. For a cycle with period P the eigenvalues are expressed by

$$\lambda_p = \rho e^{j2\pi p/P}; p = 0, 1, \ldots, P-1 \tag{3.15}$$

where ρ is given by the geomteric mean of the scalings of the cycle, i.e.

$$\rho = \left[\prod_{l=0}^{P-1} \alpha(\mu^l(m_0)) \right]^{\frac{1}{P}} . \tag{3.16}$$

Here m_0 is an arbitrary point in the cycle. Thus, eigenvalues belonging to the same cycle will be equally spaced on a circle about the origin of radius ρ. Eigenvalues not belonging to a cycle in this way are all zero.

3.10 Conclusion

Any fractal coding scheme can be modified by exchanging averaging for subsampling when domains are decimated. If this is done a decoding method is available corresponding to a computational load of one conventional iteration. The cost is a lower PSNR. In many cases the loss in PSNR is negligible, and in many applications where fast decoding is needed, subsampling is therefore to be preferred.

Acknowledgements

I am very grateful to Yuval Fisher for revealing his C code for the Quadtree method in [162] and making it available through WWW. This enabled me to make direct comparisons with a publicly "standard" reference method, and saved a lot of work.

Parts of this work was carried out at CERN, The European Laboratory for Particle Physics during a sabbatical leave from Sør-Trøndelag College. My thanks go to both institutions.

Chapter 4

On the Benefits of Basis Orthogonalization in Fractal Compression

Geir E. Øien and Skjalg Lepsøy

4.1 Introduction

This paper discusses the beneficial effects of incorporating an *orthogonalization operator* into the affine mapping which results from collage optimization in a typical fractal encoder, and which is used for attractor generation in the fractal decoder. We consider the coding problem for a general 2-dimensional signal, which is partitioned into *range blocks*, each of which to be approximated by a linear combination of a *decimated domain block*, taken from some part of the signal itself, and some *offset* block, which in general may itself lie in a signal subspace spanned by a set of *signal-independent* blocks, typically chosen to be mutually orthogonal for optimum computational and coding efficiency.

The introduction of an orthogonalization operator means that each decimated domain block is *orthogonalized with respect to the offset signal subspace.* This is to be done both during the collage optimization performed in the encoder, and in each iteration in the fractal decoding (attractor construction) process. We show that this mapping modification, under certain mild restrictions on the choice of parameters, has the following benefits:

- For a given signal resolution, exact decoder convergence towards the attractor is obtained in a minimum number of iterations. This number is signal independent and only depends on the sizes of the domain and range blocks.

- The decoder may be given an effective "pyramid" structure in which successively finer resolutions are decoded by using only one iteration to go from one resolution to the next.
- A noniterative decoder is included as a special case.
- The collage and attractor remain unchanged under the modification done to the mapping.
- The domain block coefficients, the values of which in most papers are constrained due to anticipated convergence problems (most typically to $|\alpha| < 1$), can be quantized in an unconstrained fashion, e.g. by means of Lloyd-Max quantizers based on the statistics of truly l_2-optimal values. The coefficient optimization is also performed in a computationally simpler way.
- The coefficients of the offset subspace and the orthogonalized domain block can be made to become *mutually uncorrelated* if the signal is a sample from a stationary random process (and empirically almost uncorrelated for practical signals). This is beneficial when using scalar quantization.
- Efficient clustering methods reducing the complexity of the search through the set of possible domain blocks may be designed.
- It becomes easy to show a much more restrictive collage theorem than the ones which have been traditionally used.
- It becomes easier to discuss good bases for the offset subspace, as well as to draw direct analogies between fractal compression and more traditional coding methods such as vector quantization.

4.2 Introducing Basis Orthogonalization in the Fractal Signal Model

Many discrete-time signal processing applications involve splitting a signal into subvectors, and then decomposing each subvector x into separate components in a linear space. In such cases it is mostly desirable — e.g. for reasons of computational complexity or coding efficiency — to use an *orthogonal* basis for this space. By this is of course meant that the *inner product* $\langle b_i, b_j \rangle$ between basis vector i and j, b_i and b_j, is *zero* for any $i \neq j$. In the N-dimensional Euclidean space, the inner product is commonly defined as $\langle x, y \rangle = x^T y = \sum_{n=1}^{N} x_n y_n$ where x_n and y_n are the nth samples of x and y respectively. This yields an inner product space where the l_2 *distance* $d(x, y) = \|x - y\|^{1/2} = \langle x - y, x - y \rangle^{1/2}$ is the natural metric.

For any given linear space, and an *arbitrary* given basis for this space, an orthogonal basis may be derived by applying the well known *Gram-Schmidt orthogonalization procedure* [145, 235]. This procedure both serves as a proof for the fact that every inner product space has an orthogonal basis (this is called *Gram-Schmidt's theorem* [93]), and as an algorithm for constructing it from an arbitrary basis.

In a typical fractal coding scheme (as originally suggested by Jacquin [123]), the signal to be coded is split into nonoverlapping blocks (either 1- or 2-dimensional), socalled *range* blocks. Each range block is then subject to a *collage approximation*, in which a linear combination of a decimated and amplitude scaled *domain* block, and a signal-optimized but invariant *offset* block, is formed. Forming this approximation (termed the *collage*) for all range blocks, we define an *affine* mapping $T : X \to X$ on the signal space X. This mapping has to be (strictly or eventually) *contractive* [145], since decoding of the signal is done by constructing the *attractor* of the mapping T.

The offset block subspace we, for the sake of generality, shall assume to be of dimension $K_n - 1$ for range block n. Here $K_n \geq 2$ and is assumed fixed for each given range block size. We may thus compute and store an ortho*normal* basis for the offset subspace once and for all. But the decimated domain block, which is properly regarded as a K_nth basis vector, is blockwise dependent on the signal we perform the mapping on. Therefore, the bases employed in the class of mappings T introduced by Jacquin are not completely orthogonal. We proceed to modify the mapping T to obtain orthogonality.

The linear term in the class of mappings introduced by Jacquin can be decomposed as

$$L = \sum_{n=1}^{N_r} \beta_{K_n}^{(n)} P_n D_n F_n \tag{4.1}$$

where

- N_r is the number of range blocks we divide the signal into;
- $F_n : X \to R^{D_n^2}$ fetches the best domain block (of length (number of samples) D_n^2);
- $D_n :\to R^{D_n^2} \to R^{B_n^2}$ decimates this domain block to length B_n^2, which is the length of range block n;
- $P_n : R^{B_n^2} \to X$ places the sum of the offset block and the transformed domain block in the position of range block n;
- $\beta_{K_n}^{(n)}$ is the amplitude scaling factor for the transformed domain block.

  We now want to make all the decimated domain blocks orthogonal to the offset subspace basis blocks. If the decimated domain block does not lie *within* the offset subspace (in which case it should be excluded from the domain pool, as it provides no information that is not contained within the offset subspace), this can always be obtained by means of the Gram-Schmidt procedure. In terms of linear algebra this corresponds to pre-multiplying each $D_n F_n x$ with an *orthogonalizing matrix*, $O_n : R^{B_n^2} \to R^{B_n^2}$, given as

$$O_n = I - \sum_{k=1}^{K_n - 1} b_k^{(n)} b_k^{(n)T} \tag{4.2}$$

where $b_1^{(n)}, \ldots, b_{K_n-1}^{(n)}$ are the orthonormal offset basis vectors chosen. The modified linear mapping can thus be written

$$L_O = \sum_{n=1}^{N_r} \alpha_{K_n}^{(n)} P_n O_n D_n F_n \tag{4.3}$$

and the total mapping is expressible as

$$T_O x = L_O x + t_O \tag{4.4}$$

where t_O is the modified offset term, which can be expanded as

$$t_O = \sum_{n=1}^{N_r} P_n \left(\sum_{k=1}^{K_n-1} \alpha_k^{(n)} \cdot b_k^{(n)} \right) \tag{4.5}$$

O_n can be interpreted as a *projection* onto the *orthogonal complement* of span$\{b_1^{(n)}, \ldots, b_{K_n-1}^{(n)}\}$, and hence possess the following useful properties [235]:

1. It is *symmetric*, i.e.

$$O_n^T = O_n \tag{4.6}$$

2. It is *idempotent*, i.e.

$$O_n^2 = O_n \tag{4.7}$$

3. It has *unit matrix norm*, i.e.

$$\|O_n\|_2 \stackrel{\text{def}}{=} \max_{\|x\|_2=1} \|O_n x\|_2 = 1 \tag{4.8}$$

4. For all vectors $v \in R^{B_n^2}$,

$$v^T O_n b_k^{(n)} = 0 \text{ for } k = 1, \ldots, K_n - 1 \tag{4.9}$$

The last property is exactly the orthogonality condition.

To sum up, the collage of x with respect to T_O can be written

$$\begin{aligned} T_O x &= L_O x + t_O \\ &= \left(\sum_{n=1}^{N_r} \alpha_{K_n}^{(n)} P_n O_n D_n F_n\right) x + \sum_{n=1}^{N_r} \left(\sum_{k=1}^{K_n-1} \alpha_k^{(n)} P_n b_k^{(n)} \right), \end{aligned} \tag{4.10}$$

where

- $O_n : R^{B_n^2} \to R^{B_n^2}$ orthogonalizes the decimated domain block with respect to the subspace spanned by $b_1^{(n)}, \cdots, b_{K-1}^{(n)}$, the basis vectors for the offset subspace;
- $\alpha_1^{(n)}, \cdots \alpha_{K-1}^{(n)}$ and $\alpha_K^{(n)}$ are scaling factors for the offset block and transformed domain block respectively.
- The other operators are the same as in Equation 4.1.

4.3 Decoder Convergence

Perhaps the most important effect of our introducing the orthogonalization operator is that it *secures and speeds up the decoder convergence.* It will be shown that relative to the type of mappings normally used, neither the collage, nor the attractor is changed as long as l_2-optimization is used, but *exact decoder convergence in a finite, signal-independent number of iterations* is obtained. To prove these results we introduce and prove two important properties of the new mapping T_O, under the following constraints:

1. We assume a *quadtree* range partition [87] consisting of blocks of sizes $2^{b_1} \times 2^{b_1}, \ldots, 2^{b_I} \times 2^{b_I}$, and that the corresponding domain block sizes are $2^{d_1} \times 2^{d_1}, \ldots, 2^{d_I} \times 2^{d_I}$. We assume the ordering $b_1 = 2b_2 = \cdots = 2b_I$. The decimation factors used are $2^{d_1-b_1}, \ldots, 2^{d_I-b_I}$ in both the x- and y-direction, with $d_i > b_i$ for all i[1].

2. The range partition and domain pool construction is such that every domain block in the signal (before the $2^{d_i-b_i}$-decimation) is made up of a integer number of range blocks. In practice this holds for the quadtree partition if every allowed domain address is also a position for a range block on the top level, and if $\min_i d_i \geq \max_i b_i$.

3. The $2^{d_i-b_i}$-decimation in the two spatial directions is to be done by *pure averaging* over a mask of $2^{d_i-b_i} \times 2^{d_i-b_i}$ samples.

We do not claim that the above constraints are the most general the properties to follow can be proved under. In fact, Lepsøy has proved that the same convergence properties can be attained under more general conditions [151]. However, they provide us with sufficient freedom in design, while making for simple proofs.

4.3.1 Convergence Speed

The first important property of T_O is concerned with its rate of convergence towards the attractor:

Property 1 *x_T can always be reached in a minimum number of iterations of T_O from an arbitrary initial signal if the DC component[2] is a basis block for the offset subspace.*
Proof. The attractor x_T can be written [123]

$$x_T = \sum_{j=0}^{\infty} L_O^j t_O, \tag{4.11}$$

provided this series converges. Since L_O orthogonalizes all range-sized blocks with respect to the DC component before placing them in their proper position, $L_O t_O$ has range-sized blocks of zero DC. Hence, due to the second of the constraints stated above, the domain blocks in this signal also have zero-valued DC component. Since L_O also

[1] This is necessary for the mapping T_O to possess a nontrivial attractor [163].

[2] DC = "direct current", i.e. a block with all samples equal-valued.

shrinks $2^{d_i} \times 2^{d_i}$ blocks to size $2^{b_i} \times 2^{b_i}$, $L_O^2 t_O$ has zero DC blocks of size $\frac{2^{d_i}}{2^{d_i}/2^{b_i}} \times \frac{2^{d_i}}{2^{d_i}/2^{b_i}}$, and generally $L_O^j t_O$ consists of zero DC blocks of size $\frac{2^{b_i}}{(2^{d_i}/2^{b_i})^{j-1}} \times \frac{2^{b_i}}{(2^{d_i}/2^{b_i})^{j-1}}$ samples. When this size becomes less than 1×1 samples for all i, then *every sample in the signal* $L_O^j t_O$ *must be zero.* After some manipulation we thus obtain that L_O^j is zero for all j greater than or equal to

$$J = 1 + \max_i \left\lceil \frac{b_i}{d_i - b_i} \right\rceil \tag{4.12}$$

where $\lceil u \rceil$ denotes "the smallest integer larger than u".

This means that the attractor is simply given as

$$x_T = \sum_{j=0}^{J-1} L_O^j t_O \tag{4.13}$$

with J as given in Equation 4.12. Hence, we have proved that the introduction of orthogonalization secures decoder convergence in a finite number of steps that is *only a function of the factors* d_i *and* b_i. □

Note that the above conclusion holds regardless of how the scaling coefficients are computed or quantized. This is because the elements of the orthogonalizing matrices O_n are signal-independent, and need not be quantized.

If the domain blocks are large enough compared to the range blocks, the decoding in fact becomes *noniterative*; it is simply performed by applying the mapping T_O once to the signal t_O. By inspecting the expression for J, it can be seen that this occurs when $d_i = 2b_i$, which makes $J = 2$, i.e. $x_T = L_O t_O + t_O$.

We also note that if a vector of initial length $\max B_n^2$ is to be shrunk to length 1 (i.e. a scalar sample) in a number of steps, and by a factor $2^{2d_i/b_i}$ in each step, it is impossible to do this in less than J steps, where J is as given in Equation 4.12. Thus there do not exist cases in which iteration of the T without orthogonalization converges even faster than iteration of T_O — and in practice the iteration converges much *faster* for most cases when orthogonalization is used. In image decoding experiments [88] it has been found that use of T may give wildly image-varying convergence rates; up to 50 iterations for some images. In general the tendency seems to be that signals with high activity (more small details) need more iterations than images with low activity. Jacquin and Fisher have both reported that 8 to 10 iterations are needed for most real world images if no orthogonalization is used [123, 84].

Choice of initial signal

Iteration from an *arbitary* initial signal x_0 will produce the iterate sequence $\{x_0,\ x_1 = L_O x_0 + t_O,\ x_2 = L_O^2 x_0 + L_O t_O + t_O,\ \cdots,\ x_L = \sum_{j=0}^{L-1} L_O t_O + L_O^L x_0, \cdots\}$. The last term in the $(L+1)$th signal in the sequence, x_L, is here zero for $L \geq J$ as given by Equation 4.12. In the special case of $x_0 = t_O$, however, this occurs already in the Lth signal in the sequence. Hence we need *one additional iteration* if we start from another initial signal than t_O. Therefore t_O must be deemed the most intelligent choice of initial signal in a practical implementation.

4.3.2 Invariance of Attractor

Up until now, we have only discussed how fast we can reach the attractor of T_O — not how this attractor actually looks. The positive convergence properties would be of little use if the signal modelling ability of T were reduced. In this subsection, we explore this aspect; more specifically, we prove the following property, under the extra assumption that the collage is l_2-optimized:

Property 2 *If the offset subspace consists of only the DC component, then T_O and T have the same attractor x_T.*

Proof. Let x be the original signal to be coded. The *collage* of x — which is the same with respect to both T and T_O[3]— with respect to T is simply

$$c = Lx + t = L_O x + t_O \tag{4.14}$$

We now define the error signal vectors

$$e = x - x_T \tag{4.15}$$

where x_T is the attractor of T, given by

$$x_T = \sum_{j=0}^{\infty} Lt, \tag{4.16}$$

and

$$e_c = x - c \tag{4.17}$$

Using Equations 4.16, 4.14, 4.15 and 4.17 it is easy to show that e and e_c are related by the equation

$$e = e_c + Le \tag{4.18}$$

By expanding this equation we obtain

$$e = e_c + Le_c + L^2 e_c + L^3 e_c + \cdots \tag{4.19}$$

If T_O is used instead of T then we have the attractor error

$$e_O = e_c + L_O e_c + L_O^2 e_c + L_O^3 e_c + \cdots \tag{4.20}$$

Now, since the collage is assumed to be l_2-optimized, each range-sized block in e_c is orthogonal to the offset subspace. Thus it has zero DC component. Due to the second constraint above this also holds for each domain block. Applying either of the linear terms L or L_O to a domain-blockwise zero DC signal will not introduce a nonzero blockwise DC, since the averaging decimator (constraint 3) do not change the DC component in the domain blocks — it just sums and then weighs all samples by the same factor. Hence there is no DC to remove in any of the decimated domain blocks, so if the DC component is the *only* component in the offset subspace, the action of L_O and *L is the same* on the collage error signal. Thus the two series expansions above must be equal in this case, i.e. $e = e_O$, which implies that the attractors are equal. □

[3] This is trivial, since $\text{span}\{b_1, \ldots, b_{K-1}, Ob_K\} = \text{span}\{b_1, \ldots, b_K\}$. We have not changed the collage subspace itself, only its basis.

When coefficient quantization is introduced the above conclusion will have to be modified somewhat, but not dramatically. As long as the DC quantization is "fine enough", there in practice is no change. It has been established that 8 bit uniform quantization of the DC component has no discernible influence on the attractor [198]. Also, if more offset dimensions are included, the attractor quality can never get worse than when only one is used. Thus the attractors we obtain with orthogonalization and $K \geq 2$ is always at least as good as the ones obtained without orthogonalization and $K = 2$, as long as the DC is the first offset component in both cases.

4.3.3 Similarity Between Collage and Attractor

For a T_O with orthogonalization, we can directly obtain a new collage theorem. First, note that e_c, and hence e, are DC-less over all range-sized blocks even before we start applying L_O. Hence, for these particular signals the attractor series expansion will have only $J-1$ terms instead of J. With this in mind, we get

$$\begin{aligned} d_B(x, x_T) &= \|x - x_T\| \\ &\leq \|x - T_O x\| + \|T_O x - x_T\| \\ &= \|x - T_O x\| + \|e - e_c\| \\ &= \|e_c\| + \|\sum_{j=0}^{J-2} L_O^j e_c\| \\ &\leq (\sum_{j=0}^{J-2} \|L_O\|^j) \cdot d(x, T_O x) \end{aligned} \tag{4.21}$$

which is the new collage theorem bound. In fact, also this bound can be strengthened considerably by taking into account *multiresolution* aspects of the theory — as shown by Øien, Baharav *et al.* in [200] and discussed by Baharav elsewhere in these proceedings [15]. In proving these strengthened bounds, the orthogonalization properties are also very useful.

In practice, though, the bounds given by all the available collage theorems are still very pessimistic for many cases of practical interest. Hence these theorems should be viewed mainly as an initial motivation. It is desirable to find a better justification of why the collage quality is used as an optimization criterion. We shall now give a *statistically* based argument on the relationship between the attractor and the collage.

Consider the expression for the attractor error, Equation 4.18. In sample n this equation gives

$$e_n = e_{c,n} + \sum_{j \in \mathcal{I}_n} l_{nj} e_j \tag{4.22}$$

where $\mathcal{I}_\backslash$ is the set of domain block samples used to construct range block sample n. Equation 4.22 shows that the attractor error in sample n is a sum of the *collage* error in the same sample, plus a weighted sum of the *attractor* error in the samples used by the mapping to construct sample n.

We may model the collage error as a stationary and ergodic random process[4], with expected value equal to zero:

$$m_c = \mathrm{E}[e_{c,n}] = 0 \tag{4.23}$$

By inspection of difference signals resulting from image modelling experiments we have performed this seems to be a valid assumption: Generally, the sample values in the difference signal have a narrow Gaussian- or Laplacian-like distribution, and most of the difference signal has a "noiselike" character which does not change noticeably from region to region, although some basic structures of the original signal can still be perceived.

From the above assumption it readily follows — by recursion — that the *attractor* error has shiftinvariant expected value $m = 0$, too. Furthermore, the last term in Equation 4.22,

$$\hat{m}_n = \sum_{j \in \mathcal{I}_n} l_{nj} \cdot e_j \tag{4.24}$$

can be interpreted as a *weighted estimator* of this expected value, with the samples in the region $\mathcal{I}_n$ being used to compute the estimate. If the $\frac{D}{B}$-decimation is performed by *averaging* over $\mathcal{I}_n$ we have

$$l_{nj} = \frac{\alpha_{P,n}}{I} \tag{4.25}$$

where $\alpha_{P,n}$ is the grey tone scaling coefficient for the range block of which sample n is a part, and I is the number of elements in the set $\mathcal{I}_n$ (the *averaging mask size*). Thus,

$$\hat{m}_n = \alpha_{P,n} \cdot \left(\frac{1}{I} \sum_{j \in \mathcal{I}_n} e_j \right) = \alpha_{P,n} \cdot \hat{m} \tag{4.26}$$

The estimator $\hat{m} = \frac{1}{I} \sum_{j \in \mathcal{I}_n} e_n$ is known to be an *unbiased* estimator for m. I.e.

$$\mathrm{E}[\hat{m}] = m = 0 \tag{4.27}$$

If $\{e\}$ is a *white noise* process[5] this estimator is also *consistent* [156]. In this case

$$\lim_{I \to \infty} \mathrm{Var}(\hat{m}) = 0 \tag{4.28}$$

where $\mathrm{Var}(\hat{m}) \stackrel{\text{def}}{=} \mathrm{E}[(\hat{m} - m)^2]$ — the estimator variance. Generally,

$$\mathrm{Var}(\hat{m}) = \frac{1}{I^2} \cdot \sum_{i \in \mathcal{I}_n} \sum_{j \in \mathcal{I}_n} \mathrm{E}[e_i e_j] \tag{4.29}$$

which also will decrease as I increases, unless the process $\{e\}$ is fully correlated.

Thus, when an averaging decimator is used, the attractor error converges towards the collage error as we increase the size of the averaging mask. This does of course not say anything about the quality of the collage, but it indicates that using this quality as

[4] By this we mean constant mean and shift-invariant correlation properties, with ensemble expectations being estimated by sample averages.

[5] We denote a random process of which e_n is a sample by $\{e\}$.

an optimization criterion in the encoder is more valid the larger I is. This in turn implies that if equality between attractor and collage were the only concern, one should use a large ratio $\frac{D}{B}$ (the maximal possible I is $\frac{D^2}{B^2}$). The above analysis also illustrates why the use of $\frac{D}{B} = 1$, or a pure subsampling decimator, are "worst cases" when it comes to the resemblance between attractor and collage[6]. In these cases, $I = 1$, and the estimator $\hat{m}$ employs only one sample to estimate the expected value for each region $\mathcal{I}_n$. This gives highly unpredictable results (large variance in the estimate), which corresponds to larger deviations between attractor and collage than those obtained with any other reasonable decimator. This is supported by experimental results [163, 115, 37]. In fact the tendency towards equality between attractor and collage is even stronger than what we could predict by means of statistics.

One should also note that when $D = B$, the attractor is simply $x_T = t_O$. In this case every domain block in t_O is also a block in the offset subspace, and is thus set to zero by O.

Direct attractor optimization

There also exists a special case in which both the various collage theorems and the statistical analysis in the previous subsection become redundant. This solution, first discovered by Lepsøy [149, 152], provides *direct attractor optimization* in the encoder. This is a simple consequence of the fact that the encoding algorithm produces a collage error which is blockwise orthogonal to the offset subspace, and of specific constraints made on the domain-to-range decimation factor and the domain-to-range decimator filter coefficients. We illustrate by a simple example.

Let the constraints stated in the beginning of this chapter be fulfilled. Consider Equation 4.22, the expression for the attractor error in sample n. By substituting this expression back into itself once, we obtain

$$e_n = e_{c,n} + \sum_{j \in \mathcal{I}_n} l_{nj} e_{c,j} + \sum_{j \in \mathcal{I}_n} l_{nj} \left(\sum_{k \in \mathcal{I}_j} l_{jk} e_k \right) \tag{4.30}$$

Doing similar back-substitutions over and over again will produce a series that contains only $e_{c,n}$ plus linear combinations of terms of the form $\sum_{k \in \mathcal{I}_j} l_{jk} e_{c,k}$ for various j. Now, suppose the averaging mask for each sample *is itself a range block* in the signal. In this case, $I = D = B \times B$ and

$$\sum_{k \in \mathcal{I}_j} l_{jk} e_{c,k} = \langle l_j, e_c(\mathcal{I}_j) \rangle \tag{4.31}$$

where l_j is the set of decimator filter coefficients ordered as a B^2-vector, and $e_c(\mathcal{I}_j)$ is the collage error restricted to the $B \times B$ filter mask considered. Furthermore, let

[6] There exist certain waveforms, notably *Weierstrass functions*, for which subsampling has been shown to be the correct decimation strategy [163]. However, these are mainly of academical/theoretical interest.

$l_{jk} = \frac{\alpha_{K,j}}{B^2}$, i.e. let the decimator be a pure averaging filter. In this case,

$$\langle l_j, e_c(\mathcal{I}_j)\rangle = \frac{\alpha_{K,j}}{B^2} \cdot \sum_{k \in \mathcal{I}_j} e_{c,k} \tag{4.32}$$

But, since we have orthogonalized the error with respect to the offset subspace, for which the DC component is assumed to be a basis vector, $\sum_{k\in\mathcal{I}_j} e_{c,k} = 0$ in this case. Hence $e_n = e_{c,n}$ for all samples n, which implies that the attractor and the collage are identical.

Incidentally, the above result does *not* assume a linear part with orthogonalization, but *if* we choose the orthogonal basis solution, we also obtain a *noniterative* decoder. We may also point out that the collage theorem bound given by Equation 4.21 is in this case fulfilled with *equality*: $d_B(x, x_T) = d_B(x, T_O x)$. We should note, though, that when *quantized* coefficients are used, the above proof does no longer hold exactly, since it exploits the fact that the linear term in this case operates in the same way on the original signal x and on its exact projection onto the offset subspace. After quantization, the offset component t_O no longer is this exact projection. But the attractor may still be written as $x_T = L_O t_O + t_O$, and since the fully quantized t_O is available when we start optimizing L_O, we may perform the optimization directly on this signal instead of on the original x. If we do that, we still optimize the same signal in the encoder that is synthesized in the decoder. However, the term "collage of x" is no longer an accurate name for this signal — it is instead the "collage of t_O".

4.3.4 Pyramid Decoder Structure

The decoder complexity is of vital importance because, in many applications the encoding is done once and for all, while the decoding is to be repeated many times, often in real — or close to real — time. This is typical in generation and use of compressed image data bases, and in processing of video sequences. In such situations it is of the utmost importance that the decoding can be done fast. Although we have seen that fractal coding can be done noniteratively in a special case, we have no guarantee that the restrictions inherent in that particular implementation (very large domain blocks compared to range blocks) will always give satisfactory coding results. In Figure 4.1, PSNRs for encodings of four different images at three different decimation factors are displayed. 8×8 range blocks and 2000 domain blocks were used in all cases.

For three of these four images, the noniterative case of $\log_2 \frac{D}{B} = 3$ yields the worst modelling result (although the difference is seen to be quite insignificant in at least two of those cases). We therefore seek a fast implementation of the decoding algorithm within the *most general* restrictions stated in this chapter.

First, remember that if the offset subspace contains only the DC component, the decoding algorithm produces iterates $t_O, T_O t_O, \ldots, T_O^{J-1} t_O$ that consist of blocks that are *constant-valued* over $2^{b_i} \times 2^{b_i}$ samples, $\frac{2^{b_i}}{2^{d_i}/2^{b_i}} \times \frac{2^{b_i}}{2^{d_i}/2^{b_i}}$ samples, $\ldots$, 1×1 samples respectively. *But this implies that the iterates may be subsampled — by factors* $\min_i 2^{b_i} \times 2^{b_i}$, $\max\left\{\min_i \frac{2^{b_i}}{2^{d_i}/2^{b_i}} \times \frac{2^{b_i}}{2^{d_i}/2^{b_i}}, 1 \times 1\right\}, \ldots, 1 \times 1$, *— without loss of information.* Hence the jth iteration may be performed on a signal of $\min\left\{\max_i \frac{(2^{2d_i}/2^{2b_i})^{j-1}}{2^{2b_i}}, 1 \times 1\right\}$ of the

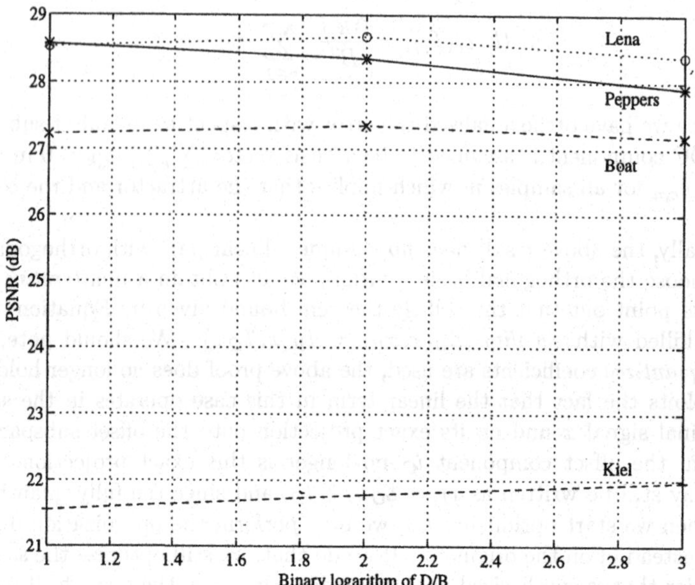

Figure 4.1: PSNRs from modelling of four different images at three different $\frac{D}{B}$ ratios.

original size. The resulting decoder structure is depicted in Figure 4.2, for the example of an image with uniform range partition and $D = 2B$. This structure is similar to that suggested by Baharav *et al.* [15], save for the fact that in our case, the attractor on the lowest resolution (which is the subsampled t_O) is obtained without iterations[7].

In the above structure, it is implied that the attractor is to be decoded at the same resolution as the original. If we want the signal decoded at a resolution which is an integer number R times the domain-to-range decimation factor less or more than the original, we only have to respectively decrease or extend the number of steps in the figure accordingly — i.e. by the same integer number of steps R — in order to obtain this.

Table 4.1 shows the number of arithmetic operations in the pyramid decoder structure, when the signal is decoded at original resolution, for a uniform range partition, and for various choices of b and d. It is seen that the decoder complexity in these examples is comparable to that of the fastest established compression techniques for all choices.

[7]Note that the use of the name "T_O" for the mappings applied in the pyramid structure in Figure 4.2 is not really mathematically correct,as these mappings operate on signals of different sizes than the one T_O operates on. However, they perform exactly the *same type of operations* on the signals on which they operate. With these reservations, there hopefully should not occur any confusion because of our lack of mathematical precision in the naming convention.

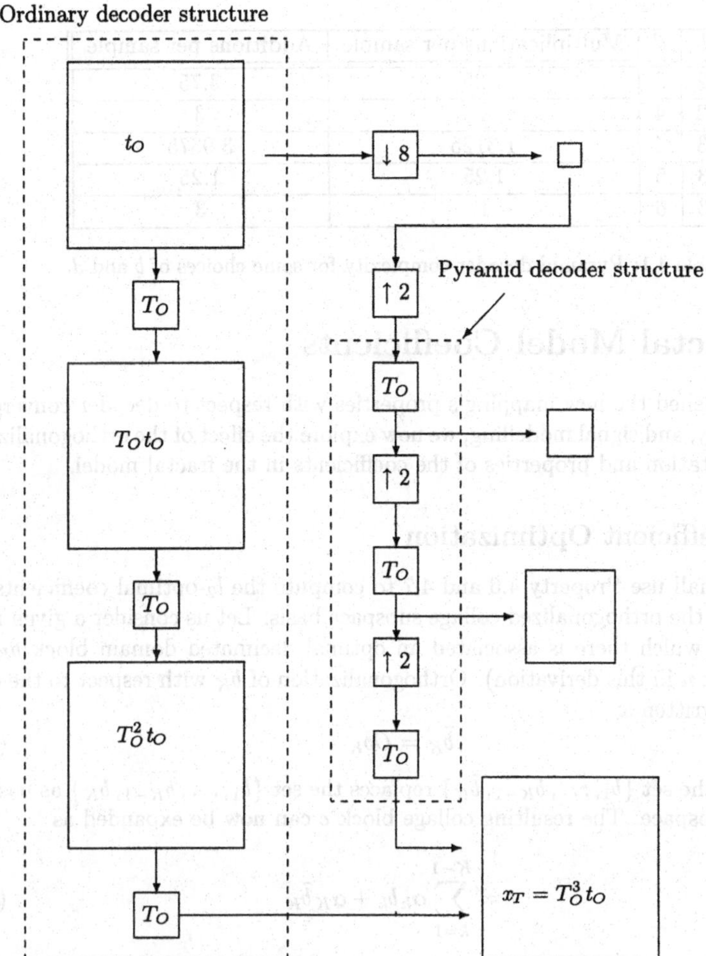

Figure 4.2: Ordinary and "pyramid" decoder structure.

b	d	Multiplications per sample	Additions per sample
2	3	1.25	3.75
2	4	1	3
3	4	1.3125	3.9375
3	5	1.25	1.25
3	6	1	3

Table 4.1: Pyramid decoder complexity for some choices of b and d.

4.4 Fractal Model Coefficients

Having established the new mapping's properties with respect to decoder convergence and complexity, and signal modelling, we now explore the effect of the orthogonalization on the computation and properties of the coefficients in the fractal model.

4.4.1 Coefficient Optimization

Initially, we shall use Property 4.6 and 4.7 to compute the l_2-optimal coefficients corresponding to the orthogonalized collage subspace basis. Let us consider a given range block r, with which there is associated an optimal decimated domain block b_K (we omit the index n in this derivation). Orthogonalization of b_K with respect to the offset basis can be written as

$$\tilde{b}_K = Ob_K \tag{4.33}$$

such that the set $\{b_1, \ldots, b_{K-1}, \tilde{b}_K\}$ replaces the set $\{b_1, \ldots, b_{K-1}, b_K\}$ as basis for the collage subspace. The resulting collage block c can now be expanded as

$$c = \sum_{k=1}^{K-1} \alpha_k b_k + \alpha_K \tilde{b}_K \tag{4.34}$$

Assuming that $b_1, \ldots, b_{K-1}$ are chosen orthonormal, the modified normal equations that must be solved in order to find the l_2-optimal coefficient values can be written [198]

$$\begin{bmatrix} 1 & 0 & \cdots & \cdots & 0 \\ 0 & 1 & 0 & \cdots & 0 \\ \vdots & \ddots & \ddots & \ddots & \vdots \\ 0 & \cdots & 0 & 1 & 0 \\ 0 & \cdots & \cdots & 0 & \|\tilde{b}_K\|^2 \end{bmatrix} \begin{bmatrix} \alpha_1 \\ \alpha_2 \\ \vdots \\ \alpha_{K-1} \\ \alpha_K \end{bmatrix} = \begin{bmatrix} \langle b_1, r\rangle \\ \langle b_2, r\rangle \\ \vdots \\ \langle b_{K-1}, r\rangle \\ \langle \tilde{b}_K, r\rangle \end{bmatrix} \tag{4.35}$$

which directly yields

$$\alpha_k = \langle r, b_k\rangle \text{ for } k = 1, \ldots, K-1 \tag{4.36}$$

and

$$\alpha_K = \frac{\langle r, \tilde{b}_K\rangle}{\|\tilde{b}_K\|^2} = \frac{r^T O b_K}{b_K^T O^T O b_K} = \frac{r^T O b_K}{b_K^T O b_K} \tag{4.37}$$

Expression 4.37 may be expanded as

$$\alpha_K = \frac{\langle r, b_K\rangle - \sum_{k=1}^{K-1} \alpha_k \langle b_k, b_K\rangle}{\|b_K\|^2 - \sum_{k=1}^{K-1} \langle b_k, b_K\rangle} \tag{4.38}$$

which shows that there is no need to *explicitly* perform the orthogonalization of the decimated domain blocks. However, *if* we do that, the solution can be simplified somewhat more — at the cost of increased storage needs in the encoder, due to the fact that the orthogonalized domain blocks must be buffered during the encoding.

All inner products in the expression 4.38 for α_K, except for $\langle r, b_K\rangle$, can be computed before the domain search commences. Furthermore, $\alpha_1, \ldots, \alpha_{K-1}$ are independent of the choice of domain block. This represents a complexity reduction relative to if we were to use the nonorthogonal basis. It can be shown [198] that the coefficients for that case are related to $\alpha_1, \ldots, \alpha_K$ through the equations

$$\beta_K = \alpha_K \tag{4.39}$$

and

$$\beta_k = \alpha_k - \langle b_k, b_K\rangle \beta_K \text{ for } k = 1, \ldots, K-1 \tag{4.40}$$

All β_k vary with both range and domain block. Hence we would have had to compute all K coefficients — not just the Kth one, as is now the case — for each combination of range and domain block, had we not introduced orthogonalization.

Another important aspect of the mapping T_O is that the convergence analysis we have performed does not involve the scaling coefficients $\{\alpha_{K_n}^{(n)}\}$ at all. No matter how big their absolute values are, their contributions are averaged out in the higher order terms of x_T. Thus T_O is always *eventually contractive*, regardless of the coefficient values. The coefficients may therefore be as optimally quantized as we wish, without concern for the convergence properties. This is contrast to previous methods, where one has been forced to constrain $\{\beta_{K_n}^{(n)}\}$ to be less than some maximum value, most usually 1.

4.4.2 Coefficient and Offset Subspace Properties

Another advantage of replacing L by L_O is that it becomes easier to see how the offset subspace should be chosen. We will consider this problem in the light of two well known concepts from source coding and quantization theory: *energy packing* and *coefficient decorrelation.*

Energy packing

The task of the offset term is not only to provide a nonlinear component such that a nontrivial attractor can be obtained. It can also be thought of as a "first approximation" to the signal in each block, while the residual from this approximation is the signal we are trying to code by means of the linear term of the mapping. Thus an important task for the offset term is to secure that important and "typical" data features are preserved in a satisfactory way, even if the linear term should fail to do a good modelling job. Using traditional compression language, the offset subspace should therefore be chosen to maximize the *energy packing* into the offset coefficients for a given dimension K. If we assume that our signal blocks are *sample vectors from a stationary random process*, it is well known that this can be achieved by letting the offset subspace basis be equal to the $K-1$ *eigenvectors* corresponding to the $K-1$ largest *eigenvalues* of the autocorrelation matrix of the range block process [133]. This set of *principal components* are the first $K-1$ basis blocks of the *Karhunen-Loève* transform associated with the process. For any K, these components form the set of basis blocks we ideally should use for the offset subspace if coding efficiency were the only concern — it simplifies the job for the linear term as much as possible, by minimizing the expected residual power, and as will be shown it also makes the setting for *coefficient quantization* as optimal as possible.

Decorrelation of the coefficients

If the offset subspace is spanned by the $K-1$ first principal components of the range block process, the coefficients $\alpha_1, \ldots, \alpha_K$ become *decorrelated.* This is easily shown: Letting $b_1, \ldots, b_{K-1}, \tilde{b}_K$ be the orthogonalized collage subspace basis and r be a range block, the coefficient vector can be expressed as

$$a = \underbrace{\begin{bmatrix} b_1^T \\ \vdots \\ b_{K-1}^T \\ \frac{\tilde{b}_K^T}{\|\tilde{b}_K\|^2} \end{bmatrix}}_{=B^T} \cdot \begin{bmatrix} r_1 \\ \vdots \\ r_{B^2} \end{bmatrix} = B^T r \tag{4.41}$$

and the corresponding correlation matrix is

$$R_{aa} = \mathrm{E}[aa^T] = B^T R_{rr} B \tag{4.42}$$

or, written out explicitly,

$$R_{aa} = \begin{bmatrix} b_1^T R_{rr} b_1 & \cdots & b_1^T R_{rr} b_{K-1} & b_1^T R_{rr} \frac{\tilde{b}_K}{\|\tilde{b}_K\|^2} \\ \vdots & \ddots & & \vdots \\ b_1^T R_{rr} \frac{\tilde{b}_K}{\|\tilde{b}_K\|^2} & b_2^T R_{rr} \frac{\tilde{b}_K}{\|\tilde{b}_K\|^2} & \cdots & \frac{\tilde{b}_K^T R_{rr} \tilde{b}_K}{\|\tilde{b}_K\|^4} \end{bmatrix} \tag{4.43}$$

where

$$R_{rr} \overset{\text{def}}{=} \mathrm{E}[rr^T] \tag{4.44}$$

Since $b_1, \ldots, b_{K-1}$ per definition are *orthonormal eigenvectors* for R_{rr}, they do not change direction (except for possibly by 180°) when multiplied by R_{rr}. Hence R_{aa} is a *diagonal* matrix, which was what we wanted to prove.

This property has positive consequences when it comes to quantization of the coefficients. It is well known that if scalar quantization of the elements in a vector process is to be used, the only signal properties that can be exploited are inter-element correlation (linear dependency) and the shape of the probability density function (pdf) of each individual element [164]. It is optimal to have uncorrelated parameters, and we may use quantization levels whose placement is optimized with respect to the pdf shape. For this, the iterative *Lloyd-Max* algorithm may be used [137]. Results from such experiments may be found in [199, 198].

4.4.3 Image Coding: Practical Considerations and Compromises

If we look at *the first* principal component of an image-like stationary random process, it is typically very close to a uniformly grey block, i.e. the DC component. This fact is demonstrated in several standard references such as [137] or [133], and is reflected in the choice of the DC component as the first basis vector in fixed transforms used for image compression, most notably the discrete cosine transform [49]. We have also shown experimentally that there is almost no correlation between the grey tone scaling coefficient and the DC component. As an example, for the 512×512 image "Boat", encoded with 8×8 range blocks and $K = 2$, we estimated the normalized autocorrelation matrix for the coefficients to be

$$\hat{R}_{aa} = \begin{bmatrix} 1.00 & 0.04 \\ 0.04 & 1.00 \end{bmatrix} \tag{4.45}$$

which shows that the intercoefficient crosscorrelation for a given range block is negligible. Since in addition the convergence properties of T_O have been proved under the assumption that the DC component is part of the offset subspace, we choose the DC component as the first offset component in all image experiments from now on.

Another question we should answer is which *dimension* $K-1$ one should choose for the offset subspace. Adding components will certainly not degrade the model quality, but it will increase the bit rate if the number of domain blocks is kept constant[8].

It is also demonstrated in the above-mentioned references that the first principal component of highly correlated, image-like processes clearly dominates over the others when it comes to signal power. The special treatment the DC component receives in practical transform coder quantization schemes is in part a consequence of this. Hence, if we have any faith in the signal modelling abilities of the fractal coder domain pool at all, it seems sensible that we should limit ourselves to a one-dimensional offset subspace for real-world images. Thus we have essentially arrived at the conclusion that, apart from the strict contractivity constraint, Jacquin's original image model [123] is in fact sufficiently general. However, this is *not* necessarily true for other types of data, whose statistical properties might be entirely different — cf. for example Jayant & Noll's example of the KLT basis for an AR(8) model of speech [137, p. 520]. If it ever comes to the point where fractal coding becomes a viable technique for compression of various kinds of data, it is nice to have a general theory available.

[8]It is possible, however, that the size of the domain pool — and hence the number of bits used to address it and the time used to search it — can be decreased as more offset components are added. Also, we may get away with larger block sizes in some image areas, which implies a *lower* rate in these areas. This should be investigated.

4.5 Encoder Complexity Reduction

In this section we discuss some aspects of encoder complexity reduction, which is made easier when orthogonalization is used.

4.5.1 Simplified Coefficient Computation

We have in fact already discussed one example of complexity reduction which is a direct consequence of the orthogonalization: Only one — instead of K — coefficient per range block is now domain block dependent. I.e. $K-1$ of the K coefficients can now be computed *once* per range block, instead of N_d times — where N_d is the number of domain blocks.

4.5.2 Simplified Distortion Minimization

Also, the actual operation of distortion minimization for each range-domain combination is simplified considerably with the new mapping structure. To find the original-to-collage block distance we only need to maximize $\alpha_K^2 \cdot \|\tilde{b}_K\|^2$ inside the domain search loop, as the squared distortion is now given by

$$d_B^2(r,c) = \|r\|^2 - \sum_{k=1}^{K-1} \alpha_k^2 - \alpha_K^2 \cdot \|\tilde{b}_K\|^2 \tag{4.46}$$

Here $\|r\|^2$ and the squared offset coefficients are independent of the outcome of the domain pool search. $\|\tilde{b}_K\|^2$ is simply the denominator in the expression 4.38 for α_K, and can be prestored for each domain block. Thus only two extra multiplications per range-domain-block combination are needed to make the decision that minimizes distortion[9].

4.5.3 Directed Domain Pool Search

As has been previously published in [198, 149, 201, 151], the orthogonalization makes it easy to introduce an algorithm for *directed* domain pool search, much along the principles of tree-structured vector quantization [97]. Space does not permit reproducing all details of the algorithm here, but the main idea is as follows: The pre-orthogonalized domain and range blocks are buffered, then the range blocks are classified into a chosen number of classes or *clusters.* With each cluster there is associated a *cluster centroid.* Each domain block is then compared to each cluster centroid and classified according to maximum similarity to one of the centroids. *Within each class*, each range-domain pair is then compared, and for each range block, the most similar domain block within the class is kept.

The reason why this algorithm is made easy through orthogonalization, is mainly that the criterion for similarity becomes so simple: The best range-domain match is simply the one for which the two blocks are *maximally parallel* in the block space

[9] Again, this holds for the *modelling* problem — a real compression system uses quantized coefficients, which modifies this conclusion somewhat, albeit not dramatically.

[149, 151]. By combining a simple centroid-finding algorithm developed by Lepsøy [149, 151] with the well-known *LBG* (Linde-Buzo-Gray) clustering algorithm [158], the range block clustering and subsequent domain block comparisons can then be done very effectively.

It has also been shown [198, 201] that it is in fact beneficial to further *decimate* all orthogonalized range and D/B-decimated domain blocks before clustering and thereafter comparing them. Typically an additional decimation factor of 2 in each spatial direction (which means that the block space dimension is reduced by a factor 4) should be used – plus a final comparison between a very restricted list (say, 20 candidates) of nondecimated "best candidates" from the decimated search [198].

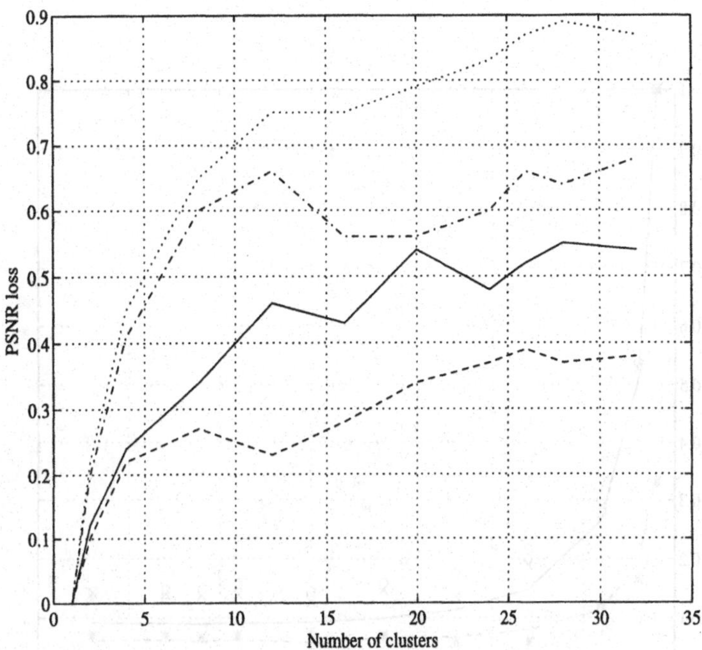

Figure 4.3: Observed PSNR loss relative to encoding by exhaustive search, as a function of the number of clusters. "···" = "Lena" without decimation. "–·–" = "Lena" with 2-decimation. "——" = "Kiel harbour" without decimation. "- - - -" = "Kiel harbour" with 2-decimation. Domain pool size = 2000 blocks, range size = 8 × 8, length of candidate list = 20.

Figure 4.3 indicates how much of a PSNR loss we may expect in coding of real-world images (resolution 512×512 pixels), relative to using exhaustive domain pool search, when the clustering method just described is used. It is seen that when a tree-structured search on 2-decimated range/domain blocks is used, the loss stabilizes at values about 0.15 – 0.20 dB *lower* than what is experienced without decimation.

It is tempting to attribute this effect to the known suboptimality of the clustering algorithm, which typically is more serious the higher the block space dimension is. In other words, it is easier for the algorithm to converge towards a "truer" optimum when the dimension of the training blocks is lower — and this effect seems to more than make up for the loss of information inherent in the decimation process.

Typically, the loss to expect for a given image is in the range of 0.3 – 0.6 dB. This drop has corresponded to a graceful visual degradation for all images we have encountered.

In Figure 4.4 we compare the theoretical and the observed (as measured in seconds computer simulation time, then normalized with respect to the complexity of an exhaustive search) encoder complexity as a function of the number of clusters, with and without decimation. (The theoretical expressions may be found in [198, 201].)

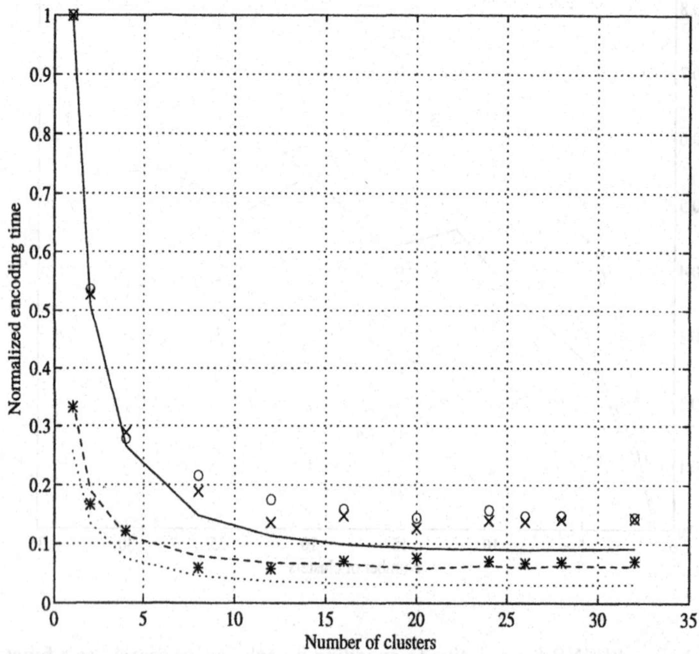

Figure 4.4: Observed and theoretical encoder complexity as a function of the number of clusters. Solid line = theoretical complexity without decimation in the classification. "..." = theoretical complexity with 2-decimation. "o" = observed complexity for "Lena" without decimation. "×" = observed complexity for "Kiel harbour" without decimation. "*" = observed complexity for "Lena" with 2-decimation. "+" = observed complexity for "Kiel Harbour" with 2-decimation. Domain pool size = 2000 blocks, range size = 8×8, length of candidate list = 20. All curves are normalized with respect to the complexity of the exhaustive search case.

It is seen that the shapes of the observed curves approximately follow those of the theoretical ones, but stabilize at a somewhat higher level. This may be attributed both to software program overhead (memory access/allocation etc.), and the fact that one of the theoretical assumptions — all clusters equal-sized — does not hold exactly in practice [149, 151].

Finally, we mention that the encoding times can be further reduced by ca. 50 % by using *image invariant* cluster centroids. Experiments have shown that this in fact gives only a negligible extra loss of quality compared to the image-specific clustering [255].

4.6 Comparisons to Other Techniques

It is helpful for our understanding of the fractal coding method to consider it in the light of other, more fully understood compression algorithms. Through such comparisons, it may be easier to uncover both the potential and the limitations of fractal coding, as well as to develop techniques for fully exploiting the former, and overcoming the latter. This comparison is also much easier to make when T is replaced by T_O.

4.6.1 Product Code Vector Quantization

The most striking comparisons can be made between fractal coding and *vector quantization (VQ)*. Image compression using VQ techniques has been a research topic since about 1980. An excellent survey of VQ can be found in [97]. The connection between VQ and fractal coding was tentatively suggested by Jacquin [123], and has been studied in more detail by Lepsøy, who has also performed empirical studies of the relative merits of the two techniques (see Chapter 2). The central idea is that during the encoding, the domain pool can be regarded as a *codebook* for the signal. However, it is a special and novel kind of codebook:

- it is signal-specific;
- it is derived from information contained only within the signal itself;
- it can be constructed without extensive training, and is thus not a result of optimization with respect to some training set representative of the source to be coded;
- it is not explicitly stored in the decoder;
- it can be approximately regenerated in the decoder, through iteration.

These are all novel features of fractal coding, but the similarity to VQ is still very strong, as is especially apparent when orthogonalization is employed. The closest link is found between fractal coding and some *product code vector quantization (PCVQ)* techniques. PCVQ is a collective name for all VQ algorithms where the vectors to be coded are broken down into several subcomponents (features) before they are coded. In the simplest PCVQ variants, these features are quantized independently of each other, each by its own subcodebook. A better, but slightly more complex structure is obtained by taking the interdependency of the features into account, in order to secure that the overall code vector always is the one giving minimum distortion [97]. Using such techniques, the quality-complexity ratio of unconstrained VQ can be significantly improved for signals of practical interest — also images [211, 97]. The name "product code" VQ reflects the fact that the number of distinct reproduction vectors (the size of the implicit representation codebook) is equal to the *product* of the subcodebook sizes used for the different features. However, the *complexity* is proportional to the *sum* of these sizes (at least if the features are independently quantized). This is typically a much smaller number than the product, and this is the reason why a better quality is achieved for the same complexity, relative to unconstrained VQ where the complexity is proportional to the codebook size (whereas in PCVQ it is a *logarithmic* function of the codebook size). PCVQ is generally suboptimal with respect to quality versus *bit rate*, however: In practice the features chosen are not always statistically independent, and the structural constraints of PCVQ prevent us from fully exploiting the dimensionality of the vector space [137, 164].

The types of PCVQ that most resemble fractal coding are:

Mean-removed VQ (MRVQ). In MRVQ, the DC component (most usually referred to as the *mean*) is removed from each vector to be coded, i.e. the vectors are orthogonalized with respect to the DC. The DC component and the *mean-removed* vector is thereafter quantized separately — the DC by a scalar quantizer and the mean-removed vector by a codebook. This procedure is strongly linked to our splitting of each block into a *offset* part — which can be regarded as a generalized mean value, and in practice, as has been discussed, often *is* the mean — and a *domain block* part.

Shape-gain VQ (SGVQ) In SGVQ each vector is normalized to l_2-norm 1, and the normalization factor (the *gain*) and the normalized vector (the *shape*) is quantized with two separate codebooks, which may be simultaneously optimized. This has a close resemblance to the workings of the *linear term* of our signal mapping, where each domain (codebook) block is allowed to be multiplied by some scalar.

In [190] MRVQ and SGVQ have been combined in a *mean-removed shape-gain* VQ (MRSGVQ) scheme. This provides the closest analogy to fractal coding known from the VQ litterature. Lepsøy has explored the merits of this method relative to the fractal method (see Chapter 2).

4.6.2 Other Lossy Compression Techniques

Most lossy compression methods can be interpreted as a combination of a *signal decomposition* onto a set of basis vectors, and a *quantization* of the weighting coefficients in this decomposition. This is perhaps most clearly seen in *transform coding*, where nonoverlapping signal blocks are projected onto a fixed, orthonormal basis for the whole block space. The projection coefficients are then quantized by one of a variety of available schemes [211, 49, 47].

Subband coding [212] obeys similar principles. Indeed, transform coding can be seen as a special case of subband coding. The signal decomposition in a subband coder is done with a linear analysis filter bank, which generally corresponds to *overlapping* basis functions in the "time" (spatial) domain, and therefore no discernible block structure. The splitting is followed by an effective quantization scheme for the subband samples.

Signal decomposition onto linear subspaces is properly regarded as a unifying principle in lossy compression. L_2-optimal fractal coding is based on this principle, too. Each block is projected onto a number of *noncomplete* bases, since each of the collage subspaces tried is of lower dimension than the range block space. It follows from this that fractal coding — as opposed to e.g. unitary transform coding and perfect reconstruction subband coders — is what we might term *fundamentally* lossy (or *model-based*): Even when infinite precision is used in the representation of the signal code, it is not an invertible operation for an arbitrary signal, so the possibility of *perfect reconstruction* is not as a rule present in a fractal coding system. This is also a feature shared with VQ. However, for many applications, this need not be of great importance. At low rates, a fundamentally lossy method need not be inferior to a method providing numerically perfect reconstruction in the case of no parameter quantization, as the perfect reconstruction property is not directly related to robustness against quantization.

4.7 Conclusion

We have modified the fractal signal mapping by introducing orthogonalization of the domain blocks with respect to the offset subspace basis. This has been shown to have the following consequences:

- the decoding algorithm is made to always converge exactly in a finite number of iterations.
- the number of iterations is signal independent, and only a function of the domain and range block sizes used.
- the collage and the attractor of the modified mapping are the same as before.
- it is not necessary to constrain the domain block scaling factors to obtain convergence.
- for a stationary random process, the model coefficients corresponding to the orthogonal collage subspace basis are uncorrelated if the offset subspace is spanned by principal components of the process. This is the optimal situation if we are to use scalar quantization.
- a "pyramid" structure may be used in the decoder, in which we build signals of successively higher resolution as the iteration proceeds, until the original size is reached. This significantly reduces the number of arithmetic operations necessary in the decoder.
- a noniterative decoder is included as a special case.

The above consequences have been shown to hold under the following constraints:

- quadtree range partition.
- domain-to-range decimation by averaging.
- each domain block consisting of an integer number of complete range blocks.
- the DC component being a part of the offset subspace (for the attractor to be unchanged relative to previous implementations it must be the *only* component).

Some of the properties also needs l_2-optimal coefficients to hold.

We have also introduced a novel argument for why it makes sense to optimize the collage instead of the attractor. It has been shown that the additional reconstruction error introduced in each attractor sample is a weighted estimate of the *expected value* of the attractor error, which is zero due to the l_2-optimization procedure.

Finally, we have used the orthogonalization to discuss good choices for the offset subspace, to lay the foundations for an improved strategy for coefficient quantization, to design an effective encoder compexity reduction scheme, and to draw analogies between fractal coding and other compression techniques, most notably mean-removed shape-gain VQ.

Chapter 5

On the Dimension of Fractally Encoded Images

Yuval Fisher, Frank Dudbridge, Ben Bielefeld

Abstract: We analytically derive the box-counting dimension of the attrator of a class of fractally encoded images. This method can be used to quickly partition an attractor derived from an image into regions of differing fractal dimension. It's important to note, however, that the dimension is computed for the attractor, not the original image.

5.1 Introduction

We discuss the fractal (box-counting) dimension (see [74]) of the attractor of a fractally encoded image. We restrict ourself to "Dudbridge-Monro"-style fractal encodings (see [67]). In these type of encodings, the image is partitioned into square blocks, each of which is encoded separately by an iterated function system containing four maps. We restrict ourselves to computing the dimension of this specific kind of IFS.

We begin with the one-dimensional case for clarity. We consider the IFS $w_1, \ldots, w_n$ give by

$$w_i(x,y) = (\frac{x}{n} + \frac{i-1}{n}, s_i y + o_i).$$

This IFS can also be written as a transformation τ on L^p functions on the unit interval $f : I \to \mathcal{R}$ given by

$$\tau(f)(x) = s(x) f(nx \bmod 1) + o(x)$$

where $s(x)$ and $o(x)$ are piecewise constant and equal to s_i and o_i, respectively, on the n subintervals of I with length $1/n$. When τ is contractive, it defines a unique attractor denoted $|\tau| \in L^p(I)$ satisfying $\tau(|\tau|) = |\tau|$. The result is then,

Theorem 5.1 ([32]) *For $f : I \to \mathcal{R}$, define $\tau(f)(x) = s(x)f(nx) + o(x)$, where $s(x)$ and $o(x)$ are piecewise constant on the n subintervals of I with length $1/n$. Let* $0 \leq$

$s_1, \ldots, s_n \leq 1$ be the values of $s(x)$. Then τ has a fixed point whose graph A has box-counting dimension

$$\dim_B A = \max\left\{1, d + \frac{\log \sum_{i=1}^{n} |s_i|^d}{\log n}\right\}$$

where d is the fractal dimension of the range of $|\tau|$.

We omit the proof, which is similar to the 2-dimensional case given later.

5.2 The Fractal Dimension of Image Blocks

We consider an image block to be an L^p function f over the unit square $I^2 = [0,1] \times [0,1]$. In this case, we define a transformation of f for each $x \in I^2$ by

$$\tau_{m,s,o}(f) : L^p(I^2) \to L^p(I^2)$$

using a scaling map $s : I^2 \to \mathcal{R}$, an offset map $o : I^2 \to \mathcal{R}$ and a spatial transformation $m : I^2 \to I^2$ as follows,

$$\tau_{m,s,o}(f)(x) = s(x) f(m(x)) + o(x).$$

In our application, we restrict $m(x)$ to be multiplication by 2 modulo 1 in each coordinate. We also restrict $s(x)$ and $o(x)$ to be piecewise constant and equal to s_1, s_2, s_3, s_4 and o_1, o_2, o_3, o_4, respectively, over the four quadrants of I^2. This becomes equivalent to four affine transformation of the unit cube, in which the xy coordinates are similitudes that map the square onto each of its quadrants and the z coordinate are transformed by affine transformations with scaling s_i and offset o_i, for $i = 1, 2, 3, 4$.

When τ is contractive, it defines a unique attractor denoted $|\tau|$. The range of $|\tau|$ is a subset of $\mathcal{R}$; it can be thought of as the projection of the graph of $|\tau|$ on the z-axis.

In practice, this type of fractal compression algorithm yields less-than-optimal results, but modifications of this algorithm, for example, those incorporating sophisticated partitioning schemes or adding more degrees of freedom, may yield good results.

Our main theorem is:

Theorem 5.2 *Let τ to be as described, and assume that it is contractive with attractor $|\tau|$ whose graph is A. Then the box-counting dimension of A is*

$$\dim_B(A) = max(2, d + \log(\sum_{i=1}^{4} |s_i|^d)/\log 2)$$

where d is the fractal dimension of the range of $|\tau|$.

Proof:

Let A be the graph of the attractor of $|\tau|$, and let E be the range of $|\tau|$. Then E is the attractor of an IFS of (one-dimensional) similtudes, and hence

$$0 \leq \overline{\dim}_B E = \underline{\dim}_B E = \dim_B E = d \leq 1.$$

Let $N_r(E)$ and $M_r(A)$ denote the smallest number of intervals and cubes, respectively, of diameter r required to cover the sets E and A. Then for every $\epsilon > 0$ (with $d > \epsilon$) there are c_1, c_2 such that

$$c_1 \max\left\{1, \left(\frac{1}{r}\right)^{d-\epsilon}\right\} \leq N_r(E) \leq c_2 \max\left\{1, \left(\frac{1}{r}\right)^{d+\epsilon}\right\}, \tag{5.1}$$

with a similar condition holding for $M_r(A)$. Also, we write λE to denote the scaling of the set E by λ, in which case

$$N_r(\lambda E) = N_{r/\lambda}(E). \tag{5.2}$$

Consider the kth iteration of τ. For any index sequence $i_1, i_2, \ldots, i_k$, $(i_j \in \{1, 2, 3, 4\})$, consider the cuboid R that lies on the grid defined by the iterates of τ with height $|s_{i_1} s_{i_2} \cdots s_{i_k}|$ and a square base with side length 2^{-k}.

The projection of $R \cap A$ on the range of $|\tau|$ is a copy of E scaled by $|s_{i_1} s_{i_2} \cdots s_{i_k}|$. So

$$\begin{aligned} M_{2^{-k}}(R \cap A) &= N_{2^{-k}}(|s_{i_1} s_{i_2} \cdots s_{i_k}| E) \\ &= N_{(2^k |s_{i_1} s_{i_2} \cdots s_{i_k}|)^{-1}}(E). \end{aligned}$$

So, using Eq. (5.2) and Eq. (5.1),

$$c_1 \max\left\{1, (2^k |s_{i_1} s_{i_2} \cdots s_{i_k}|)^{d-\epsilon}\right\} \leq M_{2^{-k}}(R \cap A) \leq$$
$$c_2 \max\left\{1, (2^k |s_{i_1} s_{i_2} \cdots s_{i_k}|)^{d+\epsilon}\right\}$$

Summing on all possible indices gives, for some constants c_3 and c_4,

$$c_3 \sum_{i_1,\ldots,i_k} \max\left\{1, (2^k |s_{i_1} s_{i_2} \cdots s_{i_k}|)^{d-\epsilon}\right\} \leq M_{2^{-k}}(A) \leq$$
$$c_4 \sum_{i_1,\ldots,i_k} \max\left\{1, (2^k |s_{i_1} s_{i_2} \cdots s_{i_k}|)^{d+\epsilon}\right\}$$

which implies that

$$c_3 \max\left\{\sum_{i_1,\ldots,i_k} 1, \sum_{i_1,\ldots,i_k} (2^k |s_{i_1} s_{i_2} \cdots s_{i_k}|)^{d-\epsilon}\right\} \leq M_{2^{-k}}(A) \leq$$
$$c_4 \left[\sum_{i_1,\ldots,i_k} 1 + \sum_{i_1,\ldots,i_k} (2^k |s_{i_1} s_{i_2} \cdots s_{i_k}|)^{d+\epsilon}\right]$$

which implies that

$$c_3 \max\left\{4^k, 2^{k(d-\epsilon)}\left(\sum_{i=1}^{4}|s_i|^{d-\epsilon}\right)^k\right\} \leq M_{2^{-k}}(A) \leq$$
$$c_4\left[4^k + 2^{k(d+\epsilon)}\left(\sum_{i=1}^{4}|s_i|^{d+\epsilon}\right)^k\right]$$

So that

$$\dim_B A = \lim_{k\to\infty}\frac{\log M_{2^{-k}}(A)}{\log 2^k} = \max\left\{2, d + \frac{\log \sum_{i=1}^{4}|s_i|^d}{\log 2}\right\}$$

■

5.3 Results

The action of τ induces an IFS consisting of (one-dimensional) similitudes in its range. Thus, it is possible to determine the dimension of the range of $|\tau|$ almost surely with respect to Lebesgue measure using only the scaling values of the IFS. That is, except for specific parameter values falling in a set with measure zero, there are only the following two cases (see [74], pg. 131):

- If $\sum |s_i| < 1$ then $\sum |s_i|^d = 1$ almost surely, in which case the fractal dimension of A is $\dim_B(A) = 2$.
- If $\sum |s_i| \geq 1$ then $d = 1$ almost surely, in which case the fractal dimension of A is the maximum of 2 and $\dim_B(A) = 1 + \log \sum |s_i| / \log 2$.

In our implementation, we disregarded the measure zero cases. We thus avoid computing d altogether, resulting in some possible error.

Fig. 5.1 shows several examples of encoded images and the resulting dimension of the encoding. In these examples, the images were encoded to achieve good fidelity, since it is clearly desirable to find an attractor that is as close as possible to the original image. Each image was partitioned into a quadtree based on a thresholding criterion for the fit between the collage for the partition and the original image. Each partition was encoded using a map of the type described in Section 2. Thus, each partition of the quadtree has its own fractal dimension.

The methods employed here are sensitive to block size and encoding fidelity. For example, it does not make sense to encode very small blocks, since data about variation within the block is lost during the decimation. Conversely, large blocks are typically not well encoded and so the resulting dimension would not be well correlated to the original image. The images in the dataset in Fig. 5.1 are all of size 256 × 256 with encodings that have a smallest block of size 4 × 4.

We note with interest that while the image of Lena clearly shows her hair and hat, the image of San Francisco shows the highest fractal dimension on the buildings (which are presumably actually rather flat).

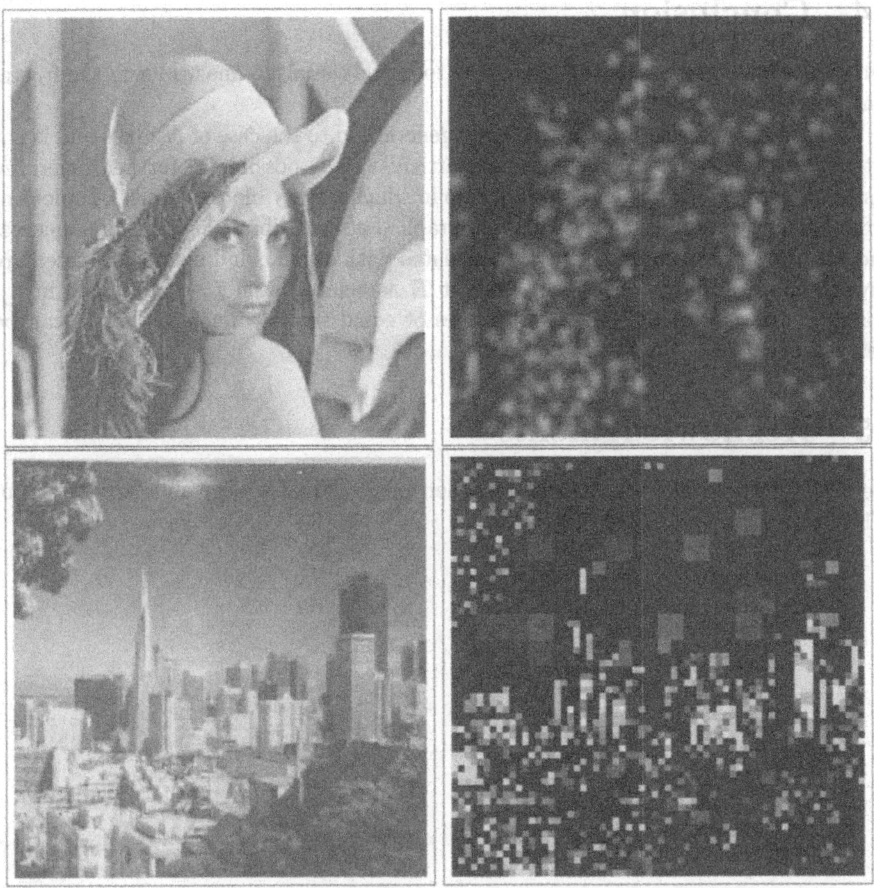

Figure 5.1: Attractors (left) and their dimension (right) using the result of Theorem 5.2. The dimension is shown with white represendng a dimension of 3 and black a dimension of 2.

5.4 Conclusion

Our results demonstrate that it is possible to quickly analyze images using the fractal dimension of their attractors.

We note that this result can be directly generalized to the case of N maps and is not restricted to the 3-dimensional case. In particular, it is possible to further generalize this result to the general fractal coding case in which library blocks consist of unions of target blocks by considering the following graph. Let each library block L_i in a general (Jacquin) fractal encoding scheme be associated with a node, and define a directed arc between node L_i and node L_j if the target T_i associated with L_i is contained in L_j. This graph partitions the image into cycles of equal fractal dimension which can be analyzed in terms of the theorem above.

5.5 Acknowledgments

The authors would like to thank K. Falconer for providing a clear and "trivial" proof of Theorem 5.2.

Chapter 6

Fractal Image Compression via Nearest Neighbor Search

Dietmar Saupe

Abstract: In fractal image compression the encoding step is computationally expensive. A large number of sequential searches through a list of domains (portions of the image) are carried out while trying to find best matches for other image portions called ranges. Our theory developed here shows that this basic procedure of fractal image compression is equivalent to multi-dimensional nearest neighbor search in a space of feature vectors. This result is useful for accelerating the encoding procedure in fractal image compression. The traditional sequential search takes linear time whereas the nearest neighbor search can be organized to require only logarithmic time. The fast search has been integrated into an existing state-of-the-art classification method thereby accelerating the searches carried out in the individual domain classes. In this case we record acceleration factors up to about 50 depending on image and domain pool size with negligible or minor degradation in both image quality and compression ratio. Furthermore, as compared to plain classification our method is demonstrated to be able to search through larger portions of the domain pool without increasing the computation time. In this way both image quality and compression ratio can be improved at reduced computation time. We also consider the application of a unitary transformation of the feature vectors which results in a reduction of the dimensionality of the search space. New results from numerical simulations are reported. Also we provide a brief overview of related work and other complexity reduction methods. This chapter is an extended version of the article [223].

6.1 Introduction

With the ever increasing demand for images, sound, video sequences, computer animations and volume visualization, data compression remains a critical issue regarding the cost of data storage and transmission times. While JPEG currently provides the industry standard for still image compression there is ongoing research in alternative methods. Fractal image compression is one of them.

6.1.1 The Search Problem in Fractal Image Encoding

Basically, a fractal code consists of three ingredients: a partitioning of the image region into portions R_k, called *ranges,* an equal number of other image regions D_k, called *domains* (which may overlap), and for each domain-range pair two *transformations,* a geometric one, $u_k : D_k \to R_k$, which maps the domain to the range, and an affine transformation, v_k, that adjusts the intensity values in the domain to those in the range. The collection of transformations may act on an arbitrary image, g, producing an output image, Tg, which is like a collage of modified copies of the domains of the image g. The iteration of the image operator T is the decoding step, i.e., it yields a sequence of images which converges to an approximation of the encoded image.

The time consuming part of the encoding step is the search for an appropriate domain for each range. The number of possible domains that theoretically may serve as candidates is prohibitively large. For example, the number of arbitrarily sized square subregions in an image of size n by n pixels is of order $O(n^3)$. Thus, one must impose certain restrictions in the specification of the allowable domains. In a simple implementation one might consider as domains, e.g., only sub-squares of a limited number of sizes and positions. This defines the so-called *domain pool.* Now for each range in the partition of the original image all elements of the domain pool are inspected: for a given range R_k and a domain D the transformations u_k and v_k are constructed such that when the domain image portion is mapped into the range the result $v_k f u_k^{-1}(x)$ for $x \in R_k$ matches the original image f as much as possible. This step (called collage coding) uses the well-known least squares method. From all domains in the pool we select the best one, i.e., the domain D that yields the best least squares approximation of the original image in the range. In other words, fractal image coding consists in approximating the image as a collage of transformed pieces of itself, which can be viewed as a collection of self-similarity properties. The better the collage fits the given image the higher the fidelity of the resulting decoded image.

In this article we cannot explain any further details and variations of fractal image compression. For introductory texts or reviews see, for example, [22, 81, 87, 126]. For a bibliographic survey of the field of fractal image compression see our paper [226].

JPEG can be termed *symmetric* in the sense that the encoding and decoding phases require about the same number of operations. On the contrary, fractal image compression allows fast decoding but suffers from long encoding times. In our papers [221, 222] we introduced and discussed a new twist for the encoding process. In [223] we demonstrated its efficiency by a series of empirical studies. In this expository article we review the material in [223] and extend the discussion of the acceleration technique.

During encoding a large pool of image subsets, the domain pool, has to be searched repeatedly many times, which by far dominates all other computations in the encoding process. If the number of domains in the pool is N, then the time spent for each search is *linear* in N, $O(N)$. Previous attempts to reduce the computation times employ *classification schemes* for the domains based on image features such as edges or bright spots. Thus, in each search only domains from a particular class need to be examined. However, this approach reduces only the factor of proportionality in the $O(N)$ complexity.

The main idea of the acceleration is the following. First we show that the fundamental searching for optimal domain-range pairs is equivalent to solving nearest-neighbor problems in a suitable Euclidean space of feature vectors of domains and ranges. The data points are given by the feature vectors (also called multi-dimensional keys) of the domains, while the query point is defined as the feature vector of a given range.

Our application of this reasoning is that we may substitute the sequential search in the domain pool (or in one of its classes) by multi-dimensional nearest neighbor search. There are well known data structures and algorithms for this task which operate in *logarithmic* time, $O(\log N)$, a definite advantage over the $O(N)$ complexity of the sequential search. Our implementation and empirical studies show that these time savings in fact do provide a considerable acceleration of the encoding and, moreover, allow an enlargement of the domain pool yielding improved image fidelity.

6.1.2 Previous Work

The problem of time complexity in fractal image compression was already very clear right from the beginning. The earliest — and some of the latest — implementations use the concept of classification as a tool for complexity reduction. The classification of ranges and domains serves the purpose of reducing the number of domains in the domain pool which need to be considered as a partner for a given range. Just like in 'real life', birds of a feather flock together. For example, if the original image contains an edge running through a range, then domains which contain only 'flat' pieces of the image can be safely discarded when searching for a good match for that range. Jacquin [124, 126] sorts ranges and domains into three classes (shade blocks, edge blocks, and midrange blocks) following a classification, well-known in image processing. The classification of Fisher, Jacobs, and Boss [122, 87] is made with a clever design of a variable number of classes (3–24–72) taking into account not only intensity values but also intensity variance across an image block.[1] Here an ordering of variances in the four sub-quadrants of an image block is used. Although successful this approach is not satisfying in the sense that a notion of neighboring classes is not available. So if the search in one class does not yield a sufficiently strong match for a domain, one cannot extend the search to any neighboring classes. The solution for this problem has been given by Caso, Obrador and Kuo in [44], where the unflexible ordering of variances of an image block has been replaced by a vector of variances. These variance vectors are strongly quantized leading to a collection of classes where each class has a neighborhood of classes which can be searched.

[1] Since we will use Fisher's implementation as a test bed for our experiments we will describe their approach in some more detail in section 6.5.

Attempts have also been made to design the set of classes adaptively, i.e. depending upon the target images. Lepsøy and Øien [151] proposed an adaptive codebook clustering algorithm and Boss and Jacobs [36] considered an archetype classification based on a set of training images. A first algorithm based on the self-organizing map (SOM) of Kohonen for codebook clustering was presented by Bogdan in [35]. In [106] the promising SOM approach for clustering has been combined with the nearest neighbor approach presented here.

In addition to the complexity reduction by using these classification schemes Bedford, Dekking, and Keane [28] suggested to use inner products with Rademacher functions to further exclude certains domains from consideration for a partner to a given range. Three multi-resolution approaches with the same goal are presented by Caso, Obrador and Kuo in [44], by Dekking in [60, 61], and by Lin and Venetsanopoulos in [157]. Another approach employing an incremental procedure at the pixel level has been given by Bani-Eqbal in [17, 16]. For a survey of some of these complexity reduction methods see the article [225].

Complexity reduction methods that are somewhat different in character are based on reducing the domain pool rigorously to a small subset of all possible domains. For example, in the work that followed Monro and Dudbridge [186] for each range the codebook block to be used to cover the range is uniquely predetermined to be a specific block that contains the range block. A similar idea has been pursued by Hürtgen and Stiller [118] where the search area for a domain is restricted to a neighborhood of the current range or additionally a few sparsely spaced domains far from the range are taken into account as an option. In [224] we have noticed that one can discard domain blocks with low variance values with a little degradation in fidelity for which we can compensate by an improved storage scheme for the domain addresses. Signes [232] and Kominek [142] pursue similar ideas for domain pool reduction. An adaptive version of spatial search based on optimizing the rate-distortion performance is presented by Barthel in [23].

The remainder of this chapter is organized as follows. In the following section we present mathematical notation and a first simple formula for the least squares error based on projections. In section 6.3 we provide the definition of the multi-dimensional keys, a theorem that establishes the mathematical foundation, and a corollary giving a necessary and sufficient condition in terms of the multi-dimensional feature vectors for a domain codebook block in order to satisfy a given tolerance criterion for the least squares error. The following sections 6.4 and 6.5 outline some practical comments on the implications of the theory and explain our implementation. Section 6.6 discusses our experiments and their results. Finally, in section 6.7, we discuss other work as far as it relates to feature vectors and then give a conclusion.

6.2 A Formula for the Least Squares Error Based on Projections

For the discussion in the paper let us assume that an image is partitioned into non-overlapping square blocks of size $N \times N$ called *range blocks*. This is not a restriction since it will be clear how the principles described carry over to more general partitions.

We consider each range block as a vector R in the linear vector space $\mathcal{R}^n$ where $n = N \times N$. The conversion from a square subimage of side length N to a vector of length $n = N^2$ can be accomplished, e.g., by scanning the block line by line. Working with vectors in place of 2D-arrays simplifies the notation considerably without losing generality.

The *domain pool* is a collection of square blocks which are typically larger than the ranges and taken also from the image, called *domain blocks*. The domain pool is enlarged by including blocks obtained after applying the eight isometrical operators to the domain blocks (i.e., rotations and reflections). Finally, by pixel averaging, the size of these blocks is reduced to the size of a range block. The resulting blocks are called *codebook blocks*.

In the encoding process for a range block a search through the codebook blocks is required. A vector representing a codebook block will be denoted by D. A small set of $p < n$ blocks independent from the image is also considered. We represent them by the vectors $B_1, B_2, \ldots, B_p \in \mathcal{R}^n$, which are chosen so as to form an orthonormal basis of a p-dimensional subspace of $\mathcal{R}^n$. They are known as *the fixed basis blocks*. The encoding problem can then be stated as the least squares problem

$$E(D,R) = \min_{a,b_1,\ldots,b_p \in \mathcal{R}} \|R - (aD + \sum_{k=1}^{p} b_k B_k)\| = \min_{x \in \mathcal{R}^{p+1}} \|R - Ax\|, \qquad (6.1)$$

where A is an n by $p+1$ matrix whose columns are $D, B_1, B_2, \ldots, B_p$ and $x = (a, b_1, \ldots, b_p) \in \mathcal{R}^{p+1}$ is a vector of coefficients.[2] This problem should be solved for all codebook blocks D and the one which gives the smallest error $\|R - (aD + \sum_1^p b_k B_k)\|$ is selected on condition that the value of the scaling factor a for the codebook block D ensures the convergence of the decoding process (e.g., by requiring $|a| < 1$). This condition on a can be removed when one uses the orthogonalized representation of Øien [198]. A basic result of linear algebra states that if the codebook block D is not in the linear span of the fixed basis blocks $B_1, \ldots, B_p$, then the minimization problem (6.1) has the unique solution

$$\bar{x} = (A^T A)^{-1} A^T R$$

where the matrix $A^+ = (A^T A)^{-1} A^T$ is also known as the pseudo-inverse of A. Thus, the range block R is approximated by the *collage* block AA^+R where AA^+ is the orthogonal projection matrix onto range(A). Now let P be the orthogonal projection operator which projects $\mathcal{R}^n$ onto the subspace $\mathcal{B}$ spanned by only the fixed basis blocks $B_1, B_2, \ldots, B_p$. Thus, by orthogonality of the fixed basis blocks we have for $R \in \mathcal{R}^n$

$$PR = \sum_{k=1}^{p} b_k B_k = \sum_{k=1}^{p} \langle R, B_k \rangle B_k.$$

Then the range block R has a unique orthogonal decomposition $R = OR + PR$ where the operator $O = I - P$ projects onto the orthogonal complement $\mathcal{B}^\perp$. For $Z =$

[2] In practise usually the root mean square error (rms) is used equivalently in place of $E(D,R)$. This is just $E(D,R)/\sqrt{n}$. Also note, that we have used the Euclidean norm instead of the squared norm. We use the notation $\langle \cdot, \cdot \rangle$ for the common inner product in $\mathcal{R}^n$, thus, $\|x\| = \sqrt{\langle x, x \rangle}$.

$(z_1, \ldots, z_n) \in \mathcal{R}^n \backslash \mathcal{B}$, we define the operator

$$\phi(Z) = \frac{OZ}{||OZ||}. \tag{6.2}$$

Now for a given domain block $D \notin \mathcal{B}$ the collage block AA^+R can be given explicitly as

$$AA^+R = \langle R, \phi(D)\rangle\phi(D) + \sum_{k=1}^{p}\langle R, B_k\rangle B_k. \tag{6.3}$$

To get the least squares error we use the orthogonality of $\phi(R), B_1, \ldots, B_p$ to express the range block R as

$$R = \langle R, \phi(R)\rangle\, \phi(R) + \sum_{k=1}^{p}\langle R, B_k\rangle B_k. \tag{6.4}$$

We insert the result for R in the first part of the collage block AA^+R in (6.3) and after three lines of computations find that

$$\langle R, \phi(D)\rangle\phi(D) = \langle R, \phi(R)\rangle\, \langle\phi(D), \phi(R)\rangle\, \phi(D).$$

Thus, the collage block can be rewritten as

$$AA^+R = \langle R, \phi(R)\rangle\, \langle\phi(D), \phi(R)\rangle\, \phi(D) + \sum_{k=1}^{p}\langle R, B_k\rangle B_k. \tag{6.5}$$

Using (6.4) and (6.5) we can now compute the least squares error

$$E(D, R) = ||R - AA^+R|| = \sqrt{\langle R - AA^+R, R - AA^+R\rangle}.$$

The result follows after a few lines of calculations, namely

$$E(D, R) = \langle R, \phi(R)\rangle\, \sqrt{1 - \langle\phi(D), \phi(R)\rangle^2}. \tag{6.6}$$

Thus, the minimization of the error $E(D, R)$ among domain codebook blocks D can be achieved using an *angle criterion:* The minimum of $E(D, R)$ occurs when the squared inner product $\langle\phi(D), \phi(R)\rangle^2$ is maximal. Since $\langle\phi(D), \phi(R)\rangle^2 = \cos^2\angle(\phi(D), \phi(R))$ this means minimizing the angle $\angle(\phi(D), \phi(R))$, or, equivalently $\angle(OD, OR)$.

In the next section we will see how the expression (6.6) for the least squares error can be seen in terms of the distances between $\phi(R)$ and $\pm\phi(D)$.

6.3 Searching in Fractal Image Compression is Nearest Neighbor Search

We consider a set of N_D codebook blocks $D_1, \ldots, D_{N_D} \in \mathcal{R}^n$ and a range block $R \in \mathcal{R}^n$. As before we let $E(D_i, R)$ denote the smallest possible least squares error of an approximation of the range data R by an affine transformation of the domain data

D_i (see (6.1) and (6.6)). Then the searching in fractal image compression consists in finding the domain codebook block index k with

$$k = \arg \min_{i=1,\dots,N_D} E(D_i, R).$$

The following theorem provides the mathematical foundation for our feature vector approach. We generalize the original version of the theorem to the case of several fixed basis blocks. We use the notation for the normalized projection operator ϕ and the linear span $\mathcal{B}$ of the (orthonormalized) fixed basis blocks $B_1, \dots, B_p$ as before.

Theorem 6.1 [221, 222].
Let $n \geq 2$ and $X = \mathcal{R}^n \backslash \mathcal{B}$. Define the function $\Delta : X \times X \to [0, \sqrt{2}]$ by

$$\Delta(D, R) = \min\left(\|\phi(R) + \phi(D)\|, \|\phi(R) - \phi(D)\|\right).$$

For $D, R \in X$ the least squares error

$$E(D, R) = \min_{a, b_1, \dots, b_p \in \mathcal{R}} \|R - (aD + \sum_{k=1}^{p} b_k B_k)\|$$

is given by

$$E(D, R) = \langle R, \phi(R)\rangle \, g(\Delta(D, R))$$

where

$$g(\Delta) = \Delta \sqrt{1 - \frac{\Delta^2}{4}}.$$

Proof. Since

$$\begin{aligned} \|\phi(R) \pm \phi(D)\| &= \sqrt{\langle \phi(R) \pm \phi(D), \phi(R) \pm \phi(D)\rangle} \\ &= \sqrt{\langle \phi(R), \phi(R)\rangle \pm 2\langle \phi(D), \phi(R)\rangle + \langle \phi(D), \phi(D)\rangle} \\ &= \sqrt{2(1 \pm \langle \phi(D), \phi(R)\rangle)} \end{aligned}$$

we have

$$\Delta(D, R) = \min\left(\|\phi(R) \pm \phi(D)\|\right) = \sqrt{2(1 - |\langle \phi(D), \phi(R)\rangle|)}.$$

This last equation yields $|\langle \phi(D), \phi(R)\rangle| = 1 - \Delta(D, R)^2/2$ and inserting the square of that in the formula (6.6) for $E(D, R)$ completes the proof.

The theorem states that the least squares error $E(D_i, R)$ is proportional to the simple function g of the Euclidean distance Δ between the normalized projections $\phi(D_i)$ and $\phi(R)$ (or $-\phi(D_i)$ and $\phi(R)$). The value of the result is not in terms of a speed-up of the calculation of the least squares error, but of a more fundamental nature. Since $g(\Delta)$ is a monotonically increasing function for $0 \leq \Delta \leq \sqrt{2}$ we conclude that *the minimization of the least squares errors $E(D_i, R)$ for $i = 1, \dots, N_D$ is equivalent*

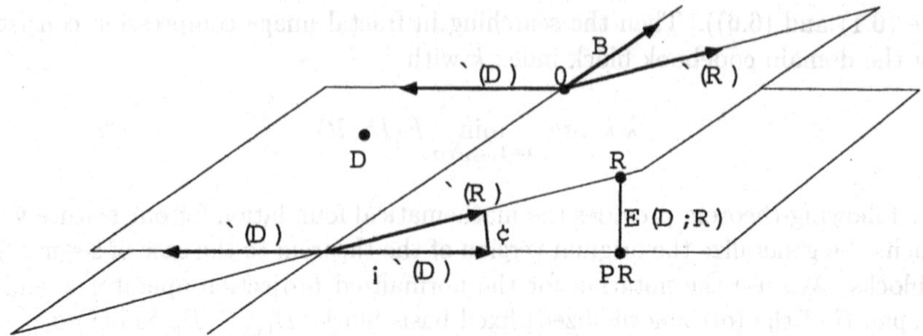

Figure 6.1: Illustration of the geometry underlying the theorem.

to the minimization of the distance expressions $\Delta(D_i, R)$. Formally, the sought index k is

$$\begin{aligned} k = \arg \min_{i=1,\dots,N_D} E(D_i, R) &= \arg \min_{i=1,\dots,N_D} \langle R, \phi(R)\rangle\, g(\Delta(D_i, R)) \\ &= \arg \min_{i=1,\dots,N_D} g(\Delta(D_i, R)) \\ &= \arg \min_{i=1,\dots,N_D} \Delta(D_i, R) \end{aligned}$$

Thus, we may replace the computation and minimization of N_D least squares errors $E(D_i, R)$ by the search for the nearest neighbor of $\phi(R) \in \mathcal{R}^n$ in the set of $2N_D$ vectors $\pm\phi(D_i) \in \mathcal{R}^n$.

For an interpretation we note that for given $D \in \mathcal{R}^n$ all vectors of the form $a\phi(D) + \sum_{k=1}^{p} b_k B_k$ can be exactly represented as a linear combination of D and $B_1, \dots, B_p$ (with zero least squares error). These vectors form a $(p+1)$-dimensional subspace of $\mathcal{R}^n$, an orthonormal basis of which is given by $\phi(D)$ and $B_1, \dots, B_p$. For all D in this space $\phi(D)$ is unique up to choice of the sign. Thus, $\phi(D)$ may serve as a 'representative' of this space and will become the multi-dimensional key for searching.

The problem of finding closest neighbors in Euclidean spaces has been thoroughly studied in computer science. For example, a method using kd-trees that runs in expected logarithmic time is presented in [94] together with pseudo code. After a preprocessing step to set up the required kd-tree, which takes $O(N \log N)$ steps, the search for the nearest neighbor of a query point can be completed in expected logarithmic time, $O(\log N)$. However, as the dimension d increases, the performance may suffer. A method that is more efficient in that respect, presented in [10], produces a so-called approximate nearest neighbor. For domain pools that are not large other methods, that are not based on space-partitioning trees, may perform better. For example, the modified equal average nearest neighbor search (ENNS) in [148] seems to be one of the best.

Before we turn to practical issues, we remark, that we can use the result of the Theorem 6.1 in order to identify all codebook blocks D_i that satisfy a given tolerance criterion $E(R, D_i) \leq \delta$. In other words, solving the equality for Δ in the expression for the error $E(D,R)$ in the theorem yields a necessary and sufficient condition for a codebook block D to fulfill the tolerance criterion.

Corollary 6.1 (A necessary and sufficient condition)
Let $\delta > 0$ and $n \geq 2$. Let R and D be in $\mathcal{R}^n \backslash \mathcal{B}$ with $\langle R, \phi(R) \rangle \geq \delta$. Then $E(D,R) = \min_{a,b_1,\ldots,b_p \in \mathcal{R}} \|R - (aD + \sum_{k=1}^p b_k B_k)\| \leq \delta$ if and only if

$$\Delta(D,R) \leq \sqrt{2 - 2\sqrt{1 - \frac{\delta^2}{\langle R, \phi(R) \rangle^2}}},$$

where $\Delta(D,R)$ is defined as in Theorem 6.1.

Proof. From Theorem 6.1, $E(D,R) = \langle R, \phi(R) \rangle\, g(\Delta(D,R))$ with $g(\Delta) = \Delta\sqrt{1 - \frac{\Delta^2}{4}}$. Thus, for $0 \leq \Delta \leq \sqrt{2}$ we have $E(D,R) \leq \delta$ if and only if $\Delta^4 - 4\Delta^2 + 4\delta^2/\langle R, \phi(R) \rangle^2 \geq 0$. From this the assertion easily follows.

The condition $\langle R, \phi(R) \rangle \geq \delta$ does not impose any restrictions. To see this, observe that in the case of $\langle R, \phi(R) \rangle < \delta$ we already have

$$E(D,R) = \langle R, \phi(R) \rangle \sqrt{1 - \langle \phi(D), \phi(R) \rangle^2} \leq \langle R, \phi(R) \rangle < \delta.$$

for *any* codebook block D. Thus, it suffices to encode R only using the fixed basis blocks, i.e., by $\sum_{k=1}^p b_k B_k$.

6.4 Practical Considerations

We continue with some remarks on generalizations and implications of the theory from the last section.

In practice, there is a limit in terms of storage for the feature vectors of domains and ranges. For example, the keys for ranges of size of 8 by 8 pixels require 64 floating point numbers each. Thus, 32K domains from a domain pool would already fill 8 MB of memory on a typical workstation, while we would like to work with pools of a hundred thousand and more domains. To cope with this difficulty, we settle for a compromise and proceed as follows. We *down-filter* all ranges and domains to some prescribed dimension of moderate size, e.g., $d = 4 \times 4 = 16$. Moreover, each of the d components of a feature vector is *quantized* (8 bits/component suffice). This allows the processing of an increased number of domains and ranges, however, with the implication that the formula of the theorem is no longer exact but only approximate. This, however, is not a severe disadvantage as pointed out in the following remark and as demonstrated in the experiments later on.

The approach of pixel averaging in order to reduce the dimensionality of the domains and ranges (64 and higher is typical) to a more feasible number (here $d = 16$) may be improved by better concentrating relevant subimage information in the d components. Based on our report [221] Barthel et al [23] have suggested and implemented an alternative reduction of dimension. They have used a two-dimensional discrete cosine transformation (DCT) of the projected codebook blocks $\pm\phi(D_i)$. The distance preserving property of the unitary transform carries over the result of our Theorem to the frequency domain and nearest neighbors of DCT coefficient vectors will yield the smallest least squares errors. In practise one computes the DCT for all domains and ranges. Then, from the resulting coefficients, the DC component is ignored and the next d coefficients are normalized and make up the feature vector.

Because of the downfiltering and the quantization of both the feature vectors and the coefficients $a, b_1, \ldots, b_p$, it can happen that the nearest neighbor in feature vector space is not the codebook block with the minimum least squares error using quantized coefficients. Moreover, it could yield a scaling factor a being too large to be allowed. To take that into consideration, we search the codebook not only for the nearest neighbor of the given query point but also for, say, the next 5 or 10 nearest neighbors (this can still be accomplished in logarithmic time using a priority queue). From this set of neighbors the non-admissible domains are discarded and the remaining domains are compared using the ordinary least squares approach. This also takes care of the problem from the previous remark, namely that the estimate by the theorem is only approximate. While the domain corresponding to the closest point found may not be the optimal one, there are usually near-optimum alternatives among the other candidates.

We make two technical remarks concerning *memory requirements* for the kd-tree. Firstly, it is not necessary to create the tree for the full set of $2N_D$ keys in the domain pool. We need to keep only one multi-dimensional key per domain, e.g., by keeping only the key which has a non-negative first component (multiply key by -1 if necessary). In this set-up a kd-tree of all $2N_D$ vectors has two symmetric main branches (separated by a coordinate hyperplane), thus, it suffices to store only one of them. Secondly, there is some freedom in the choice of the *geometric transformation* that maps a domain onto a range coming from the 8 possible rotations and reflections of a square subimage. This will create a total of 8 entries per domain in the kd-tree, enlarging the size of the tree. However, we can get away without this tree expansion. To see this, just note that we may instead consider the 8 transformations of the *range* and search the original tree for nearest neighbors of each one of them.

The *preprocessing time* to create the data structure for the multi-dimensional search is not a limitation of the method as demonstrated by our experiments.

6.5 Implementation

Our implementation is based on Fisher's adaptive quadtree algorithm [87]. The image region is subdivided into squares using a quadtree data structure. The leaf nodes of the tree correspond to the ranges while the domain pool for a given range consists of image subsquares of twice the size. The number of domains in the pool can be adjusted by parameters of the method. The quadtree construction is adaptive in the sense that for each node the corresponding domain pool is searched for a matching domain. That node then becomes a leaf node if the search yielded a match satisfying a given tolerance criterion. If the tolerance is not met, the square image region corresponding to the node is broken up into four subsquares which become the child nodes of the given node (see [81, 87] for details). Although the performance of this algorithm in terms of compression and fidelity is not the best possible it serves as a good test bed for our experiments in which we are aiming at evaluating the capabilities of the multi-dimensional nearest neighbor search in comparison to traditional complexity reduction attempts. In that respect Fisher's code is excellent because it contains an advanced classification method.

This classification works as follows. A square range or domain is subdivided into its four quadrants (upper left, upper right, lower left, and lower right). In the quadrants the average pixel intensities A_i and the corresponding variances V_i are computed ($i = 1, \ldots, 4$). It is easy to see that one can always orient (rotate and flip) the range or domain such that the average intensities are ordered in one of the three following canonical orientations:

$$\begin{aligned} &\text{major class 1: } A_1 \geq A_2 \geq A_3 \geq A_4, \\ &\text{major class 2: } A_1 \geq A_2 \geq A_4 \geq A_3, \\ &\text{major class 3: } A_1 \geq A_4 \geq A_2 \geq A_3. \end{aligned}$$

Once the orientation of the range or domain has been fixed accordingly, there are 24 different possible orderings of the variances which define 24 subclasses for each major class. If the scaling factor a is negative then the orderings in the classes must be modified accordingly. Thus, for a given range two subclasses need to be searched in order to accommodate positive and negative scaling factors.

Our implementation of the multi-dimensional nearest neighbor search is based on a code which was kindly provided by Arya and Mount [10]. The critical advances of this algorithm in comparison to the classical kd-tree approach of Friedman et al [94] are twofold. After a space partitioning tree (such as a kd-tree) has been set up for the data a priority list of nodes is maintained during nearest neighbor searching such that nodes corresponding to data volumes that are closest to the query point are searched first. Secondly, the requirement of finding the closest neighbor is relaxed to that of finding an $(1 + \epsilon)$-approximate nearest neighbor. This is a data point whose distance to the query point is not larger that $1 + \epsilon$ times the distance of the query point to its exact closest neighbor. Arya et al report that even with seemingly large parameters such as $\epsilon = 3$ in practise data points are found which have a 50% chance of being the true nearest neighbor and which are on the average only 1.05 times as far the closest neighbor. On the other hand a speed up of a factor of up to 50 over the exact ($\epsilon = 0$) case can be enjoyed. Using a second priority list we have modified the code so that a given number of approximate nearest neighbors can be returned.

We have joined the above two programs providing the following two options used for our test series:

1. The parameter ϵ and the number M of approximate neighbors requested in each search can be specified. Note that the actual number of neighbors returned for a given range is twice as large since two searches must be carried out for one range (one for positive and one for negative scaling factors).

2. The dimension of the keys for the ranges and domains can be chosen. We have worked with dimension 4 and 16 corresponding to a downfiltering of an image block to a 2 by 2 or 4 by 4 image block. This fits well with the compression code since the domain and range sizes are of the form 2^k by 2^k, thus, the filtering can conveniently be done by pixel averaging.

We also implemented the two-dimensional discrete cosine transformation (DCT) of the projected codebook blocks $\pm\phi(D_i)$ and ranges $\phi(R)$ in order to search for nearest neighbors in the frequency domain as suggested by Barthel et al. In our study we vary the number of frequency components retained and compare the performance to the direct approach using pixel averaging.

6.6 Results

In our first series of experiments we make use of the classification and replace the linear search in a class by the approximate nearest neighbor search. Here we cannot yet expect the method to come up with better quality encodings than the plain classification method, since we are searching through the same pool of codebook blocks. Thus, the goal is to show that significant computation time can be saved while providing image fidelity and compression without or with only very minor degradation.

6.6.1 Choosing Parameters for the Approximate Nearest Neighbor Search

Before the computations we need to decide which parameters of the fast search to use: $\epsilon \geq 0$, the number M of $(1+\epsilon)$-approximate nearest neighbors returned, and the dimension d of the key space. Figure 6.2 shows a study for the first two of these parameters. We consider encodings of a 512 by 512 test image (Lena) with a fixed range size (4×4 pixels), i.e., the adaptive quadtree is trivial having the identical minimal and maximal depths. The classification is active, and only one out of 72 classes is searched using the fast search. The dimension of the key space is $d = 16$, while the parameters ϵ and M are varied. The times reported in this subsection are measured on an SGI Indigo2 running an R4000 processor. The result shows that a lower quality but faster search (high value of ϵ) is beneficial in terms of quality obtained for given computation time. We settle for $\epsilon = 3$ and $M = 5$ (35.21 dB, 16 secs). When comparing this to the performance using the linear search with classification (35.56 dB, 52 secs) we obtain in this case threefold speed in exchange for a 0.3 dB drop in peak-signal-to-noise ratio (PSNR).

We have studied the performance regarding the dimension of the key space using images of various sizes and dimensions $d = 4$ and 16. We find that with the smaller dimension $d = 4$ we gain a little processing time, however, we lose up to 2 dB in image fidelity without providing a gain in compression (see also figure 6.3). Thus, in the following we downfilter domains and ranges to 4 by 4 pixels.

6.6.2 Fast Nearest Neighbor Search with Classification

We have used a few popular test images of differing resolutions: 256 by 256 Collie, 512 by 512 Baboon and Lena, and 1024 by 1024 Composite consisting of the following four subimages of resolution 512 by 512: Kiel Harbor, Peppers, Baboon, Lena. Also we vary the domain pool size by factors of 4. This way the scaling behavior of the pure

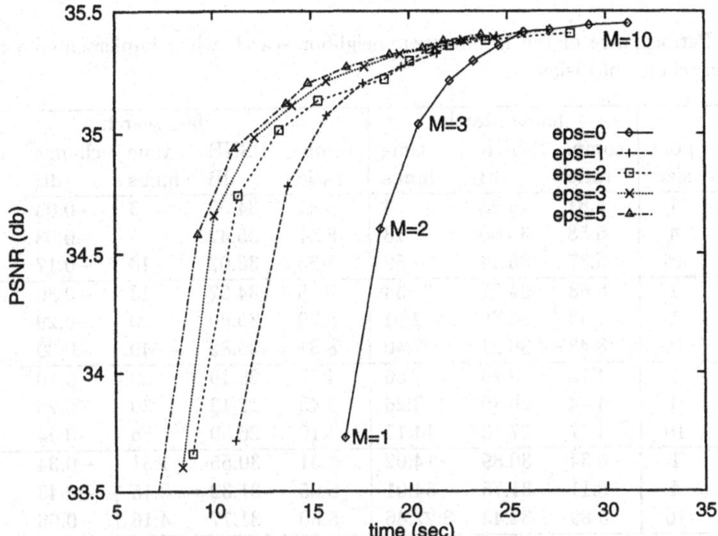

Figure 6.2: For the parameters $\epsilon = 0, 1, 2, 3, 5$ and with $M = 1, ..., 10$ neighbors returned per search we show the PSNR versus computation time. The maximal attainable PSNR is 35.56 dB (obtained in 52 secs with linear search using classification).

classification method can be compared to that of the method using nearest neighbor search.

The parameters used for the adaptive quadtree method are the default parameters [87], except for the tolerance (set at the value 4) and the minimal block size in the range partition, which is set to 4×4 pixels in all cases. Quadrupling the domain pool size is obtained by halving the domain spacing in both image x- and y-directions. The data in table 6.1 summarize the findings.[3] The results are as follows:

- The compression ratios obtained with the fast search versus the linear search differ on the average by less than 0.03. Thus, for all practical purposes the fast search does not significantly worsen the compression ratios.
- As expected a slight loss in image quality is traded in for higher speed in the encoding with the fast search (see the last two columns for the loss in PSNR versus the speed up factor).
- It is interesting to note that although different images may require much different encoding times (compare Lena with Baboon) the speed up factor scales nicely with image size and domain pool size. When the domain pool size is quadrupled or the image size is doubled (which also results in a quadrupled domain pool size) the speed up factor roughly doubles. This is reflected in the computation times.

[3]The timings in the table are improved w.r.t. the data in [223], due to a technical improvement in our code. The times reported in this subsection and the following are measured on an SGI Indy running an R4600 processor.

Table 6.1: Performance of the fast nearest neighbor search when implemented within the classification scheme of Fisher.

		linear search			fast search				
image size	pool size	comp. ratio	PSNR dB	time h:m:s	comp. ratio	PSNR dB	time h:m:s	change dB	speed up
Collie	1	5.81	34.55	4	5.81	34.52	3	−0.03	1.3
256	4	5.58	35.50	13	5.57	35.42	5	−0.08	2.6
	16	5.37	36.19	52	5.35	36.02	10	−0.17	5.2
Lena	1	8.48	34.87	34	8.45	34.57	13	−0.30	2.6
512	4	8.45	35.69	2:01	8.39	35.40	20	−0.29	6.1
	16	8.43	36.21	7:40	8.31	35.82	40	−0.39	11.5
Baboon	1	4.75	25.29	56	4.75	25.19	20	−0.10	2.8
512	4	4.44	26.39	3:26	4.43	26.13	30	−0.26	6.9
	16	4.17	27.13	14:17	4.16	26.69	56	−0.44	15.3
Comp.	1	6.34	30.89	14:02	6.31	30.55	1:31	−0.34	9.3
1024	4	6.11	31.75	54:01	6.05	31.32	2:15	−0.43	24.0
	16	5.89	32.43	3:22:46	5.80	31.77	4:16	−0.66	47.5

While quadrupled domain pool size requires also fourfold computation time with the linear search, only 1.5 to 2 times as much time is needed with the fast search.

The slight loss in image quality can be further reduced by enlarging the number M of neighbors computed in each search or by reducing ϵ. For example, with $M = 10$ and $\epsilon = 3$ we obtained (for pool size 1 and tolerance 8) PSNR degradations of only 0.02, 0.10, 0.06, 0.20 dB with speed up factors of 0.9, 1.7, 1.6, 5.2 for the four images respectively.

6.6.3 Fast Nearest Neighbor Search without Loss in Fidelity

It is also possible to accelerate the encoding of fractal image compression without having to pay the price of a slightly degraded image quality. The idea is to include more classes in the search tree instead of searching through only one class. In a first try we ignore the classification altogether and thus search the entire domain pool for each range. In fact, since all 8 orientations of all domains in all of the 72 subclasses belong to the domain pool, we estimate that about 576 times as many domains are considered in each search as compared to the method with classification (where one out of 72 classes is searched). Thus, the searching covers a lot more domains and may result in a better matching domain. Although this global attack of the problem is clearly overkill, it produces the desired qualities in some of the 12 cases corresponding to the table above. In all cases both image fidelity and compression ratio improve (by 0.88 dB PSNR and 0.30 ratio on the average). However, due to the vast search a significant speed up occurs only for the larger domain pools and images. A more sensible approach would be to enlarge the search by a smaller factor. For example, Fisher [87, page 69] notes that the 24 class search gives almost the same quality results as the 72 class search. Thus, we may expect similar improvements in PSNR and compression ratio

Table 6.2: Performance of the fast nearest neighbor search when implemented within the major classes of the classification scheme of Fisher.

image size	pool size	fast search comp. ratio	PSNR dB	time m:s	change dB	speed up
Collie	1	6.18	35.62	6	+1.10	0.7
256	4	5.88	36.15	9	+0.73	1.4
	16	5.62	36.52	17	+0.50	3.1
Lena	1	9.02	35.59	23	+1.02	1.5
512	4	8.86	36.02	35	+0.62	3.5
	16	8.59	36.27	1:10	+0.47	6.6
Baboon	1	4.76	26.34	41	+1.15	1.4
512	4	4.45	26.96	1:07	+0.83	3.1
	16	4.18	27.43	2:00	+0.75	7.1
Comp.	1	6.48	31.41	3:04	+0.86	4.6
1024	4	6.17	31.89	4:49	+0.57	11.2
	16	5.89	32.20	8:56	+0.43	22.7

but with notably reduced computation times when searching only through one of the major classes for each range.

The results are given in table 6.2 and as expected; instead of losing fidelity we gain up to over 1 dB in PSNR, while the execution times are still less than for the classification method. As compared to the fast search with classification the run times are only doubled although the number of codebook blocks considered in each search is increased by a factor of 24. For example, when we consider the image Lena with the domain pool such that domains are unions of ranges (i.e., pool size 4), the method accelerates fractal image compression using the already fast classification scheme by a factor of 3.5 and in addition improves fidelity by 0.62 dB. In this case even the compression ratio is also improved by 0.41.

6.6.4 On the Use of Orthonormally Transformed Feature Space

In the following experiment we investigate the alternative representation of the feature vectors $\pm\phi(D)$ in an orthonormally transformed space. We use the two-dimensional discrete cosine transformation for this purpose. As mentioned further above, from the DCT coefficients of D, the DC component is deleted and the next d coefficients are normalized and make up the feature vector. We fix the sizes of the ranges to 8×8 pixels and encode the 512×512 image Lena several times using the classification as in section 6.6.2 and with the following variations:

- Straight search in the classes. This gives an upper bound for the quality of all encodings in this study.

- Using the fast nearest neighbor search with feature vectors computed from pixel averaging down to size 2×2, i.e., $d = 4$.
- Using the fast nearest neighbor search with feature vectors computed from pixel averaging down to size 4×4, i.e., $d = 16$.
- Using the fast nearest neighbor search with feature vectors computed from a variable number of DCT coefficients, i.e., $d = 1, 2, \ldots, 15$.

We record the corresponding PSNR and repeat the experiment with fourfold and sixteenfold expanded domain pools. Figure 6.3 displays the results. For an interpretation of the graph we note:

- We already know that the achievable quality with pixel averaging and $d = 2 \times 2$ is poor. However, when we take 4-dimensional feature vectors from the DCT coefficients, the results are much better (0.2, 0.6, 1.0 dB for domain pool size 1x, 4x, 16x). This confirms the idea of exploiting the energy compaction in the first components of the transform.
- The fidelity obtained with pixel averaging and $d = 4 \times 4$ can be achieved with about 10 DCT coefficients. Thus, we can reduce the dimension of the feature space by 6 without penalty. This is a useful insight as the nearest neighbor search is more efficient in lower dimensions. However, the time for the additional DCT must also be taken into account.[4]

6.6.5 Quantization Issues

In the last experiment we study the effects of quantization. Quantization occurs at two places, namely for the feature vectors used in the nearest neighbor search, and then also for the coefficients $a, b_1, \ldots, b_p$. Thus, the nearest neighbor in feature vector space is not necessarily corresponding to the codebook block with the minimum least squares error using quantized coefficients.

We have compared all 4×4 range blocks R with all 8×8 domain blocks, which were first downfiltered codebook blocks D of size 4×4. For a given range the codebook blocks can be ordered in one of four ways, namely according to

1. root mean square error, which (for 4×4 image blocks) is 0.25 times the minimum least squares error $E(R, D)$, without quantization of the coefficients
2. root mean square error, but with quantization of the coefficients (uniform quantization with 5 bits for the scaling coefficient a and 7 bits for the offset b_1, only the constant fixed basis block is used)
3. distance measure $\Delta(\phi(D), \phi(R))$, treating the components of the feature vectors as unquantized real numbers

[4] Wohlberg and de Jager [258] report that the DCT representation reveals an improvement in computation time by a factor of 1.5 to 2.0 but give only few details about the parameters of their experiment.

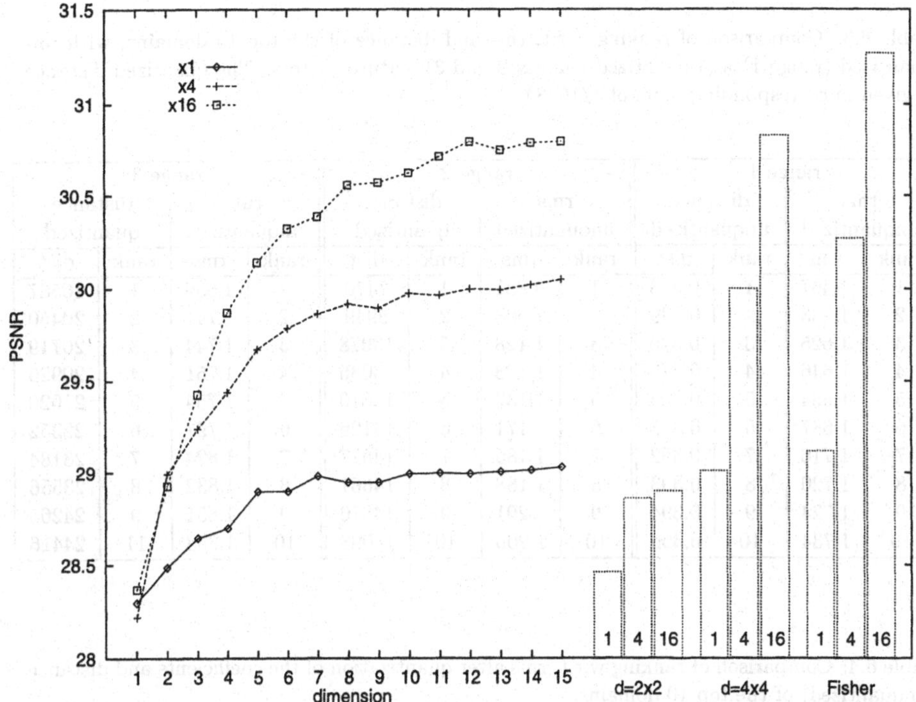

Figure 6.3: Using DCT coefficients for feature vectors. Results are shown for three domain pool sizes (1, 4, 16). On the left the fidelity in dB when using transformed feature vectors of dimension 1 to 15. The three groups of bars on the right correspond to the use of untransformed feature vectors (pixel averaging down to 2×2 or 4×4 pixels) and to the full search in each class as in the unaltered program of Fisher.

4. distance measure $\Delta(\phi(D), \phi(R))$, but with quantization of the components of the feature vectors (uniform quantization with 8 bits/component)

When we compare the ordering of the codebook blocks with respect to criteria 1 and 3 we expect a one-to-one correspondence, because that is just the statement of the theorem in section 6.3. This is exactly what happens as the table 6.3 (first 4 columns) confirms.

To investigate the losses due to the 8 bit quantization of the feature vector components we similarly compare the rankings of rms versus distance based on quantized vectors (table 6.3, columns 5 to 12) for two typical ranges. The table shows that there seems to be almost no negative quantization effect. The rankings are much the same. Thus, our quantization to one byte per feature vector component is appropriate.

In the last study in this series we examine the quantization effect that occurs with the finite resolution of the collage coefficients. In table 6.4 we display the rankings for three ranges using the (unquantized) distance in feature space relative to that given by the rms after coefficient quantization. Here we see that the quantization

Table 6.3: Comparison of ranking w.r.t. rms and distance of the top 10 domains, with unquantized (range 1) and quantized (ranges 2 and 3) feature vectors. The quantized distance is given in corresponding units of 1/16384.

range 1				range 2				range 3			
rms unquantized		distance unquantized		rms unquantized		distance quantized		rms unquantized		distance quantized	
rank	rms	rank	dist	rank	rms	rank	dist	rank	rms	rank	dist
1	1.467	1	0.333	1	0.881	1	7470	1	1.596	1	16567
2	1.533	2	0.349	2	1.008	2	9949	2	1.741	2	20450
3	1.625	3	0.370	3	1.128	5	13078	3	1.744	3	20719
4	1.646	4	0.375	4	1.128	4	13035	4	1.751	4	20926
5	1.684	5	0.384	5	1.132	3	13010	5	1.754	5	21020
6	1.687	6	0.385	6	1.171	6	14126	6	1.797	6	22352
7	1.716	7	0.392	7	1.185	7	14577	7	1.824	7	23184
8	1.720	8	0.393	8	1.188	8	14661	8	1.832	8	23556
9	1.724	9	0.394	9	1.201	9	14870	9	1.851	9	24265
10	1.734	10	0.396	10	1.205	10	15098	10	1.858	11	24416

Table 6.4: Comparison of ranking w.r.t. rms after quantization of the coefficients and distance (unquantized) of the top 10 domains.

rms (quantized)		1	2	3	4	5	6	7	8	9	10
distance	range 4	2	7	9	11	6	10	18	22	8	1
(unquantized)	range 5	27	9	48	39	46	63	56	66	84	41
	range 6	1	3	2	9	6	8	11	14	4	22

of the coefficients does have a significant effect. In some cases (range 5) we lose the best domain when we restrict the search to the ten best ones according to distance. However, most cases are similar to range 6, which is not quite so bad. This shows, that one cannot expect to achieve an overall lossless result when applying the nearest neighbor search in feature space with the restriction on the best 10 codebook blocks (compare also figure 6.2).

6.7 Other Work Related to Feature Vectors

In this section we briefly discuss other work as it relates to our feature vector approach. The material can be grouped into three parts: the straight usage of the projected and normalized codebook blocks as in this paper, the attempt of using invariant moments of ranges and domains, and tree structured methods.

6.7.1 Straight Feature Vectors

A forerunner of feature vectors as used in this paper has been presented by Hürtgen and Stiller [118]. As in the classification of Fisher, Jacobs, and Boss an image block is partitioned into its four quadrants and their mean intensities are computed. Then a vector consisting of four bits is constructed as follows: the i-th bit is 1 if the mean of the i-th quadrant is above the overall mean, and 0 otherwise. Thus, in the terminology of this paper, this is our feature vector after downsampling to size $d = 2 \times 2$ and quantizing to 1 bit per component. Due to these strict limitations a nearest neighbor search is not practical, rather, these vectors serve as a means for classification into 16 classes. Then a range is compared only with codebook blocks from the same class. We remark, that in order to accommodate negative scaling factors a a different class ought to be searched, a fact, that had been overlooked in [118].

Kominek in [143, 142] follows the straight feature vector approach with the difference that another type of data structure for the multi-dimensional feature vectors (r-trees) is adopted. However, the method previously presented in [221, 222] is not completely realized: the necessary pairing of positive/negative feature vectors to support negative scaling factors is ignored (as in [118]) and the search for a nearest neighbor in the r-tree is unnecessarily suboptimal (only a single bucket is inspected).

6.7.2 Invariant Moments

In [195], Novak assigns a 4-dimensional feature vector to each block. The components of the feature vector are certain moment invariants defined from the grey level distribution within the block. A useful property of these moment invariants is that they are invariant with respect to the geometric transformation, i.e., one feature vector suffices for each domain. The isometric versions of that domain block then have the same moment vector. However, the moments are not invariant w.r.t. the affine transformation regarding the luminance. To cope with this problem Novak proposed a normalization procedure. There are several problems with this approach: Again, the negative intensity blocks are omitted from consideration causing the loss of some of the possible fidelity. The values of the invariant moments range over several orders of magnitude, thus, a logarithmic rescaling becomes necessary before nearest neighbor search becomes feasible. And then, most importantly, the method is intuitive in the sense that no supporting theory is given to the goal that closeness in the feature space ensures good approximations in the least squares sense. The fact is, that such a theory cannot exist. Novak worked with triangular partitioning, and Frigaard continued the work in [95] using a quadtree partitioning. However, Frigaard does not normalize feature vectors with respect to mean and variance in order to make the moments invariant relative to the affine luminance transformation. He reports that normalizing would in fact degrade the overall quality of an encoding of an image, which apparently documents the weakness of the method.

Götting, Ibenthal, and Grigat [98] and Popescu and Yan [210] also pursue complexity reduction using invariant moments of different types.

6.7.3 Tree Structured Methods

Besides the dimensional reduction and the variance based classification mentioned above Caso, Obrador and Kuo propose a tree structured search in [44]. The pool of codebook blocks is recursively organized in a binary tree. Initially two (parent) blocks are chosen randomly from the pool. Then all blocks are sorted into one of two bins depending on by which of the two parent blocks the given block can be covered best in the least squares sense. This results in a partitioning of the entire pool into two subsets. The procedure is recursively repeated for each one of them until a prescribed bucket size is reached. Given a range one can then compare this block with the blocks at the nodes of the binary tree until a bucket is encountered at which point all of the codebook blocks in it are checked. This does not necessarily yield the globally best match. However, the best one (or a good approximate solution) can be obtained by extending the search to some nearby buckets. A numerical test based on the angle criterion is given for that purpose. The procedure is related to the nearest neighbor approach since the least squares criterion (minimize $E(D, R)$) is equivalent to the distance criterion (minimize $\Delta(\phi(D), \phi(R))$. Thus, the underlying binary tree can be considered to be randomized version of the kd-tree structure we have used here.

Van der Walle [246] worked on a wavelet representation of fractal image compression, where similarly to ordinary fractal image compression, range vectors (corresponding to subtrees of the tree of wavelet coefficients) have to be matched with domain vectors (also corresponding to nodes of the wavelet tree), which may be scaled by an arbitrary scaling factor. For each node a feature vector is generated based on angles between the coefficient vectors and axes in the wavelet coefficient space. These vectors are then sorted into a multi-dimensional space-partitioning data structure within which the fast search is organized. In terms of distances of feature vectors $\pm\phi(D)$ the interpretation is as follows: We define a small set of anchor points in feature space (e.g., at the positions of the main principal components of the set of all feature vectors). For each (projected and normalized) codebook block as well as for each range block we compute the distances Δ to the anchor points. Then a point in feature space that is close to a given range feature vector must necessarily have distances to the anchor points that are near those of the range. To facilitate the search for such codebook blocks, the blocks can be organized in a tree structure. The core of this method for finding nearest neighbors is reminiscent of the annulus testing known in vector quantization [117].

Of all the methods discussed in this section only the tree structured ones correctly consider positive and negative feature vector corresponding to positive and negative scaling coefficients in the collage.

6.8 Conclusion

We have reviewed the theory which links the domain-range comparison fundamental to fractal image compression with nearest neighbor search. This result leads to a new technique at the core of the encoding process. It consists of the fast approximate nearest neighbor search in a multi-dimensional feature space and can easily be integrated into existing implementations of classification methods. The approach reduces the time complexity of the encoding step thereby creating faster fractal image compression. The speed up can be adjusted so that it comes with only minor degradation in image quality and compression ratio or with improvements in both fidelity and compression.

Acknowledgements

The author thanks Klaus Bayer, Kai Uwe Barthel, Amitava Datta, Raouf Hamzaoui, Thomas Ottmann, and Sven Schuierer for fruitful discussions, Sunil Arya and Dave Mount for the fast nearest neighbor search code and Yuval Fisher for the adaptive quadtree code. Matthias Ruhl expertly adapted and merged the two programs for our studies.

Chapter 7

Fractal Image Coding: Some Mathematical Remarks on Its Limits and Its Prospects

F.M. Dekking

Abstract: We discuss both black and white and grey value image encoding with the classical schemes proposed by M. Barnsley, respectively A. Jacquin. For the black and white case we prove a Decollage Theorem from which one might deduce that the fractal coding of binary images is not feasible. For grey value images we take a close look at the different metrics in which collages are made. From this one might deduce that the mathematical basis for this type of coding is somewhat smaller than usually assumed. Finally we propose a time reduction of the domain pool search by an analysis of the grey value pyramid associated to an image. This naturally leads to the introduction of martingales, which offer a new viewpoint to multiresolution analysis.

7.1 Introduction

In the field of fractal image compression we have seen contributions from pure mathematicians, applied mathematicians and engineers. As a caricature of their different attitudes one might say that the pure mathematicians only care about quality (and then merely in a limiting sense), the applied mathematicians care about quality and compression, and the engineers care about quality, compression *and* cost. It will be our goal in the following to pay attention to some aspects of all three of these quantities. In the next section we recall some of the basic ingredients of fractal image compression to set out notation. Then we discuss in Section 3 the coding of 1-D binary images. Here we consider the perturbation of the attractor by perturbation of the parameters of the associated IFS. In the next section we first give a general context for the study of Jacquin-type algorithms, the Read-Bajraktarevic operator, and then proceed to the question how the collage is made in practice, and what the right mathematical model for this should be. In Section 4 we associate to an image its grey value pyramid and discuss the benefit of making collages at different levels of the pyramid. A useful mathematical way to describe this multiresolution analysis is given by the concept of a martingale. In the last section we give a leisurely introduction to this notion, ending by showing its connection with the hierarchical interpretation of fractal image coding of Baharov et al. ([15])

Most of the work discussed here originated from a cooperation project of Delft University of Technology and Philips Eindhoven starting at the beginning of 1990. The papers [28] and [27] reflect the results of this cooperation.

7.2 Basics

The basic idea ([19]) of fractal image coding is that of not transmitting an image x, but instead a contractive operator T which has x as its unique fixed point. The decoder than iterates T on any initial image x_0 and will recover x as $\lim T^n x_0$ as $n \to \infty$. Because of finiteness of resolution the number of iterates will also always be finite in practice. The mathematical context is thus that we have a space X (the space of images) which is complete with respect to a metric d on X, and the problem is to find an operator $T : X \to X$ which is a contraction w.r.t. d, i.e., there exists $a < 1$ such that

$$d(Tx, Ty) \leq ad(x, y) \quad \text{for all } x, y \in X.$$

Barnsley proposes for black and white images to take $X = \mathcal{K}(\mathcal{R}^d)$, the space of all non-empty compact subsets of $\mathcal{R}^d$ with the Hausdorff metric defined by

$$d_H(K, K') = \max\{\max_{x \in K} \min_{y \in K'} \| x - y \|, \max_{x \in K'} \min_{y \in K} \| x - y \|\}$$

where $\| \cdot \|$ is the Euclidean distance in $\mathcal{R}^d$.

The operator T is defined in this case by

$$T(K) = \bigcup_{i=1}^{m} f_i(K)$$

where $f_1, \ldots, f_m$ are contractions for the Euclidean metric on $\mathcal{R}^d$. The set $\{f_1, f_2, \ldots, f_m\}$ is called an iterated function system or IFS.

Both Barnsley and Jacquin propose to model grey value images by measures. We shall rather consider functions. So here $X = L_2([0,1]^d)$ the space of square integrable functions on the unit cube. The operator T is much more complicated here, see Section 4.

In both the black and white and the grey value images case M. Barnsley has formulated the Collage Theorem to find the operator T.

Theorem 7.1 *(Collage Theorem) Let (X, d) be a metric space with a contraction operator $T : X \to X$ with contraction factor a, and fixed point x^*. Then for all $x \in X$*

$$d(x, x^*) \leq \frac{1}{1-a} d(x, Tx).$$

Proof: We have

$$\begin{aligned} d(x, x^*) &= d(x, Tx^*) \\ &\leq d(x, Tx) + d(Tx, Tx^*) \\ &\leq d(x, Tx) + ad(x, x^*), \end{aligned}$$

and the statement of the theorem follows after a simple manipulation □.

The Collage Theorem is used as follows: if we can find T such that Tx is close to our image x, then the fixed point x^* of T will be close to x, provided a is not too close to 1. For more details we refer to [19] or [83].

7.3 Coding of 1-D Black and White Images

In this section we consider the coding of black and white images, represented by a subset of the unit interval [0,1] as attractors of affine IFS's $f = \{f_i : 1 \leq i \leq m\}$, where for each i

$$f_i(x) = a_i x + b_i.$$

Here the scaling factors a_i and the offsets b_i are real numbers such that

$$a^* = \max_i |a_i| < 1.$$

It is sometimes useful to write f_i in an equivalent form, parametrized by a_i and the fixed point π_i of f_i defined by $f(\pi_i) = \pi_i$ and given by

$$\pi_i = \frac{b_i}{1 - a_i}.$$

We then have

$$f_i(x) = a_i x + (1 - a_i)\pi_i.$$

The associated operator is T defined by

$$T(K) = \bigcup_{i=1}^{m} f_i(K),$$

and we call the unique non-empty compact set such that $T(A) = A$ the attractor (of the IFS).

We shall assume throughout that $A \subset [0, 1]$ (this is no real loss of generality).

The following simple result shows that under certain conditions the extremal points of the attractor are fixed points.

Let us denote the smallest and largest fixed point by

$$\pi_* = \min_i \pi_i \ , \ \pi^* = \max_i \pi_i .$$

Proposition 7.1 *Suppose all the scaling factors satisfy $a_i \geq 0$.*
Then $\min A = \pi_*$ *and* $\max A = \pi^*$.

Proof: Note first that $\pi_i \leq \pi^*$ implies $a_i\pi^* + b_i \leq \pi^*$, i.e., $f_i(\pi^*) \leq \pi^*$.
Similarly $f_i(\pi_*) \geq \pi_*$ for all i. Hence

$$\begin{array}{rcll} f_i([\pi_*, \pi^*]) & \subset & [\pi_*, \pi^*] & \text{for all } i, \text{ so} \\ \bigcup_{i=1}^m f_i([\pi_*, \pi^*]) & \subset & [\pi_*, \pi^*] & \text{i.e.,} \\ T([\pi_*, \pi^*]) & \subset & [\pi_*, \pi^*] & \text{so} \\ T^n([\pi_*, \pi^*]) & \subset & [\pi_*, \pi^*] & \text{so (letting } n \to \infty) \\ A & \subset & [\pi_*, \pi^*]. & \end{array}$$

But $\pi_* \in A$ and $\pi^* \in A$, since the attractor will certainly contain the fixed points of all the affine maps. Combined with the previous formula this leads to $\pi_* = \min A$ and $\pi^* = \max A$. □

The following example shows that the non-negativity condition on the scaling coefficients is important.

Example. Let $m = 2$, $f_1(x) = -\frac{1}{3}x + \frac{1}{3}, f_2(x) = -\frac{1}{3}x + 1$. Then A is the Cantor set (as can be seen either geometrically or by looking at the second order IFS consisting of the 4 maps f_1f_1, f_1f_2, f_2f_1 and f_2f_2). So $\min A = 0$ and $\max A = 1$. But $\pi_1 = \frac{1}{4}$ and $\pi_2 = \frac{3}{4}$.

We remark that a similar result holds for general dimensions (the a_i should be diagonal matrices with non-negative entries). This is useful in computer graphics ([107]), and in doing collages, because the number of parameters is reduced (Evelyne Lutton, private communication).

In the sequel we shall compare the attractors of *two* IFS's

$$f = \{f_i(x) = a_i x + b_i : 1 \leq i \leq m\} \text{ and } \tilde{f} = \{\tilde{f}_i(x) = \tilde{a}_i x + \tilde{b}_i : 1 \leq i \leq m\}$$

where both $a_i \geq 0$ and $\tilde{a}_i \geq 0$. The associated operators and attractors will be denoted by T and A respectively $\tilde{T}$ and $\tilde{A}$.

We now formulate an upperbound for the distance between A and $\tilde{A}$ in terms of the perturbations $|\tilde{a}_i - a_i|$ and $|\tilde{b}_i - b_i|$.

Theorem 7.2 *Let f and $\tilde{f}$ be two IFS as above. Then*

$$d_H(A, \tilde{A}) \leq \frac{1}{1 - \tilde{a}^*} \max_i \left[|\tilde{a}_i - a_i|\pi^* + |\tilde{b}_i - b_i| \right] .$$

Proof: From the collage Theorem we deduce directly that

$$d_H(A,\tilde{A}) \leq \frac{1}{1-\tilde{a}^*} d_H(A,\tilde{T}A).$$

By a well known property of the Hausdorff metric (see e.g. [19], Lemma 4, p.81), we have

$$d_H(A,\tilde{T}A) = d_H(TA,\tilde{T}A) = d_H\left(\bigcup_{i=1}^m f_i(A), \bigcup_{i=1}^m \tilde{f}_i(A)\right) \leq \max_i d_H(f_i(A),\tilde{f}_i(A)).$$

But

$$\begin{aligned} d_H(f_iA,\tilde{f}_iA) &= \max_{x\in A}\min_{y\in A}|f_i(x)-\tilde{f}_i(y)| \\ &\leq \max_{x\in A}|f_i(x)-\tilde{f}_i(x)| \\ &= \max_{x\in A}|(a_i-\tilde{a}_i)x+b_i-\tilde{b}_i| \\ &\leq |\tilde{a}_i-a_i|\max_{x\in A}|x|+|\tilde{b}_i-b_i| \\ &= |\tilde{a}_i-a_i|\pi^*+|\tilde{b}_i-b_i|, \end{aligned}$$

where in the last step we used the proposition at the beginning of this section. □

We compare this upper bound with a lower bound, which has been given a suggestive name by P. Liardet.

Theorem 7.3 *(Decollage Theorem) Let f and $\tilde{f}$ be two IFS's as above, and suppose that $\pi^* = \pi_m$ and $\tilde{\pi}^* = \tilde{\pi}_m$. Then*

$$d_H(A,\tilde{A}) \geq \frac{1}{1-\tilde{a}_*}\min_i\left[(\tilde{a}_i-a_i)\pi^*+\tilde{b}_i-b_i\right].$$

Proof: Using the proposition again, we have

$$d_H(A,\tilde{A}) \geq \tilde{\pi}^* - \pi^*$$

since π^* and $\tilde{\pi}^*$ are the largest points in A respectively $\tilde{A}$. We can rewrite this difference as

$$\begin{aligned} \tilde{\pi}^*-\pi^* &= \frac{\tilde{b}_m}{1-\tilde{a}_m}-\pi^* = \frac{1}{1-\tilde{a}_m}\left(\tilde{b}_m-\pi^*(1-\tilde{a}_m)\right) \\ &= \frac{1}{1-\tilde{a}_m}\left((\tilde{a}_m-a_m)\pi^*-(1-a_m)\pi^*+\tilde{b}_m\right) \\ &= \frac{1}{1-\tilde{a}_m}\left((\tilde{a}_m-a_m)\pi^*+\tilde{b}_m-b_m\right) \\ &\geq \frac{1}{1-\tilde{a}_*}\min_i\left[(\tilde{a}_i-a_i)\pi^*+\tilde{b}_i-b_i\right]. \square \end{aligned}$$

Combining the last two theorems we can obtain an exact perturbation result for a special type of IFS's.

Corollary 7.1 *If all $a_i = a, \tilde{a}_i = a_i + \epsilon, \tilde{b}_i = b_i + \delta$ for some $a > 0, \epsilon > 0$ and $\delta > 0$, then*

$$d_H(A, \tilde{A}) = \frac{1}{1-a-\epsilon}\left[\frac{\epsilon}{1-a}\max(b_i) + \delta\right].$$

The corollary follows since as one easily shows, the order of the fixed points is preserved in case all $a_i = a$.

Example. The Cantor set.
Let $f_1(x) = \frac{1}{3}x,\ f_2(x) = \frac{1}{3}x + \frac{2}{3}$,
$\tilde{f}_1(x) = (\frac{1}{3} + \epsilon)x,\ \tilde{f}_2(x) = (\frac{1}{3} + \epsilon)x + 2/3$.
Then $d_H(A, \tilde{A}) = \frac{3\epsilon}{2-3\epsilon}$.

Application: Quantization.
Let an IFS $f = \{f_i : 1 \le i \le m\}$ be given with $f_i(x) = a_i x + b_i$, where $a_i \ge 0$ and $b_i \in [0,1]$. Suppose the attractor A of this IFS codes an image in $[0,1]$. The $2m$ real numbers a_i and b_i have to be quantizied, say to n bits. Then the IFS f is perturbated to an IFS $\tilde{f} = \{\tilde{f}_i : 1 \le i \le m\}$, where

$$\tilde{a}_i = a_i + \epsilon_i, \qquad \tilde{b}_i = b_i + \delta_i \qquad \epsilon_i, \delta_i \in (-2^{-n}, 2^{-n}).$$

Since the image has the size of [0,1], the largest fixed point π^* of the IFS is equal to 1. Suppose the interval consists of N pixels, i.e., pixelwidth is $1/N$. Suppose we are in the worst case, where all quantization errors are of size 2^{-n}. Then the Decollage Theorem yields

$$\text{error} \ge \frac{1}{1-\tilde{a}_*}(\epsilon \cdot 1 + \delta) \ge \epsilon + \delta = 2^{-n+1}.$$

So if errors of at most 4 pixelwidths are acceptable, we have

$$4/N = 2^{-n+1} \Leftrightarrow N = 2^{n+1}.$$

If for example the image consists of $N = 4000$ pixels, we need 11 bits to code the a_i and b_i. So if we use *one* IFS with $m = 10$ maps to code the whole image, the coding of this IFS requires

$$10.22 - \log_2(10!) = 198 \text{ bits}.$$

(the 10! arises because one can ignore the order of the maps in the IFS). Consequently the compression is only 1:20, even if we happened to find a perfect collage.

7.4 Grey Value Coding: The Collage and the Attractor

The step from black and white to grey value images has led to many generalisations of the basic idea of an IFS. One will find HIFS, LIFS, PIFS, RIFS, IFZS, IFSM, and many others in the literature. Modelling grey value images with functions (rather than with measures), there is a very general operator which has been studied in mathematics as early as 1952 ([214]), which encompasses the operator associated to Jacquin's scheme.

Let X be a set, and $m : X \to X$ a map. Suppose that for each $x \in X$ maps

$$v(x, \cdot) : G \to G$$

are given on a complete metric space G. Then the Read-Bajraktarevic operator is defined by

$$Tf(x) = v(x, f(m(x))) \qquad x \in X$$

for all functions $f \in L^\infty(X, G)$.

Example ([33]). Let $X = [0,1]^2$ be the unit square $m(x) = 2x \bmod 1$ (in each coordinate) and $v(x, g) = s(x)g + o(x)$.

See the recent book "Fractal functions, fractal surfaces and wavelets" ([182]) for many more examples.

To obtain the operator associated to Jacquin's scheme, one considers a subset X of $\mathcal{R}^d$ with a partition $(R_k)_{k=1}^N$, i.e,

$$X = \bigcup_{k=1}^{N} R_k \quad , \quad R_k \cap R_j = \emptyset \quad \text{if } j \neq k.$$

Further bijections $u_k : D_k \to R_k$ are given, where the D_k are subsets of X, and grey value maps $v_k : G \to G$. The maps $m(x)$ and $v(x, g)$ are then defined for each x by

$$m(x) = u_k^{-1}(x), \quad v(x, g) = v_k(g)$$

if R_k is the unique subset of X such that $x \in R_k$.

Remark. On a discrete space this operator performs subsampling from the domains D_k rather than averaging.

We now come to the question when T will be a contraction. Let $d_G(\cdot,\cdot)$ be the metric on the grey value space G. A natural metric on the grey value functions is given by

$$d_\infty(f, h) = \sup_{x \in X} d_G(f(x), h(x))$$

for $f : X \to G$ and $h : X \to G$. Let us define the family $\{v(x, \cdot) : x \in X\}$ to be *uniformly contractive* if there exists an a^* such that

$$d_G(v(x, g), v(x, g')) \leq a^* d_G(g, g')$$

for all $x \in X$. We then have the following result.

Theorem 7.4 *If the $v(x, \cdot)$ are uniformly contractive, then T is a contraction for the d_∞-metric.*

Proof: This is straightforward: we have

$$\begin{aligned} d_\infty(Tf,Th) &= \sup_{x\in X} d_G(T(f(x)),T(h(x))) = \\ &= \sup_{x\in X} d_G(v(x,f(m(x))),v(x,h(m(x)))) \\ &\le a^* \sup_{x\in X} d_G(f(m(x)),h(m(x))) \\ &\le a^* \sup_{y\in X} d_G(f(y),h(y)) = a^* d_\infty(f,h). \quad \Box \end{aligned}$$

This result tells us that if all the v_k in the Jacquin operator have contraction coefficients smaller than 1, then there will be an attractor. But in practical work (where X is a subset of $\mathcal{R}^d$) one sees the use of the d_2-metric

$$d_2(f,h) = \left(\int_X (f(x)-h(x))^2 dx\right)^{\frac{1}{2}},$$

both for doing the collage (but see later in this section) and for measuring the quality of the approximation of the attractor to the original image. Since the collage is made in another metric than the d_∞-metric (for which we have the theorem above which implies the *existence* of the attractor), there is absolutely no guarantee that this attractor will be close to the original image. Actually the situation is even more complicated, as nobody uses the d_2-metric!

What is used by Jacquin ([123]) and many others is an image dependent metric, which we call the $d_{2,\infty}$-metric defined by

$$d_{2,\infty}(f,h) = \max_{1\le k\le N} \alpha_k \left[\int_{R_k} |f(x)-h(x)|^2 dx\right]^{\frac{1}{2}}.$$

Here the $(\alpha_k)_{k=1}^N$ are non-negative weights. There is some confusion about the choice of these weights. Jacquin states that it is natural to take the weights inversely proportional to the area (in dimension 2), i.e.

$$\alpha_k = \frac{1}{|R_k|} \qquad k = 1,\ldots,N,$$

where $|R_k|$ denotes the Lebesgue measure (i.e., area if $d = 2$) of the ranges R_k. Another choice is

$$\alpha_k = \frac{1}{\sqrt{|R_k|}} \qquad k = 1,\ldots,N.$$

This is natural because if we assume without loss of generality that $X = [0,1]^d$, the unit cube in $\mathcal{R}^d$, then one can show that

$$d_2(f,h) \le d_{2,\infty}(f,h) \le d_\infty(f,h)$$

for all bounded functions f and h.

To verify the first inequality, note that

$$\begin{aligned} d_2^2(f,h) &= \sum_{k=1}^{N} \int_{R_k} (f(x)-h(x))^2 dx = \sum_{k=1}^{N} |R_k| \frac{1}{|R_k|} \int_{R_k} (f(x)-h(x))^2 dx \\ &\leq \max_{1\leq k\leq N} \frac{1}{|R_k|} \int_{R_k} (f(x)-h(x))^2 dx = d_{2,\infty}^2(f,h) \end{aligned}$$

where the inequality holds since $\sum_{k=1}^{N} |R_k| = |X| = 1$.

For the second inequality, note that

$$d_{2,\infty}^2(f,h) \leq d_\infty^2(f,h) \max_{1\leq k\leq N} \frac{1}{|R_k|} \int_{R_k} 1 dx = d_\infty^2(f,h).$$

Another reason that the choice $\alpha_k = |R_k|^{-\frac{1}{2}}$ is natural, is that the error of least squares approximation one performs on each R_k will then have a canonical probabilistic meaning (see next section).

Now suppose we take the most commonly used Jacquin operator where all R_k have equal sizes, and the D_k all have twice the (linear) size of the R_k, and the v_k are affine contractions given by

$$v_k(g) = a_k g + b_k.$$

Then the best possible sufficient conditions for contractivity of the operator T are (see [28])

$$\begin{array}{lcl} \text{for } d_2 & : & \max_{x\in X} \sum_{k=1}^{N} 1_{D_k}(x) a_k^2 < 2^d \\ \text{for } d_{2,\infty} & : & \max_{1\leq k\leq N} |a_k| < (\frac{2}{3})^{d/2} \\ \text{for } d_\infty & : & \max_{1\leq k\leq N} |a_k| < 1. \end{array}$$

Here best possible means that for each of the three metrics there exist operators which are not contractions and where the parameters satisfy equality in the three conditions above. An example ([28]) showing this for $d_{2,\infty}$ heavily uses the fact that *one* domain can have a non-empty intersection with 3^d ranges. If one requires that domains are unions of ranges then $\max_{1\leq k\leq N} |a_k| < 1$ is sufficient for contractivity. However in the two level "parent and child" ranges employed by Jacquin, one can construct examples where T does not contract for $d_{2,\infty}$ unless the scaling coefficients satisfy $|a_k| < 1/4$ ([28]).

Considering these heavy restrictions on the scaling factors a_k it might seem quite miraculous that Jacquin's method works in practice. It must be kept in mind however, that these restrictions correspond to sufficient conditions. Moreover, global contractivity is not necessary to have convergence to a fixed point. If e.g. $Y = [a, 1]$ for some $a > 0$ and $T : Y \rightarrow Y$ is defined by $Ty = \sqrt{y}$, then $T^n y \rightarrow 1$ for all $y \in [a, 1]$, but $|Ty - Ty'| < |y - y'|$ only if $\sqrt{y} + \sqrt{y'} > 1$.

7.5 Pyramids and Martingales in Grey Value Image Coding

In this section we shall be more concrete, and replace the modelling of an image by a function over the uncountable set $X = [0,1]^d$ by a function f over a finite set

$$X = \{(i,j) : 0 \leq i,j < 2^r\}$$

taking values in the set of grey values $G = \{0,1,\ldots,255\}$.

Given a square range R and a domain D which has twice the (linear) size of R, the map $u : D \to R$ performs averaging rather than subsampling, so f over D is transformed to $g = f \circ u^{-1}$ over R by

$$g(i,j) = \tfrac{1}{4}\{f(2k,2\ell) + f(2k+1,2\ell) + f(2k,2\ell+1) + f(2k+1,2\ell+1)\}$$

if $u(2k,2\ell) = (i,j)$. The collage (on R) then consists in finding a grey value contraction $v : G \to G$ of the form $v(g) = ag + b$ such that

$$\sum_{(i,j)\in R} (f(i,j) - v(g(i,j)))^2 = \sum_{(i,j)\in R} (f(i,j) - ag(i,j) - b)^2$$

is minimal. This is a least squares problem whose solution can be written as

$$a = \frac{\text{Cov }(f,g)}{\text{Var }(g)}, \quad b = \overline{f} - a\overline{g}$$

where

$$\overline{f} = \frac{1}{|R|} \sum_{(i,j)\in R} f(i,j)$$

is the mean value or DC-component of f over R and

$$\text{Cov }(f,g) = \frac{1}{|R|} \sum_{(i,j)\in R} (f(i,j) - \overline{f})(g(i,j) - \overline{g}), \ \text{Var }(g) = \text{ Cov }(g,g).$$

The error Δ is given by

$$\Delta^2 = \text{Var } f \left[1 - \frac{\text{Cov }^2(f,g)}{\text{Var } f \text{ Var } g}\right].$$

In our work in [27] as well as in [82] this is combined with a quadtree algorithm. One fixes a quality parameter or tolerance ϵ, and if $|\Delta| > \epsilon$ the range R is subdivided into 4 smaller ranges for each of which one tries to find the best possible domain such that the least square error is smaller than ϵ. If this is not succesful one continues at most until the range has the size of one pixel, where a perfect collage is always possible.

In the algorithm in [27] very large domain pools (about 1 million domains) are used. This is partly caused by the use of domains which have the same size as the ranges (note that the contractivity of T does not depend on the scaling of domains, as shown in the previous section). This leads to excessive computation times. In the rest of this section we shall describe an idea using the grey value pyramid associated with an image (first introduced by [139]) to reduce the cost of the domain search.

The grey value pyramid associated to a $2^r \times 2^r$ image consists of the $r+1$ images $f^{(n)}$ of size $2^n \times 2^n$ $(n = 0, \ldots, r)$ defined recursively by $f^{(r)} = f$ and for $0 \leq i, j < 2^{n-1}$

$$f^{(n-1)}(i,j) = \frac{1}{4}\left[f^{(n)}(2i,2j) + f^{(n)}(2i,2j+1) + f^{(n)}(2i+1,2j) + f^{(n)}(2i+1,2j+1)\right].$$

A range $R = R^{(r)}$ in the quadtree partition of the original image induces ranges $R^{(n)}$ for $n = 0, 1, \ldots, r$ in the obvious way: $1_{R^{(n)}} = 1_R^{(n)}$. Analogously a domain D at level r induces a domain $D^{(n)}$ at level n. We then have the following result

Proposition 7.2 *(MONOTONICITY OF ERRORS) Let $f^{(n)}, g^{(n)}, R^{(n)}$ and $D^{(n)}$ be the grey value functions, the averaged grey value function, the range and the domain at level n. Let $\Delta^{(n)}$ be the associated collage error at level n. Then $|\Delta^{(n+1)}| \geq |\Delta^{(n)}|$.*

Corollary 7.2 *If no acceptable collage can be found between a range and a domain at level n, then there exists also no collage at level $n+1$.*

Since computation of the least square error at level n is four times faster than at level $n+1$ this can reduce computational cost as we will only compare these domains with a range which have not been rejected at a coarser level.

However ... there is one problem: only one quarter of the domains at level $n+1$ is represented at level n. Namely exactly those positioned at even-even positions $(2i, 2j)$. So to check all domains, one rather has a pyramid-tree, where each node branches into four nodes representing the even-even, even-odd, odd-even and odd-odd positioned domains. A code implementing this idea has been written by V. Kritchallo. Despite its very complicated data structure this code is four times faster than the original code. It is very probable that an algorithm which only considers $4 \times 4, 8 \times 8$ and 16×16 ranges with a reduced domainpool - such that e.g. the 32×32 ranges only occur at positions divisible by 4, would be much faster. Even more so if other information of the lower levels would be passed on, as e.g. the positions of the best (but still not good enough) domains.

We now pass to a proof of the proposition. As K.U. Barthel has remarked, this can be done by applying the Haar-transform, and noting that moving to a lower level corresponds to adding another basis coefficient. We shall however present a different short proof using the notion of a martingale.

Another reason to do so, is that this provides some insight in the hierarchical interpretation of fractal coding as considered in [15].

A martingale is a powerful tool in probability theory. In the next section we shall try to explain the basic ideas for a simplified form of a martingale, which is sufficient in the context of grey value pyramids.

7.6 Martingales for Electrical Engineers

Throughout this section we consider signals X and Y defined on $[0,1]$ although the same ideas apply more generally to e.g. 2-D signals.

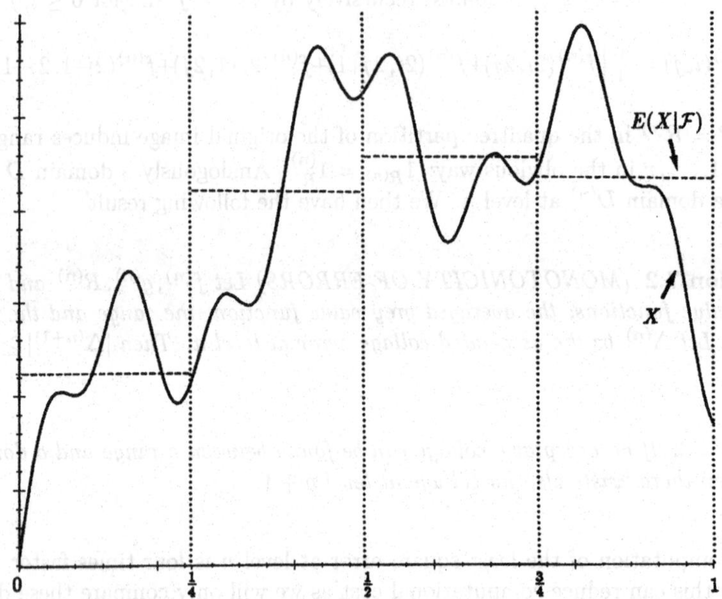

Figure 7.1: A signal X and the signal $E(X|\mathcal{F})$ obtained from filtering by a four-element partition $\mathcal{F}$.

If X is a signal we denote its mean value or DC-component as EX. By a filter $\mathcal{F}$ we understand a partition of $[0,1]$, for example

$$\mathcal{F} = \{[0, \tfrac{1}{4}), [\tfrac{1}{4}, \tfrac{1}{2}), [\tfrac{1}{2}, \tfrac{3}{4}), [\tfrac{3}{4}, 1]\}.$$

We then denote by $E(X|\mathcal{F})$ the signal filtered by $\mathcal{F}$, i.e., the signal which is constant over each element of the partition, the constant being the DC-component of the signal *on* this element (cf. Figure 1).

The operation of filtering has the following properties.

(P1) Filtering is linear:

$$E(aX + Y|\mathcal{F}) = aE(X|\mathcal{F}) + E(Y|\mathcal{F}).$$

(P2) $\tilde{\mathcal{F}}$ coarser than $\mathcal{F} \Rightarrow$

$$E(E(X|\mathcal{F})|\tilde{\mathcal{F}}) = E(X|\tilde{\mathcal{F}}),$$

in particular

$$E(E(X|\mathcal{F})) = EX.$$

(P3) Y already $\mathcal{F}$-filtered $\Rightarrow$

$$E(XY|\mathcal{F}) = YE(X|\mathcal{F}).$$

Here $\tilde{\mathcal{F}}$ coarser than $\mathcal{F}$ means that each element of $\tilde{\mathcal{F}}$ can be written as a union of elements of $\mathcal{F}$. The coarsest filter is $\tilde{\mathcal{F}} = \{[0,1]\}$, and we have of course that $E(X|\tilde{\mathcal{F}}) = EX$ for this filter. This explains the 'in particular' in (P2).

By a *filtration* we mean a sequence of refining filters $(\mathcal{F}_n)$, i.e., $\mathcal{F}_n$ is coarser than $\mathcal{F}_{n+1}$ for all $n = 0, 1, \ldots$.

Finally a *martingale* is a sequence of signals $(X_n)_{n=0}^{\infty}$ with the property that

$$E(X_{n+1}|\mathcal{F}_n) = X_n.$$

Note that each X_n is $\mathcal{F}_n$-filtered, and so $E(X_n|\mathcal{F}_n) = X_n$.

It is also useful to define a *submartingale*, which is a sequence of signals satisfying

$$E(X_{n+1}|\mathcal{F}_n) \geq X_n.$$

Lemma 7.1 *If (X_n) is a martingale, then (X_n^2) is a submartingale.*

Proof: Usually one considers this to be a special case of a so called conditional Jensen inequality. For pedagogical reasons we proceed differently. Noting first that $Y \geq 0$ implies $E(Y|\mathcal{F}) \geq 0$, then applying (P1), (P3) and the martingale property of (X_n) we have

$$\begin{aligned} 0 &\leq E((X_{n+1} - X_n)^2|\mathcal{F}_n) \\ &= E((X_{n+1} - X_n)^2|\mathcal{F}_n) \\ &= E(X_{n+1}^2|\mathcal{F}_n) - 2X_n E(X_{n+1}|\mathcal{F}_n) + X_n^2 \\ &= E(X_{n+1}^2|\mathcal{F}_n) - X_n^2. \end{aligned}$$

So $E(X_{n+1}^2|\mathcal{F}_n) \geq X_n^2$. □

We now give a simple example of a martingale which suggestively might be called the multiresolution martingale. Let the filtration $(\mathcal{F}_n)$ be defined by

$$\mathcal{F}_0 = \{[0,1)\}\ ,\ \mathcal{F}_n = \left\{\left[\frac{k}{2^n}, \frac{k+1}{2^n}\right) : 0 \leq k < n\right\}.$$

A single signal X on $[0,1)$ is given, and (X_n) is defined by

$$X_n = E(X|\mathcal{F}_n).$$

The signal X_n can be interpreted as the signal X at resolution 2^{-n}.

We now check that (X_n) has the martingale property:

$$\begin{aligned} E(X_{n+1}|\mathcal{F}_n) &= E(E(X|\mathcal{F}_{n+1})|\mathcal{F}_n) = \\ &= E(X|\mathcal{F}_n) = X_n. \end{aligned}$$

where we used (P2) in the second step.

APPLICATION 1. We return to the "Monotonicity of errors" proposition in the previous section. We see that the grey value pyramid corresponds to a 2-D multiresolution martingale, putting $X = f - \overline{f}$ and $Y = g - \overline{g}$ (substractions of the DC-component). We then have that $f^{(n)} = E(X|\mathcal{F}_n)$ and $g^{(n)} = E(Y|\mathcal{F}_n)$ where $\mathcal{F}_n$ is the filter

$$\mathcal{F}_n \left\{ \left[\frac{k}{2^n}, \frac{k+1}{2^n} \right) \times \left[\frac{\ell}{2^n}, \frac{\ell+1}{2^n} \right) : 0 \le k, \ell < 2^n \right\}$$

(or more precisely: the discrete version of that).

It is easy to check that

$$EX_n^2 = \text{ Var } (f^{(n)}), EY_n^2 = \text{ Var } (g^{(n)}) \text{ and } EX_nY_n = \text{ Cov } (f^{(n)}, g^{(n)}).$$

Hence the proposition follows directly from the following general result on pairs of martingales.

Theorem 7.5 *Let $(X_n)_{n=0}^{\infty}$ and $(Y_n)_{n=0}^{\infty}$ be square integrable martingales with respect to the same filtration $(\mathcal{F}_n)_{n=0}^{\infty}$.*

Suppose $Y_0 \not\equiv 0$. Then for each $n \ge 0$

$$EX_{n+1}^2 - \frac{(EX_{n+1}Y_{n+1})^2}{EY_{n+1}^2} \ge EX_n^2 - \frac{(EX_nY_n)^2}{EY_n^2}.$$

Proof: We have

$$\begin{aligned} EX_{n+1}^2 - \frac{(EX_{n+1}Y_{n+1})^2}{EY_n^2} &= \min_{\lambda \in \mathcal{R}} E(X_{n+1} - \lambda Y_{n+1})^2 \\ &= \min_{\lambda \in \mathcal{R}} E(E(X_{n+1} - \lambda Y_{n+1})^2 | \mathcal{F}_n) \\ &\ge \min_{\lambda \in \mathcal{R}} E(X_n - \lambda Y_n)^2 = EX_n^2 - \frac{(EX_nY_n)^2}{EY_n^2}. \end{aligned}$$

Here the first and last step use the following little trick: if $f(\lambda) = a\lambda^2 - 2b\lambda + c$, then its minimal value is given by $f(\frac{b}{a}) = c - b^2/a$. This is applied with $f(\lambda) = E(X - \lambda Y)^2 = \lambda^2 EY^2 - 2\lambda EXY + EX^2$. The second step follows by (P2), and the inequality by the Lemma (by (P1) we have that $(X_n + \lambda Y_n)$ is a martingale for each λ!). □

APPLICATION 2. We consider the hierarchical interpretation of fractal image coding of [15]. They define a family of Jacquin operators T_n, parametrized by the size of the ranges (which for fixed n all have the same size 2^{-n}). Let Z_n be the attractor of T_n. They prove that the Z_n satisfy a so called Zoom-out property. In our language this is nothing else than the martingale property:

$$Z_{n+1} = E(Z_n | \mathcal{F}_n),$$

where $(\mathcal{F}_n)$ is the multiresolution martingale.

They show next that there exists a unique function Z (called the $PIFS$ embedded function) such that $E(Z|\mathcal{F}_n) = Z_n$ for all n. Their proof is constructive. A very short non-constructive proof (with a precise convergence statement) can be given using the well known L^2-martingale convergence theorem:

Let (X_n) be a martingale w.r.t. a filtration $(\mathcal{F}_n)$. If $EX_n^2 \leq M$ for all $n \geq 0$ for some constant M, than there exists X such that $X_n \to X$ almost surely and also $E(X - X_n)^2 \to 0$ as $n \to \infty$.

Chapter 8

Linear Time Fractal Quadtree Coder

F. Dudbridge

Abstract: We describe a linear time fractal image coder which uses a quadtree construction to support a wider performance range than the standard fast coder. Image recovery is less accurate than full fractal coding and the JPEG standard, but the algorithm has the advantage of very fast compression.

8.1 Introduction

In the general fractal coding scheme introduced by Jacquin [125], an image is tiled with "range blocks", and for each range a mapping of a larger "domain block" found such that the transformed domain resembles the range. The fractal approximation image is constructed by iterating these mappings on an arbitrary initial image.

Because, for each range, the set of domain blocks must be searched for the best match, the number of possible domain blocks determines the coding speed. Methods of reducing the search are the subject of much recent research [87]. We have introduced a scheme requiring no searching [67], in which groups of four range blocks share their union as a common domain block. In this case encoding and decoding both have linear cost with image size, but the approximation quality is poor [189]. By incorporating a quadtree structure of range blocks into the scheme, the present work significantly improves coding fidelity without sacrificing linear cost.

8.2 A Model for Fractal Images

An image is modelled here by an iterated function system (IFS) together with a real-valued function defined on subsets of its attractor. An IFS is a set $W = \{w_1, \ldots, w_N\}$

of contraction mappings on $\mathcal{R}^2$, having a unique attractor set A satisfying

$$A = \bigcup_{k=1}^{N} w_k(A) = \bigcup_{k_1=1}^{N} \cdots \bigcup_{k_m=1}^{N} w_{k_1} \circ \ldots \circ w_{k_m}(A).$$

for a positive integer m. IFS theory is comprehensively described by Hutchinson [121] and Barnsley [19].

Let m be a fixed positive integer, termed the *resolution*. Let $A_{k_1 \ldots k_m}$ denote $w_{k_1} \circ \ldots \circ w_{k_m}(A)$. Then the *pixels* at resolution m are the set

$$P_m = \{A_{k_1 \ldots k_m}; k_1, \ldots, k_m = 1, \ldots N\}$$

A *grayscale function* is any function $f : P_m \to \mathcal{R}$. Fractal grayscale functions are defined by an operator M having the form

$$Mf(p) = a_{k_1} F(p) + s_{k_1} \sum_{i=1}^{N} f(A_{k_2 \ldots k_m i}) + t_{k_1}$$

for each pixel $p = A_{k_1 \ldots k_m}$, where a_k, s_k, t_k are fixed real numbers for each $k = 1, \ldots, N$ and F is a fixed grayscale function. If $|\sum s_k| < 1$, then M has a unique fixed point [67], which is the fractal function. A fractal image is therefore specified by an IFS W together with real coefficients a_k, s_k, t_k and an auxilliary function F, which will be composed of other fractal grayscale functions as described below.

8.3 Image Coding

Images are coded blockwise, with each block a power of 2 pixels square. To code a block of side 2^m pixels, the IFS W is taken to have four mappings which give the four quadrants of the block. The pixels at resolution m then correspond to actual image pixels. The problem is to find the mapping M whose fixed point f best matches the given grayscale function g. The procedure is to compute the variance of each quadrant; if the variance of quadrant k is below a preset threshold, then a_k is set to zero and s_k, t_k found by solving the normal equations obtained from least-squares minimization of the "collage error" $d(g, Mg)$ [67]:

$$\begin{bmatrix} \sum_{p \subset A_k} (v(p))^2 & \sum_{p \subset A_k} v(p) \\ \sum_{p \subset A_k} v(p) & \sum_{p \subset A_k} 1 \end{bmatrix} \begin{bmatrix} s_k \\ t_k \end{bmatrix} = \begin{bmatrix} \sum_{p \subset A_k} g(p) v(p) \\ \sum_{p \subset A_k} g(p) \end{bmatrix}$$

where $v(p) = \sum_{i=1}^{N} g(A_{k_2 \ldots k_m i})$ when $p = A_{k_1 \ldots k_m}$.

When the variance is too high, the quadrant is coded as a separate image block, with the fractal approximation function for this subblock used to define the auxilliary function F in the quadrant of the main block, for which a_k is set to one, and s_k, t_k to zero. Because subcoding a quadrant involves dividing it into four subtiles, the algorithm constructs a quadtree of tiles in which the variance of each is below the threshold. At the bottom level all tiles have low variance, and the coding is of the same form as in [67]. A hierarchy of fractal functions is developed, in which the functions at each level contribute to the definition of all higher-level functions. The construction has some analogy with that of IFS with condensation [20].

The variance is used to direct the quadtree because it proves a reliable predictor of the error of a tile coded by just s_k and t_k; the aim is to code the largest possible tiles while keeping the error within desired limits. With appropriate preprocessing, each variance may be computed in constant time. The image pixels $A_{k_1 \ldots k_m}$ are scanned in the order given by the rightmost subscript changing fastest, and for each pixel the cumulative sums of $g(p)$ and $g(p)^2$ are stored. The scan order ensures that the variance of any tile in the quadtree may be determined by one subtraction. If the cumulative sums of $v(p)$ and $v(p)^2$ are also pre-computed, then the only sum that must be explicitly computed for a set of normal equations is $\sum g(p)v(p)$. The preprocessing stage has linear cost with image size; the quadtree is computed in constant time and each tile is coded in linear time, so that the overall coding algorithm has linear cost.

8.4 Reconstruction

The fractal approximation image may be constructed accurately and rapidly. Let f be the fixed point of the mapping M, and for $n \leq m$ let $f_{k_1 \ldots k_n}$ denote the sum of f over all pixels in $A_{k_1 \ldots k_n}$. The sum over the whole block is given by

$$\sum_{k=1}^{N} f_k = \frac{\sum_{k=1}^{N} \left[a_k F_k + t_k 4^{m-1} \right]}{1 - \sum_{k=1}^{N} s_k}.$$

This is used to obtain each f_k, and thereafter all $f_{k_1 \ldots k_n}$ for $n = 1, \ldots, m$, according to the formula

$$f_{k_1 \ldots k_n} = a_{k_1} F_{k_1 \ldots k_n} + s_{k_1} f_{k_1 \ldots k_{n-1}} + t_{k_1} 4^{m-n}$$

thus finding all the pixel values. These quantities can only be calculated when the auxilliary function F is known, so the decoding process must start from the lowest level of the quadtree, where all $a_k = 0$ and the subblocks may be reconstructed from s and t only. The intermediate sums $f_{k_1 \ldots k_n}$ for $n < m$ in the lower levels are retained, to give the intermediate sums of F in higher levels.

The procedure may be thought of as a special case of the standard iteration for decoding fractal images, with parts of the image solved before others, and the initial image chosen so as to obtain exact convergence in finite time. Because the number of iterations is the resolution m, an operation count shows that decoding is also linear with image size.

8.5 Implementation

The algorithms as described are straightforward to implement. To estimate compression ratios, an efficient storage format is needed. Since the IFS W is inferred from the block size, there is no need to store it explicitly. Also, the set of coefficients a_k is

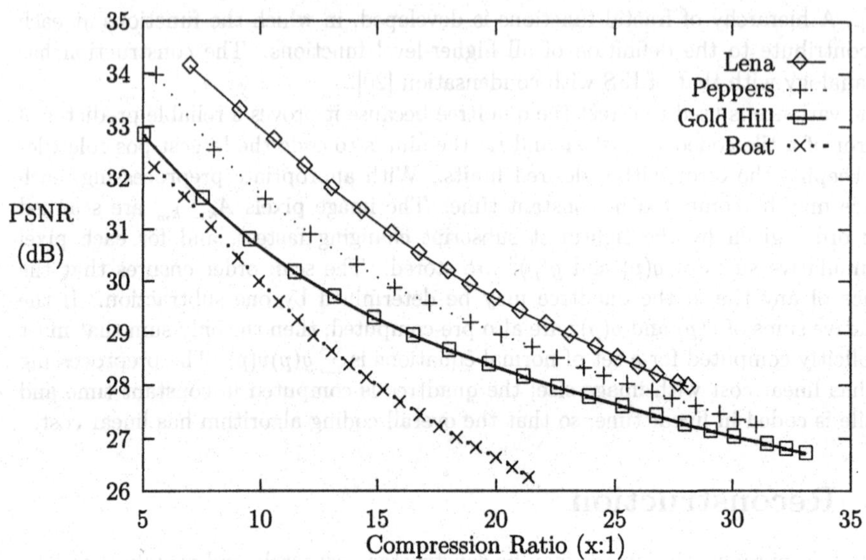

Figure 8.1: Compression versus fidelity performance for test images.

equivalent to the quadtree information, so all that need be stored for each block is its side length, and a quadtree of real values s and t.

These coefficients have no general bounds, so their quantization is not obvious. The following heuristics yield the most satisfactory results in terms of image quality, compression ratio and encoding speed, for the images used. The scaling coefficient s is clipped to the range $[-\frac{7}{32}, \frac{7}{32}]$, quantized uniformly to four bits, and the translation coefficient t recomputed from the normal equations. Note that restricting s to this range ensures that M is always contractive. t is then quantized uniformly to seven bits within the range $[-63, 192]$. If t falls outside this range, s is set to zero and t stored in the range [0,255]. The decoder assumes that t is in this range whenever s is zero; from the normal equations t will be the mean pixel value of the tile, so the convention is sound.

Results of this implementation applied to four standard test images are given in Figure 8.4. The images are sampled at 512 pixels square, 8 bits per pixel, and coded in blocks of sixteen pixels square with a three level quadtree. Each image is coded with the variance threshold varying from 50 to 1000 grayscales, in steps of 50. The compression ratios attained vary between about one-third to one-half of the JPEG standard, but our primary concern here is encoding speed. A developmental implementation on a Sparc-10 workstation achieved full encode in about 0.68 seconds for these images, and full decode in 0.58 seconds. Since the algorithms are fast and nearly symmetrical, applications such as real-time video coding are envisaged, for which reasonable compression may be delivered very rapidly.

Acknowledgements

This work was supported by a NATO postdoctoral fellowship. The author thanks H. D. I. Abarbanel for the use of research facilities, and Y. Fisher for constructive discussion and encouragement.

Chapter 9

Fractal Encoding of Video Sequences

Yuval Fisher

Abstract We present preliminary results from a scheme that encodes video sequences of digital image data. Based on a still-image fractal encoding method, the scheme encodes each frame using image pieces from its predecessor. We present results showing near real-time (5–12 frames/sec) software-only decoding; resolution independence; high compression ratios (25:1–244:1); and low compression times (2.4–66 sec/frame) as compared with standard still-image fractal schemes. More information on various fractal schemes can be found at the WWW site "http://inls.ucsd.edu/y/Fractals/".

9.1 Introduction and Notation

Recent publications of Jacquin [126] and others (see [87]) have demonstrated a new "fractal" method of encoding images as collections of self transformations. In this article we present preliminary results from a generalization of this scheme to video sequences of images. A thorough introduction and the latest implementation and theoretical results of the still-image scheme can be found in [87].

We consider an image as a function $f(x)$ from the unit square $I^2 = [0,1] \times [0,1]$ to the reals $\mathcal{R}$ where $f(x)$ for $x \in I^2$, represents the grey level at x. The range of $f(x)$ is I, with $f(x) = 0$ being black and $f(x) = 1$ being white. However, we consider the more general situation in which the range is $\mathcal{R}$, requiring that a function be clipped if necessary. Let $\mathcal{F} = \{f : I^2 \to \mathcal{R} \mid f \text{ is measurable}\}$ be the space of all images, and define a transformation $\tau : \mathcal{F} \to \mathcal{F}$ by defining a **mixing** function $m : I^2 \to I^2$, a **scaling** function $s : I^2 \to \mathcal{R}$, and an **offset** function $o : I^2 \to \mathcal{R}$ that are combined in the following way to form $\tau_{(m,s,o)}(f)(x) = s(x)f(m(x)) + o(x)$. That is, to get a new function $\tau(f)$, we grab the value of f at position $m(x)$, multiply it by $s(x)$ and add $o(x)$. Under various contractivity conditions,[1] $\lim \tau^{\circ n}(g)$ will converge to a unique fixed image $|\tau| = \tau(|\tau|) \in \mathcal{F}$, independent of g.

The uniqueness of $|\tau|$ implies that if we find functions m, s, and o such that $\tau_{(m,s,o)}(f) = f$, then τ will serve as an encoding of f. A non-trivial solution to this problem is not known. Moreover, since our goal is image compression, the functions $m(x), s(x)$ and $o(x)$ must be specified using as little information as possible. In practice, it is not possible to find a τ satisfying $\tau(f) = f$, but it is possible to find a $|\tau| \approx f$ satisfying $\tau(|\tau|) = |\tau|$. In this case, $f \approx |\tau| = \tau(|\tau|) \approx \tau(f)$, so that $f \approx \tau(f)$. The current state of fractal image compression relies on a theoretical weakness: τ is constructed to satisfy $f \approx \tau(f)$ with the hope that this will lead to $|\tau| \approx f$.

In applications, we define $m(x)$ by partitioning I^2 into disjoint subsets A_i on which we define $m : A_i \to m(A_i) = B_i$ by an affine map (that is, a linear term plus a translation). For historical reasons, we call the A_i **range blocks** and the B_i **domain blocks**, although this is backwards in this notation. We also use the terms to refer to the image above A_i or B_i, respectively. We require that $\cup A_i = I^2$ so that when $m(x)$ is defined on every A_i, it is defined on I^2. The B_i are found by a search procedure that minimizes[2] $d(f|_{A_i}, \tau(f)|_{A_i})$, where $d(\cdot,\cdot)$ should be some measure of visual similarity, often simply taken to be the rms metric $d(f,g) = \sqrt{\int (f-g)^2}$. Each B_i results in a different mixing function $m(x)$, and the search is taken over some set $\mathbf{B}$ of possible B_i and the resulting $m(x)$.

The functions $s(x)$ and $o(x)$ are taken to be piecewise constant and equal to s_i and o_i on A_i. They are also selected to minimize $d(f|_{A_i}, \tau(f)|_{A_i})$. In order to ensure convergence and to simplify the quantization, we take $|s_i| < 1$, though this is not necessary or optimal. With these restrictions on $m(x), s(x)$, and $o(x)$, the operator $\tau_{(m,s,o)}$ can be compactly specified.

9.2 Video Sequences

We consider a video sequence as a sequence of functions, $f_1, f_2, \ldots \in \mathcal{F}$. In this case, we seek transformations $\tau_1, \tau_2, \ldots$, where $\tau_i(f_{i-1}) = f_i$, and $f_0(x)$ must be defined. The τ_i should be thought of as general transformations that may include motion compensation and some type of encoding. In practice, it is not possible to find τ_i that satisfy the previous equality, so we seek a sequence $g_1 \approx f_1, g_2 \approx f_2, \ldots$ with $\tau_i(g_{i-1}) = g_i$ (and $g_0 = f_0$).

Since each image is encoded in terms of its predecessor, no contractivity condition is necessary, and decoding consists of an application of τ_i. It is not hard to show that using contractive transformations τ_i will cause transmission or computational errors to be dissipated, unlike expansive τ_i case with which these errors are magnified.

[1] The most restrictive contractivity condition requires $|s(x)| < 1$. This assures convergence in $\mathcal{F} = L^p(I^2)$ with the L^p metric as well as a.e. pointwise convergence. This condition also allows for simpler quantization of the $s(x)$ values (see [87]).

[2] The notation $f|_A$ denotes the restriction of the function f to the set A.

9.3 Implementation

Given the ith image f_i to be encoded and the previous encoded frame g_{i-1}, we quadtree-partition I^2 at least M_{min} times. Each element of the quadtree partition determines a range A_j. The portion of f_i over A_j is first compared with domains from g_{i-1} of the same size; this is a crude form of motion compensation without any motion. This amounts to taking $s(x) = 1$, $o(x) = 0$, and $m(x)$ a translation on A_j. If the rms difference of g_{i-1} and f_i on A_j is below a predetermined tolerance t_{mc}, then we define m, s and o to copy this piece of g_{i-1} to g_i. Otherwise, we encode the range A_j in the following way.

The encoding of A_j is done by comparing $f_i|_{A_j}$ to domain blocks B of twice the size in g_{i-1}. A comparison consists of finding s and o, to minimize the rms difference between $f_i|_{A_j}$ and $\tau_{(m,s,o)_i}(g_{i-1})|_B$, where m is affine and selected so that $m(A_j) = B$. Since we allow many possible domains B, we reduce the enormous search involved by classifying both $g_{i-1}|_{B'}$ and $f_i|_{A_j}$ into 1 of 72 classes, as described below. Only domains and ranges from the same class (or a small number n_c of classes) are compared. If the comparison yields an rms error less than a predetermined tolerance t_e or if a maximum M_{max} number of quadtree partition have already been made, the resulting transformation is stored; otherwise, A_j is quadtree-partitioned again and the process is repeated.

We now describe the classification scheme, a complete description of which can be found in Chapter 3 of [87]. The scheme computes the average luminosity of each quadrant of a domain or range block, and rotates and/or flips the block to one of the three possible canonical positions with brightest upper left quadrant. Each canonical position corresponds to one of three major classes. Each of these three classes is further resolved into 24 subclasses depending on the ordering of the variance of the four quadrants of each block. Since each major class is associated with a canonical orientation, domains and ranges are only compared in one orientation induced by their classification, as opposed to all 8 possible ways of mapping one square onto another.

Since it is possible that $s(x) < 0$ will be optimal for a particular domain-range comparison, and since multiplying the domain pixels by a negative scaling function will change the order of the quadrant luminosities, each image piece must be compared with domains blocks from two classes, one corresponding to positive scaling and one to negative scaling. Thus, when we compare a range with domains from n_c classes, we are really comparing both positively and negatively scaled domains (which corresponds to searching $2n_c$ classes).

We select domains centered on a lattice with spacing depending on the domain size. The lattice factor λ is divided into the domain size (which is twice the range size) to determine the lattice spacing. For example, for an 8×8 range, the domains are of size 16×16, and $\lambda = 4$ means that domains are selected from all 16×16 image pieces spaced by 4 pixels horizontally and vertically. A $\lambda = 2$ would give a spacing of 8 pixels, etc.

Finally, the quadtree partition is made on squares with side length equal to a power of 2, so rectangular images are partitioned by decomposing them into disjoint squares. Each image is also postprocessed to minimize artifacts along the partition boundaries. The postprocessing consists of a weighted average of pixels along the boundary of the

Table 9.1: Results for the "missa" sequence.

t_{mc}	t_e	n_c	λ	M_{max}	M_{min}	PSNR	Comp.	Dec. (fps)	Enc. (spf)
4.0	3.0	3	1	4	6	36.02	24.93:1	5.49	10.09
4.0	3.0	24	1	4	6	37.01	24.95:1	5.51	66.36
8.0	8.0	1	1	4	6	33.80	79.03:1	9.61	2.48
8.0	8.0	3	1	1	6	32.03	116.54:1	11.62	3.33
8.0	8.0	24	4	1	5	31.44	244.29:1	12.00	44.13

quadtree partition (again, see Chapter 3, [87] for a complete description).
Comments: The initial image $f_0 = 0$ we used cannot encode f_1 well. The following are possible remedies:

1. Encode f_1 using a still-image fractal scheme or some other method.

2. Use some fixed f_0 that contains good domain blocks, for example, a VQ code book.

3. Ignore the problem and do the best possible. Then f_1 will be composed of piecewise constant blocks. The next frame f_2 will also be composed of piecewise constant blocks, but of half the size. This way, after a small number of frames (depending on the decoding size), the blocking problem disappears.

The first scheme has the advantage that the first several frames are well encoded, but it is more complicated. The last two schemes are easier to implement, since all frames are handled the same way. Also, decoding times for these frames are roughly the same for each frame. We present results using the third scheme. We have found that by the third frame there are almost no noticeable artifacts due to selecting $f_0 = 0$.

9.4 Results

We present results for the 360 × 288 grey-scale "missa" sequence. In all of the runs, 5 bits were used to store s_i and 7 bits were used to store o_i, using linear quantization. The maximum and minimum range side lengths for the quadtree partitions based on $M_{max} = 6$ and $M_{min} = 4$ were 4×4 and 16 × 16. When $n_c = 3$ the three major classes were searched for a fixed subclass; when $n_c = 24$ the subclasses were searched for a fixed major classes.

Table 9.1 shows parameter, PSNR, and encoding/decoding time results. The PSNR values are averaged over all the frames in the sequence, as are the encoding and decoding times. These times are given in CPU seconds on a Silicon Graphics Personal Iris 4D/35. The decoding time is given in average frames per second and includes all disk access and display functions. The encoding time is given in seconds per frame (spf).

9.5 Discussion

The decoding times are low, even in our crude implementation. It seems highly likely that a significant reduction in decoding time can be achieved by code and algorithmic optimization. We speculate that frame rates of 90fps should be attainable on current workstation hardware with such optimizations alone. Frame rates of 30fps can already be achieved on high-end workstations (e.g., 100Mhz SGI Indigo). Encoding times can be improved, but this will be moderated by the inclusion of true motion compensation and color.

We observed that the high fidelity encodings have no artifact flicker. The low fidelity encodings have a little flicker. We also observed that decoding at a larger size (than the original) created detail that was commensurate with the local image: edges remained edges and texture remained texture. We have found that, in general, the lower the compression, the better the decoded magnified data looks.

9.6 Conclusion

Even though fractal image compression is in its infancy, it is already showing promise as a viable compression technique. Its strength, aside from giving compression/fidelity results that are comparable to DCT and wavelet methods[86], is its computationally simple decoding. Its potential is strong, if it receives even a small proportion of the many thousands of man-years of work that has been invested in DCT and wavelets techniques.

Chapter 10

Theory of Generalized Fractal Transforms

Bruno Forte and Edward R. Vrscay

10.1 Introduction

The aim of this chapter is to present a unified treatment of the various *fractal transform* methods for the representation and compression of computer images which have been based, in some way, on the method of Iterated Function Systems (IFS). These methods, which include "traditional" IFS and IFS with probabilities (IFSP), Iterated Fuzzy Set Systems (IFZS), Iterated Function Systems with Maps (IFSM) and variations, have been designed following a common pattern. Let (X, d) denote a complete metric space, the "base space" which may represent the computer screen, e.g. $[0,1], [0,1]^2$ with Euclidean metric. The IFS component, consisting of N contraction maps, $w_i : X \to X$, will be written as $\mathbf{w}$. An image or target is then represented as a point in an appropriate complete metric space (Y, d_Y). The metric spaces used in the various IFS-type methods are listed below:

IFS [121, 20, 19]: $\mathcal{H}(X)$, the set of nonempty compact subsets of X.

IFZS [39]: $\mathcal{F}^*(X)$, the set of all functions $u : X \to [0,1]$ which are 1) upper semicontinuous on (X, d) and 2) normalized, i.e. for each $u \in \mathcal{F}^*(X)$ there exists an $x_0 \in X$ for which $u(x_0) = 1$.

IFSM [89, 90, 91]: $\mathcal{L}^p(X, \mu)$, the space of p-integrable functions with respect to a measure μ, $1 \le p \le \infty$. Fractal Transforms [22, 81, 87] are a special case of IFSM. The Bath Fractal Transform [187, 188] is an IFSM with place-dependent grey level maps.

IFSP [121, 20]: $\mathcal{M}(X)$, the set of probability measures on $\mathcal{B}(X)$, the σ-algebra of Borel subsets of X.

Along with the IFS maps $\mathbf{w}$ (except in the case of IFS on $\mathcal{H}(X)$) there is an associated set of functions $\Phi = \{\phi_1, \phi_2, \ldots, \phi_N\}$, $\phi_i : \mathbf{R} \to \mathbf{R}$, which satisfy conditions that depend on the particular metric space (Y, d_Y) being used. The pair of vectors $(\mathbf{w}, \Phi)$ then determine a *fractal transform operator* T which is designed to map Y into itself. It is desirable that T be contractive on (Y, d_Y) so that it possesses a unique and globally attracting fixed point $\bar{y} \in Y$, i.e. $T\bar{y} = \bar{y}$.

Given a $u \in Y$, its image Tu will be constructed for each point $x \in X$ (or each subset $S \in \mathcal{H}(X)$). Except in the case of IFSP (which is no longer considered for image representation), most practical as well as theoretical studies devise methods which either assume that the $w_i(X)$ do not overlap, or at least ignore any such overlap. (Indeed, if $(Tu)(x)$ is to exist for all $x \in X$ and X is closed and not discrete or finite, then some sets $w_i(X)$ *must* overlap with each other, if only at one point.) As a result, each point $x \in X$ is considered to have only one preimage $w^{-1}_{i(x)}(x) \in X$. The value of the fractal transform of u at $x \in w_i(X)$ is simply $(Tu)(x) = \phi_i(u(w_i^{-1}(x)))$.

In the spirit of our earlier work on IFS-type methods on function spaces, namely IFZS and IFSM, we consider the more general non-overlapping case when x has more than one preimage, i.e. $w^{-1}_{i^*_k}(x), 1 < k \leq n(x)$. There is then the question of how to combine the $n(x)$ *fractal components* $\phi_{i^*_k}(u(w^{-1}_{i^*_k}(x)))$, to form our *generalized fractal transform* $(Tu)(x)$. In this chapter, we postulate a set of common rules for combining fractal components. Some of these rules were already considered in the development of IFZS [39]. Understandably, such a method of generalized fractal transforms may not necessarily be useful in the problem of fractal image compression since the coding of any region of an image with more than one fractal component is usually viewed as redundant and contrary to the goal of data compression. Our study, however, is based on a view of fractal transform methods as viable methods of approximating functions and measures in the same spirit as Fourier series/transforms, orthogonal function expansions and, more recently, wavelet expansions.

Previously [89], we compared briefly the various IFS-type methods before outlining a solution of the inverse problem for IFSM [90]. In this chapter, we again consider all of these methods, but with the purpose of unifying them *under one common scheme.* The first step is to establish the IFS method, traditionally viewed as a method of geometrically constructing fractal-type sets in $\mathcal{H}(X)$, as a fractal transform method over an appropriate function space whose elements are *bitmap*, i.e. black and white, images. A method of representing images with varying grey levels is clearly desirable. There is a straightforward transition from this IFS approach to the method of IFZS which works with the grey level range [0,1]. In the overlapping case for IFZS, the prescription for combining several fractal components, namely the *supremum* operator, carries over from the IFS method. A modification of the IFZS method [91], motivated in part by restrictions associated with the Hausdorff metric, yields the method of IFSM on the function space $\mathcal{L}^1(X)$. It is then natural to consider IFSM on $\mathcal{L}^p(X)$.

The final step is to provide a link between the IFS, IFZS and IFSM methods on function spaces and the method of IFSP on the probability measure space $\mathcal{M}(X)$. This is accomplished by constructing an IFS-type fractal fransform on the space of distributions $\mathcal{D}'(X)$. IFSP and IFSM correspond to particular cases of this distributional IFS (IFSD). Another noteworthy result of this method is an expression for integrals of the form $\int_X f(x)\bar{u}(x)dx$, where $\bar{u}$ denotes the fixed point of an IFSM. (This is analogous to the expression for integrals of functions with respect to an IFSP invariant measure.)

Finally, we mention that the theory described in this chapter may be easily extended to the "block encoding" [125] or "Local IFS" methods [22] which are currently employed in fractal image compression. In these methods, the IFS maps are assumed to map subsets of X, the *domain* or *parent* blocks, to smaller subsets, the *range* or *child* blocks. In Chapter 11, we consider inverse problems for generalized fractal transforms.

10.2 Generalized Fractal Transforms

In this section, we formally define generalized fractal transforms of image functions and provide a set of common rules for constructing such transforms from their fractal components. The mathematical setting is provided by the following ingredients:

1. **The base space:** denoted, as above, by (X, d). The space representing the pixels; a compact subset of $\mathbf{R}^n$. Without loss of generality, $X = [0,1]^n$ with Euclidean metric.

2. **The IFS component:** For an $N \in \mathbf{N}$, let $\mathbf{w} = \{w_1, w_2, \ldots, w_N\}$. In many cases we can relax the condition that the w_i be contraction maps on X. Note that different sets $w_i(X)$ may overlap. The principal classes of IFS functions used are:

   $Con(X) = \{w : X \to X | d(w(x), w(y)) \le c d(x, y)$ for some $c \in [0, 1)$, $\forall x, y \in X\}$, the set of contraction maps on X.

   $Con_1(X) \subset Con(X)$: the set of one-to-one contraction maps on X.

   $Aff_1(X)$: the set of affine maps on X which are one-to-one. The representation of such maps in $X = [0,1]^n$ is given by

$$\mathbf{y} = \mathbf{A}\mathbf{x} + \mathbf{b}, \tag{10.2.1}$$

   where $\mathbf{A}$ is an $n \times n$ matrix with nonzero determinant and $\mathbf{b}$ is an n-vector. The Jacobian of this transformation will be denoted by $|J| = |\lambda_1\lambda_2 \ldots \lambda_n|$, where the λ_i are the eigenvalues of $\mathbf{A}$.

3. **The image function space:** $\mathcal{F}(X) = \{u : X \to R_g \subseteq \mathbf{R}^+\}$, the functions which will represent our images. The **grey level range** R_g will denote the range of a particular class of image functions used in a given fractal transform method. (In practical applications, R_g is bounded.)

4. **The grey level component:** Associated with the IFS maps $\mathbf{w}$ will be a vector of N functions $\Phi = \{\phi_1, \phi_2, \ldots, \phi_N\}$, $\phi_i : R_g \to R_g$. We may also consider $\phi_i : R_g \times X \to R_g$, i.e. "place-dependent" grey level maps. See, for example, Ref. [188, 187, 91].

5. **The fractal components of** u will be given by $f_i : X \to R_g$, $1 \le i \le N$, where
$$f_i(x) = \begin{cases} \phi_i(u(w_i^{-1}(x))), & x \in w_i(X), \\ 0, & x \notin w_i(X). \end{cases} \tag{10.2.2}$$
In other words, the fractal component $f_i(x)$ represents a modified value of the grey level of u at the ith preimage of x (if it exists).

6. **The generalized fractal transform of** $u \in \mathcal{F}(X)$: $F_k : [R_g]^k \to R_g$, $1 \le k \le N$, where
$$F_k(\mathbf{t}) = F_k(t_1, t_2, \dots, t_k), \quad t_i \in R_g, \quad 1 \le i \le k. \tag{10.2.3}$$
The F_k combine the k distinct fractal components $t_i = f_i(x)$ subject to conditions described below. The transform F_N defines an operator $T : \mathcal{F}(X) \to \mathcal{F}(X)$ that associates to each image function $u \in \mathcal{F}(X)$ the image function $v = Tu$. The way in which the F_k combine the fractal components depends on the image function space used.

The grey level ranges and fractal transforms for IFZS and IFSM are given below.

IFZS: $R_g = [0, 1]$. $F_N(x) = \sup_{1 \le i \le N}\{f_i(x)\}$.

IFSM: $R_g = \mathbf{R}^+$. $F_N(x) = \sum_{i=1}^N f_i(x)$.

We now outline the properties which are to be satisfied by the fractal transform operators F_k. This discussion follows the same pattern discussed in [39] with some additional comments.

1. $F_N(t_1, t_2, \dots, t_N) = F_N(t_{i_1}, t_{i_2}, \dots, t_{i_N})$, where $\{i_1, i_2, \dots, i_N\}$ is any permutation of the index set $\{1, 2, \dots, N\}$ (symmetry).

2. $F_N(t_1, t_2, \dots, t_N) = F_2(F_{N-1}(t_1, \dots, t_{N-1}), t_N)$ (recursivity).
Properties 1 and 2 imply that
$$F_N(t_1, t_2, \dots, t_N) = F_2(F_2(F_2(\dots F_2(t_1, t_2), t_3), t_4), \dots, t_N), \tag{10.2.4}$$
and, in particular, that $F_2(F_2(t_1, t_2), t_3) = F_2(t_1, F_2(t_2, t_3))$. Thus, F_2 is an associative binary operation on $\mathbf{R}^+ \times \mathbf{R}^+$. We shall let S denote such a binary operation and assume that it satisfies the following set of additional properties:

3. $S : [R_g]^2 \to R_g$ is continuous.

4. $S(0, y) = y$, $\forall y \in R_g$, i.e. 0 is an *identity* element; the combination of a pixel with brightness $y > 0$ with one of zero brightness yields a pixel with brightness y.

5. S is nondecreasing, i.e. $x_a < x_b$ implies that $S(x_a, y) \le S(x_b, y)$, $\forall y \in R_g$. The brighter a pixel, the brighter its combination with another pixel.

6. $S(x, x) \ge x$, $\forall x \in R_g$; the combination of two pixels of equal brightness should not result in a darker pixel.

7. For all $y \in R_g$, $S(s,y) = s$ for $s = \sup z \in R_g$ (s may be infinite), i.e. s is the *annihilator*.

There is a representation theorem for topological semigroups on $\mathbf{R}$ which will be useful for the construction of appropriate associative operators for our fractal transforms.

Theorem 10.1 *[159] If $S : [\mathbf{R}^+]^2 \to \mathbf{R}^+$ satisfies Properties (1)-(7) above, then there exist:*

1. *a discrete (finite or countably infinite) index set I,*
2. *a sequence of disjoint open intervals $\{(a_i, b_i)\} \subset \mathbf{R}^+$, $i \in I$, with $0 = a_1 < b_1 \leq a_2 < b_2 \leq \ldots$,*
3. *a sequence of numbers $s_i \in \mathbf{R}^+$, $i \in I$ and*
4. *a sequence of continuous and strictly increasing functions $f_i : [a_i, b_i] \to [0, s_i]$, $i \in I$, with $f_i(a_i) = 0$ and $f_i(b_i) = s_i$, such that*
$$S(x,y) = g_i(f_i(x) + f_i(y)), \qquad \forall (x,y) \in [a_i, b_i]^2, \tag{10.2.5}$$
where $g_i : [0,\infty] \to [0, s_i]$, the pseudoinverse *of f_i is defined as*
$$g_i(x) = \begin{cases} f_i^{-1}(x), & x \in [0, s_i], \\ b_i, & x \notin [s_i, \infty], \end{cases} \tag{10.2.6}$$
and finally,
$$S(x,y) = \sup(x,y) \quad if \quad (x,y) \in [\bar{\mathbf{R}}^+]^2 - \bigcup_{i \in I} [a_i, b_i]^2. \tag{10.2.7}$$

Examples:

1. With the condition that $S(x,x) = x$ for all $x \in R_g = [0,s] \subset \mathbf{R}^+$, we have
$$S(x,y) = \sup\{x,y\} \quad \forall (x,y) \in [0,s]^2. \tag{10.2.8}$$
This represents an extreme case where the index set $I = \{1\}$ and the sequence $\{(a_i, b_i)\}$ reduces to the interval $(0,s)$. Then all $x \in [0,s]$ are idempotents of S, with 0 being the identity and s the annihilator. This was the natural choice for the IFZS case [39].
2. $S(x,x) = x$ for only $x = 0$ or $x = s$. Again, there is only one interval $(a,b) = (0,s)$. If we choose $f(x) = x$, then
$$S(x,y) = \min\{x+y, s\} \quad \forall (x,y) \in [0,s]^2. \tag{10.2.9}$$
3. $R_g = \mathbf{R}^+$, i.e. $s = \infty$. A possible binary operation is the summation operator, i.e.
$$S(x,y) = x + y \quad \forall (x,y) \in [0,\infty)^2. \tag{10.2.10}$$
The binary operation in Eq. (10.2.10) is employed by the IFSM method [89, 90, 91].

10.3 From IFS to IFSM Fractal Transforms

In this section, we construct a scheme to unify existing IFS-type fractal transforms on function spaces, as outlined in the Introduction. At the end we review the basic properties of IFSP on the space of probability measures $\mathcal{M}(X)$, in preparation for Section 10.4, where fractal transforms of functions and IFSP are related through distributions. In what follows, we define the contraction factor of a map $w \in Con(X)$ to be

$$c := \sup_{x,y \in X, x \neq y} d(w(x), w(y))/d(x,y). \tag{10.3.1}$$

For an N-map IFS $\mathbf{w} = (w_1, w_2, \ldots, w_N)$, the contraction factors of the w_i will be denoted by c_i. We then define $c = \max_{1 \leq i \leq N}\{c_i\}$.

10.3.1 IFS

Here $Y = \mathcal{H}(X)$, the set of nonempty compact subsets of X and d_Y is the Hausdorff metric h, defined as follows. Let the distance between a point $x \in X$ and a set $S \in \mathcal{H}(X)$ be given by

$$d(x, S) = \inf_{z \in S} d(x, z). \tag{10.3.2}$$

Then for each $S_1, S_2 \in \mathcal{H}(X)$,

$$h(S_1, S_2) = \max\{\sup_{x \in S_1} d(x, S_2), \sup_{z \in S_2} d(z, S_1)\}. \tag{10.3.3}$$

Now let $\mathbf{w} = \{w_1, w_2, ..., w_N\}$, $w_i \in Con(X)$. Associated with each contraction map w_i is a set-valued mapping $\hat{w}_i : \mathcal{H}(X) \to \mathcal{H}(X)$ defined by $\hat{w}_i(S) = \{w_i(x) | x \in S\}$ for $S \in \mathcal{H}(X)$. Then the usual IFS operator $\hat{\mathbf{w}}$ associated with the N-map IFS $\mathbf{w}$ is defined as follows:

$$\hat{\mathbf{w}}(S) = \bigcup_{i=1}^{N} \hat{w}_i(S), \quad S \in \mathcal{H}(X). \tag{10.3.4}$$

The IFS operator $\hat{\mathbf{w}}$ is contractive on $(\mathcal{H}(X), h)$ [121]:

$$h(\hat{\mathbf{w}}(A), \hat{\mathbf{w}}(B) \leq ch(A, B), \quad \forall A, B \in \mathcal{H}(X). \tag{10.3.5}$$

The completeness of $(\mathcal{H}(X), h)$ guarantees the existence of a unique fixed point $\bar{y} = A \in \mathcal{H}(X)$. The set A, also called the *attractor*, is the IFS representation of an image. From Eq. (10.3.4), it satisfies the following self-tiling property,

$$A = \bigcup_{i=1}^{N} \hat{w}_i(A). \tag{10.3.6}$$

We now formulate the IFS method over an appropriate function space. First, let $w_i \in Con_1(X)$, $1 \leq i \leq N$. Let $I_S(x)$ denote the characteristic function of a set $S \in \mathcal{H}(X)$, i.e.

$$I_S(x) = \begin{cases} 1, & x \in S, \\ 0, & x \notin S. \end{cases} \tag{10.3.7}$$

Now let $A, B \in \mathcal{H}(X)$ and $C = A \cup B \in \mathcal{H}(X)$. It follows that

$$I_C(x) = \sup\{I_A(x), I_B(x)\}. \tag{10.3.8}$$

From the property that $I_{\hat{w}_i(S)}(x) = I_S(w_i^{-1}(x))$, we then have, from Eq. (10.3.8),

$$I_{\hat{\mathbf{w}}(S)}(x) = \sup_{1 \le i \le N} \{I_S(w_i^{-1}(x))\}. \tag{10.3.9}$$

Now consider the function space $\mathcal{F}_{BW}^*(X)$ (for "black and white") defined by

$$\mathcal{F}_{BW}^*(X) = \{u : X \to \{0,1\} \mid supp(u) \in \mathcal{H}(X)\}. \tag{10.3.10}$$

In this case, the support of $u \in \mathcal{F}_{BW}^*(X)$ is given by

$$\begin{aligned} supp(u) &= \{x \in X \mid u(x) = 1\} \\ &=: [u]^1, \end{aligned} \tag{10.3.11}$$

where we have introduced the IFZS notation $[u]^1$ to denote the "1-level" set of u. In other words, u represents a *bitmap* (black and white) image whose white region $[u]^1$ is nonempty and closed. In fact, $\mathcal{F}_{BW}^*(X) \subset \mathcal{F}^*(X)$, the latter being the complete metric space on which IFZS is formulated. It is thus natural to consider the following metric on $\mathcal{F}_{BW}^*(X)$:

$$d_{BW}(u,v) = h([u]^1, [v]^1), \quad \forall u, v \in \mathcal{F}_{BW}^*. \tag{10.3.12}$$

Completeness of $(\mathcal{F}_{BW}^*(X), d_{BW})$ follows from the completeness of $(\mathcal{H}(X), h)$. From Eq. (10.3.9), the fractal transform operator $T : \mathcal{F}_{BW}^*(X) \to \mathcal{F}_{BW}^*(X)$ associated with the N-map IFS $\mathbf{w}$ is given by

$$(Tu)(x) = \sup_{1 \le i \le N} \{u(w_i^{-1}(x))\}, \quad \forall x \in X. \tag{10.3.13}$$

The contractivity of T on $(\mathcal{F}_{BW}^*(X), d_{BW})$ follows immediately from the contractivity of the IFS operator $\hat{\mathbf{w}}$ on $(\mathcal{H}(X), h)$. Thus there exists a unique fixed point $\bar{u} \in \mathcal{F}_{BW}^*(X)$ of the operator T, i.e. $T\bar{u} = \bar{u}$. Moreover, $[\bar{u}]^1 = A$, the attractor of the IFS $\mathbf{w}$ so that

$$[\bar{u}]^1 = \bigcup_{i=1}^{N} \hat{w}_i([\bar{u}]^1). \tag{10.3.14}$$

In this formulation, pixels can assume only two grey level values, namely 0 and 1, or black and white (or vice versa). As such, only the geometry of an attractor is revealed. "Real" images, however, are not only black and white. Instead, their pixels can assume a range of nonnegative grey level values, e.g. $u(x) \in [0,1]$. For this reason, it would be desirable to modify the above IFS method so that such a range of grey level values could be produced. This is easily accomplished by modifying the fractal components in Eq. (10.3.13) as follows:

$$(Tu)(x) = \sup_{1 \le i \le N} \{\phi_i(u(w_i^{-1}(x)))\}, \quad \forall x \in X. \tag{10.3.15}$$

where the $\phi_i : [0,1] \to [0,1]$ are *grey level maps.* Subject to some conditions on these maps (which guarantee that $T : \mathcal{F}^*(X) \to \mathcal{F}^*(X)$) we then arrive at the method of Iterated Fuzzy Set Systems (IFZS) [39].

10.3.2 IFZS

Here, images are represented by functions $u \in \mathcal{F}^*(X)$ (cf. Section 10.1). The grey level range is $R_g = [0,1]$. The metric d_Y for the space $Y = \mathcal{F}^*(X)$ is defined as follows. We first define the α-level sets of $u \in \mathcal{F}^*(X)$ for $\alpha \in [0,1]$:

$$
\begin{aligned}
[u]^\alpha &:= \{x \in X : u(x) \geq \alpha\}, \quad \alpha \in (0,1], \\
[u]^0 &:= \overline{\{x \in X : u(x) > 0\}},
\end{aligned} \tag{10.3.16}
$$

where $\bar{S}$ denotes the closure of the set S in (X,d). Clearly, $[u]^\alpha \in \mathcal{H}(X)$ for $0 \leq \alpha \leq 1$. Then for $u, v \in \mathcal{F}^*(X)$, define

$$
d_\infty(u,v) = \sup_{0 \leq \alpha \leq 1} h([u]^\alpha, [v]^\alpha). \tag{10.3.17}
$$

The metric space $(\mathcal{F}^*(X), d_\infty)$ is complete [64].

The IFZS is defined by the following:

1. **The IFS component:** $\mathbf{w} = \{w_1, \ldots, w_N\}$, $w_i \in Con(X)$, $1 \leq i \leq N$,
2. **The grey level component:** $\Phi = \{\phi_1, \phi_2, ..., \phi_N\}$, $\phi_i : [0,1] \to [0,1]$, such that for all $i \in \{1,2,\ldots,N\}$:

   (i) ϕ_i is nondecreasing on [0,1],

   (ii) ϕ_i is right continuous on [0,1),

   (iii) $\phi_i(0) = 0$.

   In addition,

   (iv) $\phi_{i^*}(1) = 1$ for at least one $i^* \in \{1,2,...,N\}$.

The IFZS fractal transform $T : \mathcal{F}^*(X) \to \mathcal{F}^*(X)$ is defined as

$$
(Tu)(x) = \sup_{1 \leq i \leq N} \{\phi_i(\tilde{u}(w_i^{-1}(x)))\}, \quad \forall x \in X, \tag{10.3.18}
$$

where, for $B \subseteq X$, (i) $\tilde{u}(B) = \sup_{z \in B}\{u(z)\}$ if $B \neq \emptyset$ and (ii) $\tilde{u}(\emptyset) = 0$. The conditions imposed on the functions ϕ_i guarantee that T maps $(\mathcal{F}^*(X), d_\infty)$ to itself. The relation between level sets of a function u and those of its image Tu is given by

$$
[Tu]^\alpha = \bigcup_{i=1}^{N} \hat{w}_i([\phi_i \circ u]^\alpha), \quad \alpha \in [0,1]. \tag{10.3.19}
$$

The contractivity of the IFS maps w_i implies that that T is a contraction map on $(\mathcal{F}^*(X), d_\infty)$ since [39]

$$
d_\infty(Tu, Tv) \leq c d_\infty(u,v), \quad \forall u, v \in \mathcal{F}^*(X). \tag{10.3.20}
$$

The completeness of this space guarantees the existence of a unique fixed point $\bar{u} \in \mathcal{F}^*(X)$ of the operator T. From Eq. (10.3.19), the α-level sets of $\bar{u}$ obey the following generalized self-tiling property:

$$
[\bar{u}]^\alpha = \bigcup_{i=1}^{N} w_i([\phi_i \circ \bar{u}]^\alpha), \quad \alpha \in [0,1], \tag{10.3.21}
$$

which can be compared to the self tiling of $\alpha = 1$ level sets for the IFS case in Eq. (10.3.14).

We draw the reader's attention to the fact that the use of the *sup* operator in the IFZS operator T is a natural choice. The IFZS method is based on the properties of level sets of functions in $\mathcal{F}^*(X)$. Taking the *sup* of two functions $u(x)$ and $v(x)$ for all $x \in X$ corresponds to taking the union of their respective α-level sets.

The Hausdorff metric d_∞ is very restrictive, however, from both practical (i.e. image processing) as well as theoretical perspectives. In [91], two fundamental modifications were made to the IFZS approach:

1. For a $\mu \in \mathcal{M}(X)$ and $u, v \in \mathcal{F}^*(X)$, define $\forall \alpha \in [0,1]$,

$$\begin{aligned} G(u,v;\alpha) &= \mu([u]^\alpha \Delta [v]^\alpha) \\ &= \int_X |I_{[u]^\alpha}(x) - I_{[v]^\alpha}(x)| d\mu(x), \end{aligned} \tag{10.3.22}$$

   where Δ denotes the symmetric difference operator: For $A, B \subseteq X$, $A \Delta B = (A \cup B) - (A \cap B)$.

2. Now let ν be a finite measure on $\mathcal{B}(R_g)$ and define

$$g(u,v;\nu) = \int_{R_g} G(u,v;\alpha) d\nu(\alpha). \tag{10.3.23}$$

   An application of Fubini's Theorem yields

$$\begin{aligned} g(u,v;\nu) &= \nu(\{0\})\mu([u]^0 \Delta [v]^0) + \int_{X_u} \nu((v(x), u(x)]) d\mu(x) \\ &+ \int_{X_v} \nu((u(x), v(x)]) d\mu(x), \end{aligned} \tag{10.3.24}$$

   where $X_u = \{x \in X \mid u(x) > v(x)\}$ and $X_v = \{x \in X \mid v(x) > u(x)\}$.

It can be shown that $g(u,v;\nu)$ is a pseudometric on $\mathcal{L}^1(X,\mu)$. In the particular case that $\nu = m$, the Lebesgue measure on the grey level range R_g, Eq. (10.3.34) becomes

$$g(u,v;m) = \int_X |u(x) - v(x)| d\mu(x), \tag{10.3.25}$$

the $\mathcal{L}^1(X,\mu)$ distance between u and v. The restrictive Hausdorff metric d_∞ over α-level sets has been replaced by a weaker pseudometric (metric on the measure algebra) involving integrations over X and R_g. The result is a fractal transform method on the function space $\mathcal{L}^1(X,\mu)$. While it appears that only the $\mathcal{L}^1$ distance can be generated by a measure ν on $\mathcal{B}(R_g)$, it is still natural to consider fractal transforms over the general function spaces $\mathcal{L}^p(X,\mu)$.

10.3.3 IFSM

Let μ be a measure on $\mathcal{B}(X)$ and for any integer $p \geq 1$, let $\mathcal{L}^p(X,\mu)$ denote the linear space of all real–valued functions u such that u^p is integrable on $(\mathcal{B}(X),\mu)$. We choose $Y = \mathcal{L}^p(X,\mu)$. The metric d_Y is defined by the usual $\mathcal{L}^p$–norm, i.e. $d_Y(u,v) := d_p(u,v)$, where

$$d_p(u,v) = \| u - v \|_p = \left[\int_X |u(x) - v(x)|^p d\mu(x) \right]^{1/p}. \tag{10.3.26}$$

The IFSM is then defined by the following:

1. **The IFS component:** $\mathbf{w} = \{w_1, \ldots, w_N\}$, $w_i \in Aff_1(X)$, $1 \leq i \leq N$,
2. **The grey level component:** $\Phi = \{\phi_1, \phi_2, ..., \phi_N\}$, $\phi_i \in Lip(\mathbf{R};\mathbf{R})$, where

$$\begin{aligned} Lip(\mathbf{R};\mathbf{R}) &= \{\phi : \mathbf{R} \to \mathbf{R} \mid |\phi(t_1) - \phi(t_2)| \leq K|t_1 - t_2|, \\ &\quad \forall \ t_1, t_2 \in \mathbf{R} \text{ for some } K \in [0,\infty)\}. \end{aligned} \tag{10.3.27}$$

Since our function space involves integrations, it is natural to define the following fractal transform operator T corresponding to an N-map IFSM $(\mathbf{w},\Phi)$:

$$(Tu)(x) := \sum_{k=1}^{N} f_k(x), \tag{10.3.28}$$

where the fractal components f_k are defined in Eq. (10.2.2). The above conditions on the w_i and ϕ_i guarantee that $T : \mathcal{L}^p(X,m) \to \mathcal{L}^p(X,m)$ for all $p \in [1,\infty]$, where m denotes Lebesgue measure on X.

Now let $X = [0,1]^D$, $\mu = m$ and $1 \leq p \leq \infty$. Also let $u, v \in \mathcal{L}^p(X,m)$ with fractal components f_k and g_k, $1 \leq k \leq N$, respectively. Then from the relation

$$\begin{aligned} \| Tu - Tv \|_p &= \| \sum_{k=1}^{N} [f_k(x) - g_k(x)] \|_p \\ &\leq \sum_{k=1}^{N} \| f_k(x) - g_k(x) \|_p, \end{aligned} \tag{10.3.29}$$

we obtain the result

$$d_p(Tu,Tv) \leq C_p d_p(u,v), \quad C_p = \sum_{k=1}^{N} |J_k|^{1/p} K_k, \tag{10.3.30}$$

where $|J_k|$ is the Jacobian associated with the affine transformation $x = w_k(y)$. In the special μ-nonoverlapping case, i.e. where the sets $w_i(X)$ overlap only on sets of zero μ-measure (the standard assumption in practical applications in the literature), we may use the relation

$$\| Tu - Tv \|_p^p = \sum_{k=1}^{N} \| f_k(x) - g_k(x) \|_p^p \tag{10.3.31}$$

to obtain the result

$$d_p(Tu, Tv) \leq \bar{C}_p d_p(u, v), \quad \bar{C}_p = \left[\sum_{k=1}^{N} |J_k| K_k^p\right]^{1/p}. \tag{10.3.32}$$

Note that

$$\bar{C}_p \leq C_p \leq K, \quad K = \max_{1 \leq k \leq N} K_k. \tag{10.3.33}$$

In the nonoverlapping case, with $p = \infty$, we also have

$$\| Tu - Tv \|_\infty \leq K \| u - v \|_\infty, \quad \forall u, v \in \mathcal{L}^\infty(X, m). \tag{10.3.34}$$

This is the usual bound presented in the literature on fractal transforms [22, 87].

In applications, we shall be using *affine IFSM*, i.e. $w_k \in Aff_1(X)$ and

$$\phi_k(t) = \alpha_k t + \beta_k, \quad t \in \mathbf{R}, \quad 1 \leq k \leq N. \tag{10.3.35}$$

If the associated operator T is contractive on $\mathcal{L}^p(X, \mu)$, then its fixed point $\bar{u}$ satisfies the equation

$$\bar{u}(x) = \sum_{k=1}^{N} [\alpha_k \psi_k(x) + \beta_k \chi_k(x)], \tag{10.3.36}$$

where $\psi_i(x) = \bar{u}(w_k^{-1}(x))$ and $\chi_k(x) = I_{w_k(X)}$. In other words, $\bar{u}$ may be written as a linear combination of both functions $\psi_k(x)$ and piecewise constant functions $\chi_k(x)$ which are obtained by dilatations and translations of $\bar{u}$ and $I_X(x)$, respectively. This is somewhat reminiscent of the role of scaling functions in wavelet theory.

The following result guarantees that the use of affine IFSM is sufficient from a theoretical perspective.

Theorem 10.2 *Let $X = [0,1]^D$ and $\mu \in \mathcal{M}(X)$. For a $p \geq 1$ define $\mathcal{L}_A^p(X, \mu) \subset \mathcal{L}^p(X, \mu)$ to be the set of fixed points $\bar{u}$ of all contractive N-map affine IFSM* $(\mathbf{w}, \Phi)$ *for $N \geq 1$. Then $\mathcal{L}_A^p(X, \mu)$ is dense in* $(\mathcal{L}^p(X, \mu), d_p)$.

The proof of this theorem is based on the property that the set of all step functions in X is dense in $(\mathcal{L}^p(X, \mu), d_p)$.

"Place-Dependent" IFSM

The above method may be easily generalized to "place-dependent" IFSM (PDIFSM), that is, IFSM with grey level maps having the form $\phi_k : \mathbf{R} \times X \to \mathbf{R}$, $1 \leq k \leq N$. In other words, the ϕ_i are dependent both on the grey-level value at a preimage as well as the location of the preimage itself. (This is analogous to IFS with place-dependent probabilities [21].) Much of the theory developed above for IFSM easily extends to place-dependent IFSM as we outline below. This is the basis of the "Bath Fractal Transform" and its effectiveness in coding images has been discussed in the literature [187, 188, 259].

The fractal components $f_k(x)$ of a function $u \in \mathcal{L}^p(X,\mu)$ will be given by

$$f_k(x) = \begin{cases} \phi_k(u(w_k^{-1}(x)), w_k^{-1}(x)), & x \in w_k(X), \\ 0, & x \notin w_k(X). \end{cases} \tag{10.3.37}$$

The operator T associated with an N-map PDIFSM $(\mathbf{w}, \Phi)$ will have the form

$$(Tu)(x) = \sum_{k=1}^{N}{}' \phi_k(u(w_k^{-1}(x)), w_k^{-1}(x)), \tag{10.3.38}$$

We first define the following set of uniformly Lipschitz functions,

$$\begin{aligned} Lip(\mathbf{R}, X; \mathbf{R}) &= \{\phi : \mathbf{R} \times X \to \mathbf{R} : |\phi(t_1, s) - \phi(t_2, s)| \leq K|t_1 - t_2|, \\ &\quad \forall t_1, t_2 \in \mathbf{R}, \forall s \in X \text{ for some } K \in [0, \infty)\}. \end{aligned} \tag{10.3.39}$$

If $w_i \in Aff_1(X)$ and $\phi_i \in Lip(\mathbf{R}, X; \mathbf{R})$ for $1 \leq i \leq N$ then $T : \mathcal{L}^p(X, m) \to \mathcal{L}^p(X, m)$ for $1 \leq p < \infty$. Furthermore, if $X \subset \mathbf{R}^D$, $D \in \{1, 2, ...\}$, and $\mu = m$ then the relation in Eq. (10.3.30) holds.

Some possible forms for the place-dependent grey level maps ϕ are as follows:

1. $\phi(t, s) = \sum_{i=0}^{n} a_i(s) t^i$, where the $a_i : X \to \mathbf{R}$, bounded on X,
2. $\phi(t, s) = f(t) + g(s)$ ("separable") with suitable conditions on f and g, e.g. $f \in Lip(\mathbf{R}; \mathbf{R})$ and $g : X \to \mathbf{R}$ bounded on X.

It is convenient to work with ϕ maps which are only first degree in the grey-level variable t, i.e.

$$\phi(t, s) = \alpha t + \beta + g(s), \quad g : X \to \mathbf{R}, \text{ bounded on } X, \tag{10.3.40}$$

$$\phi(t, s) = \alpha(s) t + \beta(s), \quad \alpha, \beta : X \to \mathbf{R}, \text{ bounded on } X. \tag{10.3.41}$$

The action of the first set of maps can be considered as a "place-dependent" shift in grey-level value. The second set of maps produce a more direct interaction between position and grey-level value. In Figure 10.1 are presented histogram approximations of fixed points $\bar{u}$ of two rather simple affine PDIFSM in order to show the effects of place-dependence.

10.3.4 IFSP

Associated with an N-map IFS $\mathbf{w}$, is a set of probabilities $\mathbf{p} = \{p_1, p_2, ..., p_N\}$, $p_i \geq 0$, with $\sum_{i=1}^{N} p_i = 1$. Let $\mathcal{B}(X)$ denote the σ-algebra of Borel subsets of X generated by all the elements of $\mathcal{H}(X)$. Then $Y = \mathcal{M}(X)$, the set of all probability measures on $\mathcal{B}(X)$. Here, the (Markov) operator associated with the IFSP $(\mathbf{w}, \mathbf{p})$ is defined as follows: For a $\mu \in \mathcal{M}(X)$ and each $S \in \mathcal{H}(X)$,

$$(T\mu)(S) = (M\mu)(S) = \sum_{i=1}^{N} p_i \mu(w_i^{-1}(S)). \tag{10.3.42}$$

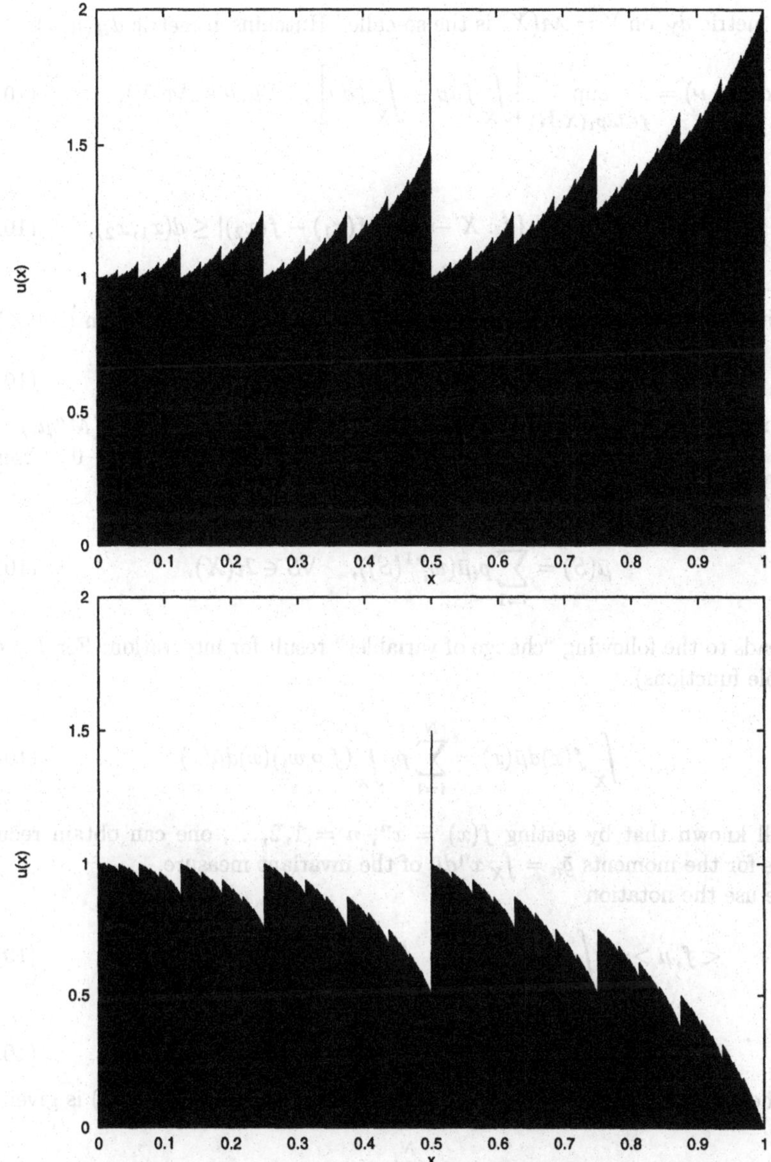

Figure 10.1: Fixed-point attractor functions $\bar{u}(x)$ of the following 2-map PDIFSM: $w_1(x) = \frac{1}{2}x$, $w_2(x) = \frac{1}{2}x + \frac{1}{2}$, $\phi_1(t,s) = \frac{1}{2}t + \frac{1}{2}$, $\phi_2(t,s) = \frac{1}{2}t + \frac{1}{2} + \gamma s$. When $\gamma = 0$, $\bar{u}(x) = 1$ (a.e.). (a) $\gamma = \frac{1}{2}$. (b) $\gamma = -\frac{1}{2}$.

The metric d_Y on $Y = \mathcal{M}(X)$ is the so-called Hutchinson metric $d_H(\mu, \nu)$:

$$d_H(\mu,\nu) = \sup_{f \in Lip_1(X;\mathbf{R})} \left[\int_X f d\mu - \int_X f d\nu \right], \quad \forall \mu, \nu \in \mathcal{M}(X), \tag{10.3.43}$$

where

$$\begin{aligned} Lip_1(X;\mathbf{R}) &= \{ f : X \to \mathbf{R} \mid |f(x_1) - f(x_2)| \leq d(x_1, x_2), \\ &\quad \forall x_1, x_2 \in X \}. \end{aligned} \tag{10.3.44}$$

The contractivity of the IFS maps w_i implies the contractivity of M on $(\mathcal{M}(X), d_H)$ [121]:

$$d_H(M\mu, M\nu) \leq c d_H(\mu, \nu), \quad \forall \mu, \nu \in \mathcal{M}(X). \tag{10.3.45}$$

There exists a unique $\bar{\mu} \in \mathcal{H}(X)$ such that (1) $M\bar{\mu} = \bar{\mu}$ and (2) $d_H(M^n\mu, \bar{\mu}) \to 0$ as $n \to \infty$. Moreover, $supp(\bar{\mu}) \subseteq A$, with the equality when all $p_i > 0$. From Eq. (10.3.42) it follows that

$$\bar{\mu}(S) = \sum_{i=1}^{N} p_i \bar{\mu}(w_i^{-1}(S)), \quad \forall S \in \mathcal{H}(X), \tag{10.3.46}$$

which leads to the following "change of variables" result for integration: For $f \in C(X)$ (or simple functions),

$$\int_X f(x) d\bar{\mu}(x) = \sum_{i=1}^{N} p_i \int_X (f \circ w_i)(x) d\bar{\mu}(x). \tag{10.3.47}$$

It is well known that by setting $f(x) = x^n$, $n = 1, 2, \ldots$, one can obtain recursion relations for the moments $\bar{g}_n = \int_X x^n d\bar{\mu}$ of the invariant measure.

If we use the notation

$$< f, \mu > := \int_X f(x) d\mu(x), \quad f \in C(X), \quad \mu \in \mathcal{M}(X), \tag{10.3.48}$$

then

$$< f, T\mu > = < T^\dagger f, \mu >, \tag{10.3.49}$$

where the adjoint operator $T^\dagger : C(X) \to C(X)$ (referred to as T in [20]) is given by

$$(T^\dagger f)(x) = \sum_{i=1}^{N} p_i (f \circ w_i)(x). \tag{10.3.50}$$

We may iterate this procedure to obtain, for $n = 1, 2, \ldots$,

$$\begin{aligned} < f, T^n\mu > &= < (T^\dagger)^n f, \mu > \\ &= \sum_{i_1,\ldots,i_n}^{N} p_{i_1} \cdots p_{i_n} \int_X (f \circ w_{i_1} \circ \ldots \circ w_{i_n})(x) d\mu. \end{aligned} \tag{10.3.51}$$

For an $x_0 \in X$, let $\mu = \delta_{x_0}$, a Dirac unit mass at x_0. Since $d_H(T^n\mu, \bar{\mu}) \to 0$, one obtains

$$\int_X f(x)d\bar{\mu}(x) = \lim_{n\to\infty} \sum_{i_1,\ldots,i_n}^{N} p_{i_1}\cdots p_{i_n}(f \circ w_{i_1} \circ \ldots \circ w_{i_n})(x_0). \tag{10.3.52}$$

This formula has been used to provide estimates for integrals involving $\bar{\mu}$ which cannot be solved recursively. The computation of the multiple sums involve the enumeration of an N-tree to n generations.

By setting $f(x) = I_S(x)$, where $S \subseteq X$, the above relation becomes

$$\bar{\mu}(S) = \lim_{n\to\infty} \sum_{i_1,\ldots,i_n}^{N} p_{i_1}\cdots p_{i_n} I_S(w_{i_1} \circ \ldots \circ w_{i_n}(x_0)). \tag{10.3.53}$$

The term involving I_S indicates whether or not the point $w_{i_1} \circ \ldots \circ w_{i_n}(x_0)$ lies in S. The quantity $p_{i_1}p_{i_2}\cdots p_{i_n}$ represents the probability of choosing the finite sequence $\{\sigma_{i_1}, \sigma_{i_2}, \ldots, \sigma_{i_n}\}$. Therefore for each $n > 0$, the sum is equal to the probability that the point x_n lies in S.

There is a connection between Eq. (10.3.53) and the Random Iteration Algorithm or "Chaos Game" [19], defined as follows: Pick an $x_0 \in X$ and define the iteration sequence

$$x_{n+1} = w_{\sigma_n}(x_n), \quad n = 0, 1, 2, \ldots, \tag{10.3.54}$$

where the σ_n are chosen randomly and independently from the set $\{1, 2, \ldots, N\}$ with probabilities $P(\sigma_n = i) = p_i$. A straightforward coding argument shows that for almost every code sequence $\sigma = \{\sigma_1, \sigma_2, \ldots)$ the orbit x_n is dense on the attractor A of the IFS **w**. As such, the Chaos Game can be used to generate computer approximations of A. However, it also provides approximations to the invariant measure $\bar{\mu}$ as a consequence of the following ergodic theorem for IFS [70]: For almost all code sequences $\sigma = (\sigma_1, \sigma_2, \ldots)$,

$$\lim_{n\to\infty} \frac{1}{n+1} \sum_{k=0}^{n} f(x_k) = \int_X f(x)\bar{\mu}(x) \tag{10.3.55}$$

for all continuous (and simple) functions $f : X \to \mathbf{R}$. By setting $f(x) = I_S(x)$ in Eq. (10.3.55) for an $S \subseteq X$, we obtain

$$\bar{\mu}(S) = \lim_{n\to\infty} \frac{1}{n+1} \sum_{k=0}^{n} I_S(x_k). \tag{10.3.56}$$

In other words, $\bar{\mu}(S)$ is the limit of the relative visitation frequency of S during the chaos game.

10.4 IFS on the Space of Distributions $\mathcal{D}'(X)$

In what follows $X = [0, 1]$ although the extension to $[0, 1]^n$ is straightforward. Distributions [220, 230, 236] are defined as linear functionals over a suitable space of "test functions", to be denoted (following the standard notation in the literature) as $\mathcal{D}(X)$. In this chapter, $\mathcal{D}(X) = C^\infty(X)$, the space of infinitely differentiable real-valued functions on X. (Note: In the literature, $\mathcal{D}(X)$ is normally taken to be $C_0^\infty(X)$, the set of $C^\infty(X)$ functions with compact support on X. With this choice, the expressions for distributional derivatives simplify due to the vanishing of boundary terms.) The space of distributions on X, to be denoted as $\mathcal{D}'(X)$, is the set of all bounded linear functionals on $\mathcal{D}(X)$, that is, $F : \mathcal{D}(X) \to \mathbf{R}$, such that

1. $|F(\psi)| < \infty$ for all $\psi \in \mathcal{D}(X)$,

2. $F(c_1\psi_1 + c_2\psi_2) = c_1F(\psi_1) + c_2F(\psi_2)$, $c_1, c_2 \in \mathbf{R}$, $\psi_1, \psi_2 \in \mathcal{D}(X)$.

The space $\mathcal{D}'(X)$ will include the following as special cases:

1. Functions $f \in \mathcal{L}^p(X, m)$, $1 \le p \le \infty$, for which the corresponding distributions are given by
$$F(\psi) = \int_X f(x)\psi(x)dx, \quad \forall\psi \in \mathcal{D}(X), \tag{10.4.1}$$

2. Probability measures $\mu \in \mathcal{M}(X)$, for which the corresponding distributions are given by
$$F(\psi) = \int_X \psi(x)d\mu(x), \quad \forall\psi \in \mathcal{D}(X), \tag{10.4.2}$$

3. The "Dirac delta function", $\delta(x-a)$, which may be defined in the distributional sense as follows: For a point $a \in X$, $F(\psi) = \psi(a)$ for all $\psi \in \mathcal{D}(X)$. This is often written symbolically as
$$F(\psi) = \int_X \psi(x)\delta(x-a)dx. \tag{10.4.3}$$

Our goal is to construct an IFS-type fractal transform operator $T : \mathcal{D}'(X) \to \mathcal{D}'(X)$ which, under suitable conditions, will be contractive with respect to a given metric on $\mathcal{D}'(X)$. In the spirit of Section 10.2, the fractal components of a distribution $u \in \mathcal{D}'(X)$ would be defined (symbolically) as $f_i(x) = (\phi_i \circ u \circ w_i^{-1})(x)$ and then combined to form T. Such a transform would serve to join the IFSM method over function spaces and the IFSP method over measure spaces under one common scheme.

The following property is very important in establishing a representation theory for distributions in $\mathcal{D}'(X)$.

Theorem 10.3 *[236] For any distribution/linear functional $F \in \mathcal{D}'(X)$, there exists a sequence of test functions $f_n \in \mathcal{D}(X)$, $n = 1, 2, \ldots$, such that for all $\psi \in \mathcal{D}(X)$,*
$$\begin{aligned}\lim_{n\to\infty} F_n(\psi) &= \lim_{n\to\infty} \int_X f_n(x)\psi(x)dx \\ &=: F(\psi).\end{aligned} \tag{10.4.4}$$

By recourse to this result, it will be convenient to express the linear functional $F \in \mathcal{D}'(X)$ symbolically as

$$\begin{aligned} F(\psi) &= \int_X f(x)\psi(x)dx \\ &= <\psi, f>, \end{aligned} \tag{10.4.5}$$

even though there may not exist a pointwise function $f(x)$ which defines the distribution F (e.g. Dirac distribution). The sequence of test functions f_n in the above theorem will then be said to converge to the distribution f "in the sense of distributions". For notational convenience we shall write that "$f \in \mathcal{D}'(X)$".

Lemma 10.1 *Let $w \in Aff_1(X)$ with Jacobian $|J| \neq 0$ and $u \in \mathcal{D}'(X)$ with associated linear functional*

$$F(\psi) = \int_X u(x)\psi(x)dx, \quad \psi \in \mathcal{D}(X). \tag{10.4.6}$$

Then the distribution $v = u \circ w^{-1} \in \mathcal{D}'(X)$ may be defined (symbolically) as

$$\begin{aligned} G(\psi) &= \int_X u(w^{-1}(x))\psi(x)dx \\ &= |J| \int_X u(x)(\psi \circ w)(x)dx, \quad \psi \in \mathcal{D}(X). \end{aligned} \tag{10.4.7}$$

Proof: Since $u \in \mathcal{D}'(X)$ there exists a sequence $u_n \in \mathcal{D}(X)$ which converges to u in the sense of distributions, i.e.

$$\lim_{n\to\infty} \int_X u_n(x)\psi(x)dx = \int_X u(x)\psi(x)dx, \quad \forall\, \psi \in \mathcal{D}(X). \tag{10.4.8}$$

For $n \geq 1$ define $v_n(x)$ as

$$v_n(x) = \begin{cases} u_n(w^{-1}(x)), & x \in \hat{w}(X), \\ 0, & \text{otherwise.} \end{cases} \tag{10.4.9}$$

By the change of variable $x = w(y)$ (with Jacobian $|J| \neq 0$),

$$\int_X v_n(y)\psi(x)dx = |J| \int_X u_n(y)(\psi \circ w)(y)dy. \tag{10.4.10}$$

Since w is affine, $\psi \circ w \in \mathcal{D}(X)$ for any $\psi \in \mathcal{D}(X)$. Therefore for each $n \geq 1$,

$$\begin{aligned} \lim_{n\to\infty} \int_X v_n(x)\psi(x)dx &= |J| \lim_{n\to\infty} \int_X u_n(y)(\psi \circ w)(y)dy \\ &= |J| \int_X u(y)(\psi \circ w)(y)dy, \end{aligned} \tag{10.4.11}$$

and the theorem is proved.□

Example: Let $w(x) = \frac{1}{2}x$ and $u(x) = \delta(x)$, the "Dirac delta function" at $a = 0$. Then $u(w^{-1}(x)) = \delta(2x) = \frac{1}{2}\delta(x)$.

Definition 10.1 *Let $f \in \mathcal{D}'(X)$ and $\{f_n\}$ any sequence in $\mathcal{D}(X)$ such that $f_n \to f$ in the sense of distributions. Now let $g : \mathbf{R} \to \mathbf{R}$ be such that*

$$L(\psi) := \lim_{n\to\infty} \int_X g(f_n(x))\psi(x)dx$$

exists for all $\psi \in \mathcal{D}(X)$ independently of the sequence $\{f_n\}$. Then we define the distribution $g \circ f \in \mathcal{D}'(X)$ in terms of the above limits, i.e.

$$\begin{aligned} < \psi, g \circ f > &= \lim_{n\to\infty} \int_X g(f_n(x))\psi(x)dx, \quad \forall\, \psi \in \mathcal{D}(X) \\ &=: (G \circ F)(\psi). \end{aligned} \tag{10.4.12}$$

If g is affine on $\mathbf{R}$, i.e. $g(x) = ax + b$, where $a, b \in \mathbf{R}$, then trivially the distribution $g \circ f$ exists for all $f \in \mathcal{D}(X)$. Note, however, that if g is Lipschitz on $\mathbf{R}$, then the distribution $g \circ f$ need not exist. For example, $\sin(\delta(x))$ is not defined.

Definition 10.2 *A function $g : \mathbf{R} \to \mathbf{R}$ will be said to satisfy a weak Lipschitz condition on $\mathcal{D}'(X)$ if there exists a $K \geq 0$ such that for all $\psi \in \mathcal{D}(X)$,*

$$\begin{aligned} \left| \int_X [(g \circ f_1)(x) - (g \circ f_2)(x)]\psi(x)dx \right| &\leq K \left| \int_X [f_1(x) - f_2(x)]\psi(x)dx \right| \\ &\quad \forall f_1, f_2 \in \mathcal{D}'(X). \end{aligned} \tag{10.4.13}$$

If g is affine on $\mathbf{R}$, then it satisfies a weak Lipschitz condition on $\mathcal{D}'(X)$.

Lemma 10.2 *Let $g : \mathbf{R} \to \mathbf{R}$ satisfy a weak Lipschitz condition on $\mathcal{D}'(X)$. Then for any $f \in \mathcal{D}'(X)$, $g \circ f \in \mathcal{D}'(X)$ exists.*

Proof: Let $f \in \mathcal{D}'(X)$ and, from Theorem 10.3, $f_n \in \mathcal{D}(X)$, $n = 1, 2, \ldots$ such that $f_n \to f$ as $n \to \infty$ in distribution. This implies that for any $\psi \in \mathcal{D}(X)$, given an $\epsilon > 0$, there exists an $M_\psi > 0$ such that

$$\left| \int_X [f_n(x) - f_m(x)]\psi(x)dx \right| < \epsilon, \quad \forall m, n \geq M_\psi. \tag{10.4.14}$$

Since

$$\begin{aligned} \left| \int_X [(g \circ f_n)(x) - (g \circ f_m)(x)]\psi(x)dx \right| &\leq K \left| \int_X [f_n(x) - f_m(x)]\psi(x)dx \right| \\ &\leq K\epsilon \quad \forall m, n \geq M_\psi, \end{aligned} \tag{10.4.15}$$

we may define, for each $\psi \in \mathcal{D}(X)$,

$$\int_X (g \circ f)(x)\psi(x)dx = \lim_{n\to\infty} \int_X (g \circ f_n)(x)\psi(x)dx. \tag{10.4.16}$$

□

We now define an N-map IFS on Distributions (IFSD) $(\mathbf{w}, \Phi)$ as follows:

1. **The IFS component:** $\mathbf{w} = \{w_1, w_2, \ldots, w_N\}$, $w_i \in Aff_1(X)$, with Jacobian $|J_i| \neq 0$,

2. **The grey level component:** $\Phi = \{\phi_1, \phi_2, \ldots, \phi_N\}$, $\phi_i : \mathbf{R} \to \mathbf{R}$ satisfies a weak Lipschitz condition on $\mathcal{D}'(X)$ (with Lipschitz constant K_i).

An operator $T : \mathcal{D}'(X) \to \mathcal{D}'(X)$ will now be associated with this N-map IFSD. For any $f \in \mathcal{D}'(X)$, the distribution $g = Tf$ will be defined by the linear functional

$$\begin{aligned} G(\psi) &= \int_X g(x)\psi(x)dx \\ &= \int_X (Tf)(x)\psi(x)dx \\ &= \sum_{i=1}^{N} \int_X (\phi_i \circ f \circ w_i^{-1})(x)\psi(x)dx. \end{aligned} \tag{10.4.17}$$

From Lemmas 1 and 2, it follows that T maps $\mathcal{D}'(X)$ into itself. We now define the following metric on $\mathcal{D}'(X)$:

$$d_{\mathcal{D}'}(f,g) = \sup_{\psi \in \mathcal{D}_1(X)} \left| \int_X f(x)\psi(x)dx - \int_X g(x)\psi(x)dx \right|, \quad \forall f,g \in \mathcal{D}'(X). \tag{10.4.18}$$

where $\mathcal{D}_1(X) = \{\psi \in C^\infty(X) \mid \|\psi\|_\infty \leq 1\}$.

Theorem 10.4 *The metric space $(\mathcal{D}'(X), d_{\mathcal{D}'})$ is complete.*

Proof: Let $\{f_n\}_{n=1}^\infty$ be a Cauchy sequence in $(\mathcal{D}'(X), d_{\mathcal{D}'})$, that is, for any $\epsilon > 0$, there exists an $\bar{N}(\epsilon)$ such that $d_{\mathcal{D}'}(f_m, f_n) < \epsilon$ for all $m, n \geq \bar{N}(\epsilon)$. From the definition of $d_{\mathcal{D}'}$ in Eq. (10.4.18), it follows that for any fixed $\psi \in \mathcal{D}_1(X)$, the sequence of real numbers $\{t_n(\psi)\}_{n=1}^\infty$, where

$$t_n(\psi) = \int_X f_n(x)\psi(x)dx, \tag{10.4.19}$$

is a Cauchy sequence on $\mathbf{R}$. Let $\bar{t}(\psi)$ denote the limit of this sequence. By setting $F(\psi) = \bar{t}(\psi)$ we define a bounded linear functional F on $\mathcal{D}_1(X)$. This procedure can easily be extended to all test functions $\psi \in C^\infty(X)$ by noting that $M^{-1}\psi \in \mathcal{D}_1(X)$ where $M = \|\psi\|_\infty$. Therefore $(\mathcal{D}'(X), d_{\mathcal{D}'})$ is complete.□

Remark: By restricting the test functions employed in the $d_{\mathcal{D}'}$ metric to $\mathcal{D}_1(X)$ (as opposed to the entire space $C^\infty(X)$), we ensure that the set of Cauchy sequences in $\mathcal{D}'(X)$ is nonempty.

Theorem 10.5 *Let $(\mathbf{w}, \Phi)$ be an N-map IFSD, $w_i \in Aff_1(X)$. Then for any $f, g \in \mathcal{D}'(X)$,*

$$d_{\mathcal{D}'}(Tf, Tg) \leq C_D d_{\mathcal{D}'}(f,g), \qquad C_D = \sum_{i=1}^{N} |J_i| K_i. \tag{10.4.20}$$

Proof: Let $\psi \in \mathcal{D}_1(X)$. Then

$$\begin{aligned}
|\int_X [(Tf)(x) & - (Tg)(x)]\psi(x)dx| \\
&= \left|\sum_{i=1}^{N} \int_{X_i} [\phi_i(f(w_i^{-1}(x))) - \phi_i(g(w_i^{-1}(x)))]\psi(x)dx\right| \\
&= \left|\sum_{i=1}^{N} |J_i| \int_X [\phi_i(f(y)) - \phi_i(g(y))](\psi \circ w_i)(y)dy\right| \\
&\leq \sum_{i=1}^{N} |J_i| \left|\int_X [\phi_i(f(y)) - \phi_i(g(y))](\psi \circ w_i)(y)dy\right| \\
&\leq \sum_{i=1}^{N} |J_i|K_i \left|\int_X [f(y) - g(y)](\psi \circ w_i)(y)dy\right|. \qquad (10.4.21)
\end{aligned}$$

The desired result follows from the fact that $\psi \circ w_i \in \mathcal{D}_1(X)$. □

Corollary 10.1 *Let* $(\mathbf{w}, \Phi)$ *be an* N*-map IFSD,* $w_i \in Aff_1(X)$*, such that*

$$C_D = \sum_{i=1}^{N} |J_i|K_i < 1. \qquad (10.4.22)$$

Then there exists a unique distribution $\bar{u} \in \mathcal{D}'(X)$ *such that* $T\bar{u} = \bar{u}$*, where* T *is the operator asssociated with the IFSD. Furthermore,* $d_{\mathcal{D}'}(T^n u, \bar{u}) \to 0$ *as* $n \to \infty$ *for all* $u \in \mathcal{D}'(X)$.

10.4.1 Affine IFSD and the Connection with IFSP and IFSM

For affine IFSD $(\mathbf{w}, \Phi)$ on $X = [0, 1]$, the IFS and grey level maps have the forms

$$w_i(x) = s_i x + a_i, \quad \phi_i(t) = \alpha_i t + \beta_i, \quad 1 \leq i \leq N, \qquad (10.4.23)$$

with $c_i = |J_i| = |s_i|$. Given an operator $T : \mathcal{D}'(X) \to \mathcal{D}'(X)$ associated with an affine IFSD, an adjoint operator $T^\dagger : \mathcal{D}(X) \to \mathcal{D}(X)$ may be defined as follows: For all $\psi \in \mathcal{D}(X)$,

$$< \psi, Tf > = < T^\dagger \psi, f >$$

$$= \sum_{i=1}^{N} \alpha_i c_i \int_X f(x)(\psi \circ w_i)(x)dx + \sum_{i=1}^{N} \beta_i c_i \int_X (\psi \circ w_i)(x)dx. \qquad (10.4.24)$$

In the special case that $\alpha_i > 0$ and $\beta_i = 0$, $1 \leq i \leq N$ and $\sum_{i=1}^{N} c_i \alpha_i = 1$, then $T^\dagger$ becomes the adjoint operator on $C(X)$ associated with the N-map IFSP $(\mathbf{w}, \mathbf{p})$ with probabilities $p_i = c_i \alpha_i$, cf. Eq. (10.3.50). The associated IFSD operator T coincides with the Markov operator M on $\mathcal{M}(X)$. However, T is not necessarily contractive in the complete metric space $(\mathcal{D}'(X), d_{\mathcal{D}'})$ since we may have $C_D = 1$. (See Example 3 below.)

However, in the subset $\mathcal{M}(X) \subset \mathcal{D}'(X)$, T is contractive with respect to the Hutchinson metric d_H. (Note that the test functions used for the d_H metric are Lip_1 functions.) By construction T maps each "shell" of measures $\mathcal{M}_K(X) = \{\mu | \mu(X) = K > 0\}$ to itself and is contractive on that shell with respect to the d_H metric. Therefore, there exists a ray of fixed point measures of T which belong to $\mathcal{D}'(X)$.

Now let the fixed point distribution $\bar{u}$ of an N-map affine IFSD be considered as a density function for a measure $\bar{\mu}$ on $\mathcal{B}(X)$, i.e.

$$\bar{\mu}(S) = \int_X I_S(x)\bar{u}(x)dx = \int_S \bar{u}(x)dx. \tag{10.4.25}$$

(Note that $\bar{\mu}$ is not necessarily a probability measure although we may suitably rescale $\bar{u}$ to make it so.) From Eq. (10.4.24),

$$\bar{\mu}(S) = \sum_{k=1}^{N} \alpha_k c_k \bar{\mu}(w_k^{-1}(S)) + \sum_{k=1}^{N} \beta_k c_k m(w_k^{-1}(S)), \tag{10.4.26}$$

where m denotes Lebesgue measure. Again in the special case that $\alpha_i > 0$ and $\beta_i = 0$, $1 \leq i \leq N$ and $\sum_{i=1}^{N} \alpha_i c_i = 1$, Eq. (10.4.26) is identical to Eq. (10.3.46) for invariant measures of IFSP.

Examples: In all cases $X = [0,1]$ and $N = 2$ with $w_1(x) = \frac{1}{2}x$ and $w_2(x) = \frac{1}{2}x + \frac{1}{2}$.

1. $\phi_1(t) = \phi_2(t) = \frac{1}{2}t + \frac{1}{2}$. The associated fractal transform operator T is contractive in $(\mathcal{D}'(X), d_{\mathcal{D}'})$ with contraction factor $C_D = \frac{1}{2}$. The fixed point $\bar{u}(x) = 1$ is an element of $\mathcal{D}'(X)$ as well as of $\mathcal{L}^p(X, m)$ for all $p \in [0, \infty]$, since T is also contractive in $(\mathcal{L}^p, d_p)$.

2. $\phi_1(t) = \phi_2(t) = \frac{1}{2}t$. T is again contractive in $(\mathcal{D}'(X), d_{\mathcal{D}'})$ with $C_D = \frac{1}{2}$. Here, $\bar{u}(x) = 0$.

3. $\phi_1(t) = \phi_2(t) = t$. T is not contractive in $(\mathcal{D}'(X), d_{\mathcal{D}'})$. Here, $C_D = 1$ and the equality in Eq. (10.4.20) holds. The functions $u(x) = c$ (a.e.), $c \in \mathbf{R}$, are fixed points of T. A fixed point attractor $\bar{u} \in \mathcal{D}'(X)$ is the Lebesgue measure on [0,1]. In order to see this, let us take $u_0(x) = \delta(x)$, the Dirac delta function at $x = 0$ and form the sequence $u_{n+1} = Tu_n$, for $n \geq 0$. From Eq. (10.4.17), $u_1(x) = \frac{1}{2}\delta(x) + \frac{1}{2}\delta(x - \frac{1}{2})$ and

$$u_n(x) = \frac{1}{2^n} \sum_{k=0}^{2^n - 1} \delta(x - \frac{k}{2^n}). \tag{10.4.27}$$

   Let F_n denote the linear functionals associated with the u_n, i.e. $F_n(\psi) = \int_X u_n(x)\psi(x)dx$. Then $\lim_{n\to\infty} F_n(\psi) = \int_X \psi(x)dx$.

   If $u_0(x) = K\delta(x)$, $K > 0$, then the sequence $u_{n+1} = Tu_n$ converges to the uniform Lebesgue measure m_K where $m_K([0,1]) = K$.

   One final remark concerning this example: It is also a 2-map IFZS on [0,1] (cf. Section 10.3.2). The associated IFZS operator T is contractive on $(\mathcal{F}^*(X), d_\infty)$ and the attractor is $\bar{u}(x) = 1$.

10.4.2 Integrals Involving Affine IFSD

Let $(\mathbf{w}, \Phi)$ be an N-map affine IFSD on $X = [0,1]$ with associated operator $T : \mathcal{D}'(X) \to \mathcal{D}'(X)$. If $f \in \mathcal{D}'(X)$ is defined by $F(\psi) = <\psi, f>$, the distribution $g = Tf$ will be given by $G(\psi)$, where

$$G(\psi) = \sum_{i=1}^{N} \alpha_i c_i \int_X f(y)(\psi \circ w_i)(y)dy + \sum_{i=1}^{N} \beta_i c_i \int_X (\psi \circ w_i)(y)dy \qquad (10.4.28)$$

By iterating this procedure, we obtain, for $n = 1, 2, \ldots$,

$$\begin{aligned} <\psi, T^n f> &= <(T^\dagger)^n \psi, f> \\ &= \sum_{i_1,\ldots,i_n}^{N} p_{i_1} \cdots p_{i_n} \int_X f(y)(\psi \circ w_{i_1} \circ \ldots \circ w_{i_n})(y)dy \\ &+ \sum_{k=1}^{n} \sum_{i_1,\ldots,i_k}^{N} p_{i_1} \cdots p_{i_{k-1}} q_{i_k} \int_X (\psi \circ w_{i_1} \circ \ldots \circ w_{i_k})(y)dy, \end{aligned} \qquad (10.4.29)$$

where $p_i = \alpha_i c_i$ and $q_i = \beta_i c_i$, $1 \le i \le N$. (This result may be compared with the IFSP case in Eq. (10.3.51).)

If T is contractive then it possesses a fixed point $\bar{u} \in \mathcal{D}'(X)$. Moreover, $T^n f \to \bar{u}$ as $n \to \infty$ in distribution for any $f \in \mathcal{D}'(X)$. Setting $f = \bar{u}$ and $n = 1$ in Eq. (10.4.29), we obtain

$$\begin{aligned} \int_X \bar{u}(x)\psi(x)dx &= \sum_{i=1}^{N} \alpha_i c_i \int_X \bar{u}(y)(\psi \circ w_i)(y)dy \\ &+ \sum_{i=1}^{N} \beta_i c_i \int_X (\psi \circ w_i)(y)dy. \end{aligned} \qquad (10.4.30)$$

For example, in the case $\psi(x) = x^n$, we obtain a set of equations which permit the recursive computation of the moments $g_n = \int_X x^n \bar{u}(x)dx$. These equations necessarily coincide with those obtained from the IFSM method [91]. In the special case $\beta_i = 0$, $1 \le i \le N$, and $\sum_{i=1}^{N} p_i = 1$, we obtain the recursion relations for moments g_n of the invariant measure of the affine IFSP $(\mathbf{w}, \mathbf{p})$.

In general, however, integrals involving the fixed point $\bar{u}$ can not be solved recursively or in closed form, e.g. $\int_X x^{1/2}\bar{u}(x)dx$. We may, however, use the fact that

$$\int_X \bar{u}(x)\psi(x)dx = \lim_{n\to\infty} \int_X (T^n f)(x)\psi(x)dx, \quad \forall\, \psi \in \mathcal{D}'(X). \qquad (10.4.31)$$

For an $x_0 \in X$, set $f(x) = \delta(x - x_0)$, the Dirac delta function at x_0, to obtain

$$\int_X (T^n f)(x)\psi(x)dx = \sum_{i_1,\ldots,i_n}^{N} p_{i_1} \cdots p_{i_n} (\psi \circ w_{i_1} \circ \ldots \circ w_{i_n})(x_0)$$

$$+\sum_{k=1}^{n}\sum_{i_1,\ldots,i_k}^{N}\alpha_{i_1}\ldots\alpha_{i_{k-1}}\beta_{i_k}\int_{X_{i_1,i_2,\ldots,i_k}}\psi(x)dx, \tag{10.4.32}$$

where $X_{i_1,i_2,\ldots,i_k} = \hat{w}_{i_1}\circ\hat{w}_{i_2}\circ\ldots\circ\hat{w}_{i_k}(X)$. This expression is somewhat more complicated than Eq. (10.3.51), its counterpart for measures. However, the integrals are generally easy to compute. As with Eq. (10.3.51), the evaluation of this expression involves the enumeration of N-trees to n generations.

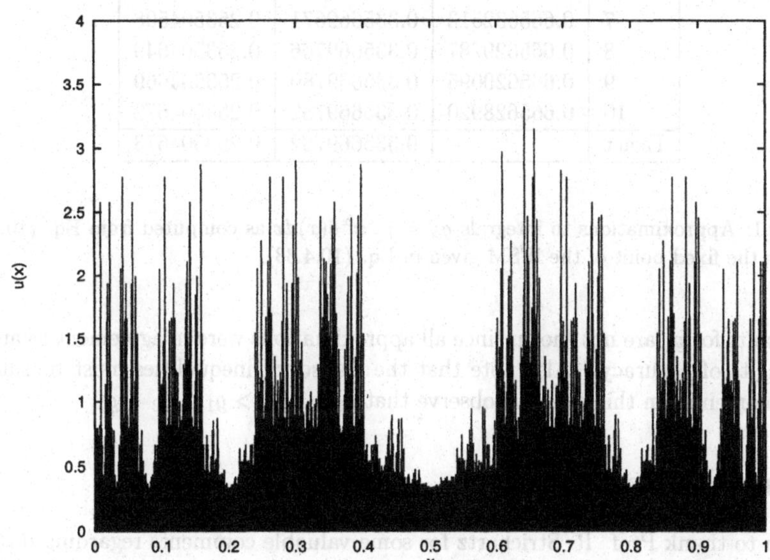

Figure 10.2: The fixed point attractor $\overline{u}(x)$ of the IFSM given in Eq. (10.4.33) of the text.

Example: We consider the following 3-map affine IFSM:

$$\begin{aligned} w_1(x) &= 0.4x, & \phi_1(t) &= t, \\ w_2(x) &= 0.4x+0.3, & \phi_2(t) &= 0.25t+0.25, \\ w_3(x) &= 0.4x+0.6, & \phi_3(t) &= t. \end{aligned} \tag{10.4.33}$$

The associated operator T is contractive in $(\mathcal{L}^p(X,m),d_p)$ for $p=1$. A histogram approximation of the attractor $\bar{u}$ of this IFSM is shown in Figure 10.2. The power moments g_n of $\bar{u}$,

$$g_n = \int_X x^n\bar{u}(x)dx, \quad n=0,1,2,\ldots, \tag{10.4.34}$$

may be computed in closed form via Eq. (10.4.30) [91]. The first four moments are:

$$g_0 = 1, \quad g_1 = \frac{1}{2}, \quad g_2 = \frac{431}{1284}, \quad g_3 = \frac{217}{856}. \tag{10.4.35}$$

In Table 10.1 are shown approximations to these moments as well as to the integral $g_{1/2} = \int_X x^{1/2}\bar{u}(x)dx$, as computed from Eq. (10.4.32). (There is no closed form expression for $g_{1/2}$.) The convergence of the approximations with increasing n is evident.

n	$g_{1/2}$	g_2	g_3
1	0.677591201	0.323333333	0.235
2	0.668876203	0.333893333	0.25084
3	0.666482494	0.335413973	0.25312096
4	0.665849446	0.335632945	0.253449418
5	0.665685332	0.335664477	0.253496716
6	0.665643243	0.335669018	0.253503527
7	0.665632513	0.335669671	0.253504508
8	0.665629787	0.335669766	0.253504649
9	0.665629095	0.335669780	0.253504669
10	0.665628920	0.335669782	0.253504672
Exact		0.335669782	0.253504673

Table 10.1: Approximations to integrals $g_\alpha = \int_X x^\alpha \bar{u}(x)dx$ as computed from Eq. (10.4.32). Here $\bar{u}$ is the fixed point of the IFSM given in Eq. (10.4.33).

(The results for g_1 are not shown since all approximations were in agreement to at least seven digits of accuracy.) Also note that the Hausdorff inequalities must be satisfied by the moments. In this case, we observe that $g_0 > g_{1/2} > g_1 > g_2 > g_3$.

Acknowledgements

We wish to thank Prof. R. Strichartz for some valuable comments regarding distributions. We also thank Dr. F. Mendivil for helpful discussions during the final preparation of this manuscript including its revision. This research was supported in part by the Natural Sciences and Engineering Research Council of Canada (NSERC) in the form of individual Operating Grants as well as a Collaborative Projects Grant (with C. Tricot and J. Lévy Véhel), all of which are gratefully acknowledged.

Chapter 11

Inverse Problem Methods for Generalized Fractal Transforms

Bruno Forte and Edward R. Vrscay

11.1 Introduction

In the previous chapter, we presented a unified treatment of IFS/Fractal Transform methods in terms of "generalized fractal transforms". A theory of Fractal Transforms on Distributions was developed so that the various IFS methods on function spaces, e.g. IFS, IFZS and IFSM (including "Fractal Transforms" and the "Bath Fractal Transform") could be united with IFSP on measure spaces.

In this chapter we examine various methods to attack the inverse problem of function/measure approximation using generalized fractal transforms, which involve the use of IFS maps whose "range blocks" may overlap. As such, these results may not appear to be greatly effective in the problem of fractal image compression since overlapping blocks are viewed as redundant in the coding of an image. However, our philosophy has been to develop a systematic theory of approximation of functions, measures and distributions using a complete "basis set" of IFS maps.

As before, we consider such target functions or images to be elements of an appropriate complete metric space (Y, d_Y). The underlying idea in fractal compression is the approximation, to some suitable accuracy, of a target $y \in Y$ by the fixed point $\bar{y}$ of a contraction mapping $f \in Con(Y)$, i.e. $f(\bar{y}) = \bar{y}$. It is then f which is stored in computer memory. By Banach's celebrated Fixed Point Theorem or Contraction Mapping Principle (CMP), the unique fixed point $\bar{y}$ may be readily generated by iteration of f, using an arbitrary "seed" image $y_0 \in Y$. We have been interested [91, 92, 89, 90] in a complete mathematical solution to the following formal inverse problem:

Given a "target" $y \in Y$ and an $\epsilon > 0$, find a map $f_\epsilon \in Con(Y)$ such that $d_Y(y, \bar{y}_\epsilon) < \epsilon$, where $f_\epsilon(\bar{y}_\epsilon) = \bar{y}_\epsilon$. As is well known, the inverse problem is somewhat simplified by the following corollary to the CMP, which is now referred to in the IFS literature as the "Collage Theorem":

Theorem 11.1 *Let (Y, d_Y) be a complete metric space. Given a $y \in Y$ suppose that there exists a map $f \in Con(Y)$ with contraction factor $c_f \in [0, 1)$ such that $d_Y(y, f(y)) < \delta$. Then*

$$d_Y(y, \bar{y}) < \frac{\delta}{1 - c_f}, \tag{11.1.1}$$

where $\bar{y} = f(\bar{y})$ is the fixed point of f.

In other words, the closer that f maps a target point y to itself, the closer that y is to the fixed point $\bar{y}$ of f. By making δ sufficiently small (if possible), $\bar{y}$ may become a suitable approximation to y. Rigorous solutions to the inverse problem of approximation using IFS-type methods (i.e. for arbitary $\delta > 0$) have been provided in [92] (for measures) and [89, 90, 91] (for functions) as well as algorithms for the construction of these approximations. Our basic strategy has been to work with an infinite set $\mathcal{W} = \{w_1, w_2, \ldots\}$ of *fixed* affine contraction maps which satisfy density conditions appropriate to the metric space Y being employed. From this set, sequences of N-map IFS $\mathbf{w}^N$ may then be chosen to produce approximations of arbitrary accuracy. As such, our formal solution establishes that the set of fixed points generated from this infinite set of IFS maps $\mathcal{W}$ is dense in (Y, d_Y).

The layout of this chapter is as follows. In Section 11.2 we briefly review our solution of the inverse problem for IFSM since it is a direct method. Section 11.3 is devoted to measure approximation using IFSP. This is an indirect method, since it is the moments of the target measure μ that we seek to approximate as closely as possible. In Section 11.4 we formulate a general fractal transform method on orthogonal function expansions. In the specific case of the local IFSM method applied to wavelet expansions, the method leads to "discrete wavelet fractal compression." In Section 11.5, a Collage Theorem for Fourier transforms of measures is given.

11.2 Inverse Problem for IFSM

The inverse problem for IFSM is a "direct method" since we may work on the target functions/images directly with IFS operators.

11.2.1 Collage Theorem for IFSM in $\mathcal{L}^2(X, \mu)$

From the Collage Theorem, the inverse problem for the approximation of functions in $\mathcal{L}^p(X, \mu)$ by IFSM may be posed as follows:

Given a target function $v \in \mathcal{L}^p(X, \mu)$ and a $\delta > 0$, find an IFSM $(\mathbf{w}, \Phi)$ with associated operator T such that $d_p(v, Tv) < \delta$.

For an N-map contractive IFSM $(\mathbf{w}, \Phi)$ on (X, d) with associated operator T, the squared $\mathcal{L}^2$ collage distance is given by

$$\begin{aligned}\Delta^2 &= \| v - Tv \|_2^2 \\ &= \int_X [\sum_{k=1}^{N}{}' \phi_k(v(w_k^{-1}(x))) - v(x)]^2 d\mu. \qquad (11.2.1)\end{aligned}$$

Following our discussion in the previous section (and our strategy in [92]), we consider the IFS maps w_i to be fixed. The problem reduces to the determination of grey level maps ϕ_i which minimize the collage distance Δ^2. In the special "μ-nonoverlapping case", i.e.,

1. $\cup_{k=1}^{N} X_k = \cup_{k=1}^{N} w_i(X) = X$, i.e. the sets $X_k = w_k(X)$ "tile" X, and
2. $\mu(w_i(X) \cap w_j(X)) = 0$ for $i \neq j$, then

the squared collage distance Δ^2 becomes

$$\begin{aligned}\Delta^2 &= \sum_{k=1}^{N} \int_{X_k} [\phi_k(v(w_k^{-1}(x)) - v(x)]^2 d\mu \\ &= \sum_{k=1}^{N} \Delta_k^2, \qquad (11.2.2)\end{aligned}$$

i.e. the sum of collage distances over the μ-nonoverlapping subsets X_k. The minimization of each integral is a continuous version of "least squares" with respect to the measure μ: For each subset X_k, find the $\phi_k : R_g \to R_g$ which provides the best $\mathcal{L}^2(X, \mu)$ approximation to the graph of $v(x)$ vs. $v(w_k^{-1}(x))$ for $x \in X_k$.

Most, if not all, applications in the literature assume the μ-nonoverlapping property, with $\mu = m$ (Lebesgue measure on X) and $w_k \in Aff_1(X)$. In the following discussion, however, we consider the more general case where the sets $w_i(X)$ can overlap on sets of nonzero μ-measure. We also assume the following:

3. $\cup_{k=1}^{N} w_k(X) = X$, i.e. the sets $w_i(X)$ "tile" X. Note that $w_i \in Aff_1(X)$ implies that $|J_i| > 0$ for $1 \leq i \leq N$, where $|J_i|$ denotes the Jacobian associated with the mapping $y = w_i(x)$.
4. affine grey level maps, i.e. $\phi_i : \mathbf{R}^+ \to \mathbf{R}^+$, where $\phi_i(t) = \alpha_i t + \beta_i, \quad t \in \mathbf{R}^+$. Thus, $\alpha_i, \beta_i \geq 0$ for $1 \leq i \leq N$.

The squared $\mathcal{L}^2$ collage distance then becomes

$$\begin{aligned}\Delta^2 &= < v - Tv, v - Tv > \\ &= \sum_{k=1}^{N}\sum_{l=1}^{N} [< \psi_k, \psi_l > \alpha_k \alpha_l + 2 < \psi_k, \chi_l > \alpha_k \beta_l + < \chi_k, \chi_l > \beta_k \beta_l] \\ &\quad - 2 \sum_{k=1}^{N} [< v, \psi_k > \alpha_k + < v, \chi_k > \beta_k] + < v, v >, \qquad (11.2.3)\end{aligned}$$

where

$$\psi_k(x) = u(w_k^{-1}(x)), \quad \chi_k(x) = I_{w_k(X)}(x), \quad x \in w_k(X). \tag{11.2.4}$$

Note that Δ^2 is a quadratic form in the ϕ-map parameters α_i and β_i, i.e.

$$\Delta^2 = \mathbf{x}^T\mathbf{A}\mathbf{x} + \mathbf{b}^T\mathbf{x} + d_0, \tag{11.2.5}$$

where $\mathbf{x}^T = (\alpha_1, ..., \alpha_N, \beta_1, ..., \beta_N) \in \mathbf{R}^{2N}$. The elements of the symmetric matrix $\mathbf{A}$ are given by

$$\begin{aligned} a_{i,j} &= <\psi_i, \psi_j> \\ a_{N+i,N+j} &= <\chi_i, \chi_j> \\ a_{i,N+j} &= <\psi_i, \chi_j>, \quad 1 \le i \le N,\ 1 \le j \le N. \end{aligned} \tag{11.2.6}$$

As well,

$$\begin{aligned} b_i &= -2 <v, \psi_i> \\ b_{N+i} &= -2 <v, \chi_i>, \quad 1 \le i \le N \end{aligned} \tag{11.2.7}$$

and $d_0 =<v, v>=\| v \|_2^2$.

For a given target $v \in \mathcal{L}^p(X, \mu)$, assuming $\| v \|_1 \neq 0$, we denote the feasible set of N-map IFSM grey-level parameters as

$$\Pi_v^{2N} = \{(\alpha_1, ..., \alpha_N, \beta_1, ..., \beta_N) \in \mathbf{R}^{2N} \mid \| Tv \|_1 \le \| v \|_1,\ \alpha_i, \beta_i \ge 0\}. \tag{11.2.8}$$

(Note that Π_v^{2N}, which is compact in the natural topology on $\mathbf{R}^{2N}$, depends on the target function v.) The condition $\| Tv \|_1 \le \| v \|_1$ is a linear inequality constraint in the grey level map parameters:

$$\sum_{k=1}^{N} (\alpha_k \| v \circ w_k^{-1} \|_1 + \beta_k \mu(X_k)) \ \le\ \| v \|_1 . \tag{11.2.9}$$

For the case $X \subset \mathbf{R}^D$, $\mu = m$ and $w_i \in Aff_1(X)$, $1 \le i \le N$, which will be used in all applications, the above linear inequality constraint becomes

$$\sum_{k=1}^{N} |J_k|(\alpha_k \| v \|_1 + m(X)\beta_k) \ \le\ \| v \|_1, \tag{11.2.10}$$

The minimization of Δ^2 may now be written as the following QP problem with linear constraints:

$$\text{minimize} \quad \mathbf{x}^T\mathbf{A}\mathbf{x} + \mathbf{b}^T\mathbf{x} + d_0, \quad \mathbf{x}^T \in \Pi_v^{2N}. \tag{11.2.11}$$

The advantages of QP problems have been discussed in [92]. Briefly,

1. QP algorithms locate an absolute minimum of the objective function Δ^2 in the feasible region Π_v^{2N} in a finite number of steps and

2. in many problems, the minimum value Δ^2_{min} is achieved on a boundary point of the feasible region. In such cases, if $(\alpha_k, \beta_k) = (0, 0)$ then $\phi_k(t) = 0$ which implies that the associated IFS map w_k is superfluous. QP (as opposed to gradient-type schemes) will locate such boundary points, essentially discarding such superfluous maps. The elimination of such maps represents an increase in the data compression factor. (This feature was observed with minimization of the collage distance involving IFS with probabilities [92].)

The following result guarantees that, with the exception of a degenerate case, the IFSM operator T corresponding to a feasible N-map IFSM grey-level parameter $\mathbf{x}^T \in \Pi_v^{2N}$ is contractive in $\mathcal{L}^1(X, \mu)$.

Proposition 11.1 *Let $X \subset \mathbf{R}^D$, $\mu = m^{(D)}$ and $v \in \mathcal{L}^1(X,\mu)$, $\| v \|_1 \neq 0$. Assume that $w_i \in Aff_1(X)$ for $1 \le i \le N$ and $\mathbf{x}^T = (\alpha_1, ..., \alpha_N, \beta_1, ..., \beta_N) \in \Pi_v^{2N}$. Then the operator T corresponding to the N-map IFSM $(\mathbf{w}, \Phi)$ is contractive in $(\mathcal{L}^1(X,\mu), d_1)$ except possibly when $\beta_1 = \beta_2 = ... = \beta_N = 0$. In this special case $\overline{u} \equiv 0$ is a fixed point of T.*

Note that $\mathbf{x}^T \in \Pi_v^{2N}$ does *not* guarantee that T is contractive in $(\mathcal{L}^2(X,\mu), d_2)$. Hence, the Collage Theorem does not apply in $\mathcal{L}^2(X, \mu)$. Nevertheless, our algorithm to approximate functions in $\mathcal{L}^2(X, \mu)$ exploits the contractivity of T in $(\mathcal{L}^1(X,\mu), d_1)$.

11.2.2 Formal Solution to the IFSM Inverse Problem

In this section, we outline the basic ideas behind a formal solution to the inverse problem posed above. Detailed proofs appear in [91]. These results would not be of as much interest to fractal compression as they would be to the approximation theory of functions using fractal transforms.

One intuitively expects that the $\mathcal{L}^2$ collage distance $\| v - Tv \|_2$ can be made arbitrarily small by adding IFS maps with increasing degrees of refinement, i.e. by increasing N. A trivial yet practical way of doing this is by simply dividing up the base space X into smaller regions with minimal overlap. This is essentially the approach adopted in image compression, e.g. quadtrees using nonoverlapping IFS maps. However, in the spirit of our earlier work on measures, we would like to consider a *complete* and infinite set of IFS maps which can provide various degrees of refinement. Therefore, in our formal solution to the inverse problem, we construct sequences of N-map IFSM, denoted as $(\mathbf{w}^N, \Phi^N)$, $N = 1, 2, \ldots$,

$$\mathbf{w}^N = (w_1, w_2, \ldots, w_N), \quad \Phi^N = (\phi_1, \phi_2, \ldots, \phi_N), \tag{11.2.12}$$

where the IFS maps w_i are chosen from a *fixed* and infinite set $\mathcal{W}$ of contraction maps. The (contractive) operators T^N associated with these N-map IFSM will play the role of f in the Collage Theorem. In order for the collage distance to become arbitrarily small with increasing N, a set of conditions will have to be imposed on the set of maps $\mathcal{W}$, according to the following definitions.

Definition 11.1 *Let (X, d) be a compact metric space and $\mu \in \mathcal{M}(X)$. A family $\mathcal{A}$ of subsets $A = \{A_i\}$ of X is "μ-dense" in a family $\mathcal{B}$ of subsets B of X if for every $\epsilon > 0$ and any $B \in \mathcal{B}$ there exists a collection $A \in \mathcal{A}$ such that $A \subseteq B$ and $\mu(B \setminus A) < \epsilon$.*

Definition 11.2 *Let $\mathcal{W} = \{w_1, w_2, ...\}$, $w_i \in Con(X)$ be an infinite set of contraction maps on X. We say that $\mathcal{W}$ generates a "μ-dense and nonoverlapping" - to be abbreviated as "μ-d-n" - family $\mathcal{A}$ of subsets of X if for every $\epsilon > 0$ and every $B \subseteq X$ there exists a finite set of integers $i_k \geq 1, 1 \leq k \leq N$, such that*

1. $A \equiv \cup_{k=1}^{N} w_{i_k}(X) \subseteq B$,

2. $\mu(B \setminus A) < \epsilon$ *and*

3. $\mu(w_{i_k}(X) \cap w_{i_l}(X)) = 0$ *if* $k \neq l$.

A useful set of affine maps satisfying such a condition on $X = [0,1]$ with respect to Lebesgue measure is given by the following "wavelet-type" functions:

$$w_{ij}(x) = 2^{-i}(x + j - 1),\ i = 1, 2, ...,\ j = 1, 2, ..., 2^i. \tag{11.2.13}$$

For each $i^* \geq 1$, the set of maps $\{w_{i^*j}, j = 1, 2, ..., 2^{i^*}\}$ provides a set of 2^{-i^*}-contractions of [0,1] which tile [0,1]. As such, the set $\mathcal{W}$ provides N-map IFS with arbitrarily small degrees of refinement on (X, d).

We now describe our algorithm. Let $\mathcal{W}$ be an infinite set of fixed one-to-one affine contraction maps on $X \subset \mathbf{R}^D$ which generates a μ-dense and nonoverlapping family of subsets of X and let $\mathbf{w}^N$ denote N-map truncations of $\mathcal{W}$. Given a target function $v \in \mathcal{L}^2(X, m)$, the region Π_v^{2N}, as defined in Eq. (11.2.8), contains all feasible points $\mathbf{x}^N = (\alpha_1, ..., \alpha_N, \beta_1, ..., \beta_N) \in \mathbf{R}^{2N}$, each of which defines a unique N-vector of affine grey level maps Φ^N,

$$\Phi^N = \{\alpha_1 t + \beta_1, ..., \alpha_N t + \beta_N\}. \tag{11.2.14}$$

For an $\mathbf{x}^N \in \Pi_v^{2N}$, let $T^N : \mathcal{L}^p(X, \mu) \to \mathcal{L}^p(X, \mu)$ be the operator associated with the N-map IFSM $(\mathbf{w}^N, \Phi^N)$ and

$$\Delta_N^2 = \| v - T^N v \|_2^2 \tag{11.2.15}$$

denote the corresponding squared $\mathcal{L}^2$ collage distance. Since $\Delta_N^2 : \Pi_v^{2N} \to \mathbf{R}^+$ is continuous in the natural topology on $\mathbf{R}^{2N}$, it attains an absolute minimum value, $\Delta_{N,\min}^2$ on Π_v^{2N}. For each N, we may determine this minimum value using QP. The following result confirms that our procedure provides a solution to the formal inverse problem posed earlier.

Theorem 11.2 $\Delta_{N,\min}^2 \to 0$ *as* $N \to \infty$.

This theorem implies that the set of all attractors $\bar{u}$ of N-map IFSM constructed from $\mathcal{W}$ is dense in $\mathcal{L}^p(X)$.

In Figure 11.1 are given some results of this method as applied to the approximation of $v(x) = \sin(\pi x)$ on $X = [0,1]$. The wavelet maps of Eq. (11.2.13) were used. The truncated IFS map vectors $\mathbf{w}^N$ in Eq. (11.2.12) were formed by arranging the wavelet w_{ij} maps in the following manner:

$$w_{1,1}, w_{1,2}, w_{2,1}, \ldots, w_{2,4}, w_{3,1}, \ldots, w_{3,8}, \ldots. \tag{11.2.16}$$

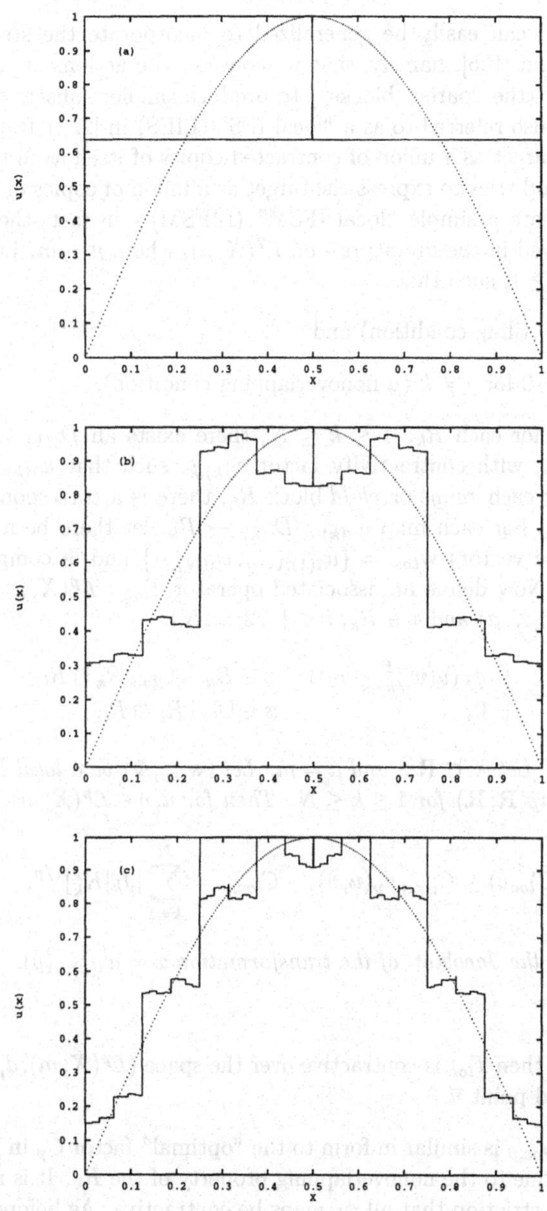

Figure 11.1: Approximations to the target set $v(x) = \sin(\pi x)$ on $X = [0,1]$ yielded by the "normal" IFSM method of Section 11.2.2, using the wavelet-type basis of Eq. (11.2.13), with $N = 2, 6$ and 14 maps, respectively.

11.2.3 "Local IFSM" on $\mathcal{L}^p(X,m)$)

Our IFSM method can easily be generalized to incorporate the strategy originally described by Jacquin [125], namely, that we consider the actions of contractive maps w_i on *subsets* of X (the "parent blocks") to produce smaller subsets of X (the "child blocks"). (This is also referred to as a "local IFS" (LIFS) in [22].) Rather than trying to approximate a target as a union of contracted copies of itself as in the IFS method, the local IFS method tries to express the target as a union of copies of *subsets* of itself.

Here we formulate a simple "local IFSM" (LIFSM) - in fact, the usual "Fractal Transform" employed in the literature - on $\mathcal{L}^2(X,\mu)$, where $\mu = m$. Let $R_k \subset X$, $k = 1, 2, ..., N$, with $N \geq 1$, such that

1. $\cup_{k=1}^N R_k = X$ (tiling condition) and
2. $\mu(R_j \cap R_k) = 0$ for $j \neq k$ (μ-nonoverlapping condition).

Also suppose that for each R_k, $1 \leq k \leq N$, there exists an $D_{j(k)} \subseteq X$ and a map $w_{i(k),k} \in Aff_1(X)$, with contractivity factor $c_{i(k),k}$, such that $w_{i(k),k}(D_{i(k)}) = R_k$. In other words, for each *range* or *child* block R_k, there is a corresponding *domain* or *parent* block $D_{i(k)}$. For each map $w_{i(k)} : D_{i(k)} \to R_k$, let there be a grey level map $\phi_k : \mathbf{R} \to \mathbf{R}$. The vectors $\mathbf{w}_{loc} = \{w_{i(1),1}, ..., w_{i(N),N}\}$ and Φ comprise an N-map LIFSM $(\mathbf{w}_{loc}, \Phi)$. Now define an associated operator $T_{loc} : \mathcal{L}^p(X,\mu) \to \mathcal{L}^p(X,\mu)$ as follows: For $u \in \mathcal{L}^p(X,\mu)$ and $x \in R_k$, $k \in \{1, 2, ..., N\}$,

$$(T_{loc}u)(x) \equiv \begin{cases} \phi_k(u(w_{i(k),k}^{-1}(x))), & x \in R_k - \cup_{l \neq k} R_k \cap R_l, \\ 0, & x \in \cup_{l \neq k} R_k \cap R_l. \end{cases} \tag{11.2.17}$$

Proposition 11.2 *Let $X \subset \mathbf{R}^D$ and $\mu = m$. Let $(\mathbf{w}_{loc}, \Phi)$ be a local IFSM defined as above, with $\phi_k \in Lip(\mathbf{R};\mathbf{R})$ for $1 \leq k \leq N$. Then for $u, v \in \mathcal{L}^p(X,m)$,*

$$d_p(T_{loc}u, T_{loc}v) \leq C_{loc,p} d_p(u,v), \quad C_{loc,p} = [\sum_{k=1}^N |J_k| K_k^p]^{1/p}, \tag{11.2.18}$$

where $|J_k|$ denotes the Jacobian of the transformation $x = w_{i(k),k}(y)$.

Remarks:

1. If $C_{loc,p} < 1$, then T_{loc} is contractive over the space $(\mathcal{L}^p(X,m), d_p)$ and possesses a unique fixed point $\bar{u}$.

2. The factor $C_{loc,p}$ is similar in form to the "optimal" factor $\bar{C}_p$ in Eq. (10.3.32) of Chapter 10, due to the nonoverlapping property of the R_k. It is not necessary to impose the restriction that all ϕ_k maps be contractive. As before, it follows that

$$C_{loc,p}(D,p) \leq K, \quad K = \max_{1 \leq k \leq N} K_k. \tag{11.2.19}$$

The weaker upper bound, K, which is independent of p, is identical to the result for the "Fractal Grey-Scale Transform" [22].

Given the above N-map LIFSM, we now compute the squared $\mathcal{L}^2$ collage distance, i.e.

$$\begin{aligned}\Delta^2 &= \| T_{loc}v - v \|_2^2 \\ &= \sum_{k=1}^{N} \int_{R_k} [\phi_i(v(w_{i(k),k}^{-1}(x)) - v(x)]^2 dx \\ &= \sum_{k=1}^{N} \Delta_k^2. \end{aligned} \tag{11.2.20}$$

Again, because the range blocks are conveniently nonoverlapping, the problem reduces to the minimization of each squared collage distance Δ_k^2 over the block R_k, a "least squares" determination of ϕ_k. In the special case that the ϕ_k maps are affine, the minimization of each Δ_k^2 is a quadratic programming problem in the two parameters α_k and β_k.

Given a target set v, a formal solution of the inverse problem for the nonoverlapping LIFSM case is straightforward, following the ideas of Section 11.2.2. The formal construction is outlined in [91].

In Figure 11.2 are some approximations to the target $v(x) = \sin(\pi x)$ using the nonoverlapping local IFSM method. As expected, the accuracy of approximation has improved. Some caution must be employed, however, as seen in Figure 11.2(c), where two range blocks and eight domain blocks have been used. The approximation is rather poor since the "halves" of the function $\sin(\pi x)$ provide poor collages of the rather straight portions R_k, $k = 1, 2, 3, 6, 7, 8$. As a result, it is necessary to employ more refined partitions for the parent cells.

11.2.4 Inverse Problem With Place-Dependent IFSM

The above methods for a formal solution to the inverse problem can be applied to place-dependent IFSM (introduced in Chapter 10). The expression for the squared $\mathcal{L}^2$ collage distance will depend upon the functional form assumed for the ϕ_k maps. We consider the following "nonoverlapping IFS" case:

1. $X \subset \mathbf{R}^D$ and $\mu = m$. We consider only the case $D = 1$ here, since the expressions involving the variable $s \in X$ become quite complicated.

2. $w_i \in Aff_1(X)$. As well, $X = \cup_{i=1}^N X_i$, where $X_i = w_i(X)$; in other words, the X_i "tile" the space X.

3. $\mu(X_i \cap X_j) = 0$ for $i \neq j$ (μ-nonoverlapping condition).

We assume that the grey-level maps ϕ_i assume the following functional form:

$$\phi_i(t, s) = \alpha_i(s)t + \beta_i(s), \quad t \in \mathbf{R}, \; s \in X, \tag{11.2.21}$$

where

$$\alpha_i(s) = \sum_{j=0}^{n_1} a_{ij}s^j, \quad \beta_i(s) = \sum_{j=0}^{n_2} b_{ij}s^j. \tag{11.2.22}$$

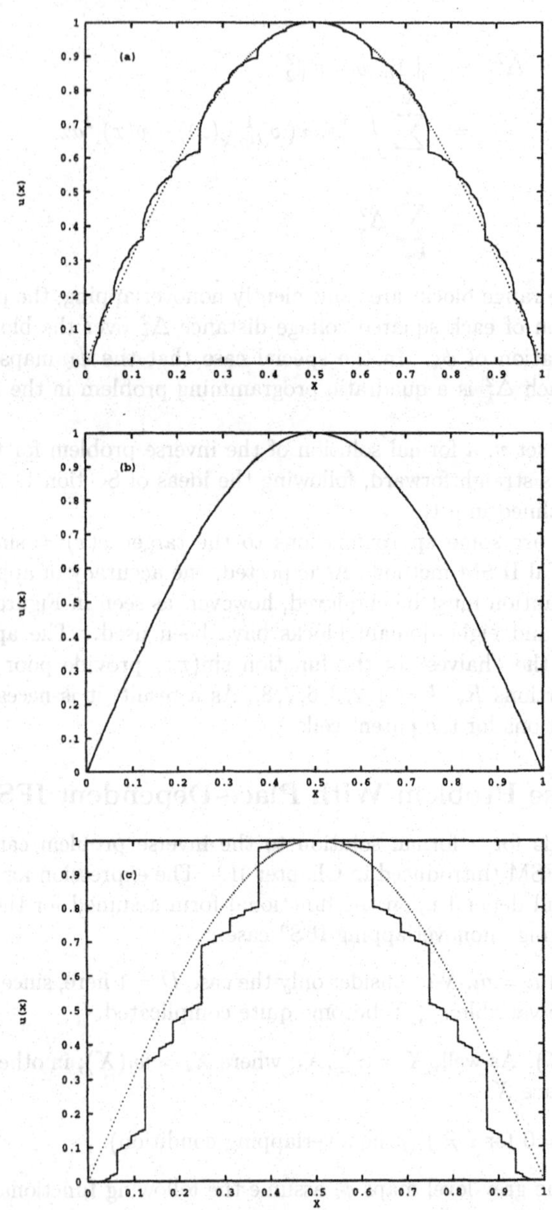

Figure 11.2: Approximations to the target set $v(x) = \sin(\pi x)$ on $X = [0,1]$ yielded by the nonoverlapping Local IFSM method of Section 11.2.3 using N_I parent intervals and N_J child intervals. (a) $(N_I, N_J) = (2,4)$. (b) $(N_I, N_J) = (4,8)$. (c) $(N_I, N_J) = (2,8)$.

(The special case $n_1 = n_2 = 0$ corresponds to the "normal" IFSM affine maps.) Each squared collage distance Δ_i^2 over X_i becomes

$$\begin{aligned}\Delta_i^2 &= \int_{X_i} [\alpha_i(w_i^{-1}(x))v((w_i^{-1}(x))) + \beta_i(w_i^{-1}(x)) - v(x)]^2 dx \\ &= c_i \int_X [\alpha_i(x)v(x) + \beta_i(x) - v(w_i(x))]^2 dx \\ &= c_i \int_X [v(x)\sum_{j=0}^{n_1} a_{ij}x^j + \sum_{k=0}^{n_2} b_{ik}x^k - v(w_i(x))]^2 dx. \qquad (11.2.23)\end{aligned}$$

Δ_i^2 is a quadratic form in the coefficients a_{ij}, b_{ij}. The coefficients of this quadratic form involve power moments of the functions v, v^2 and $v \circ w_i$ as well as moments over X. The minimization of Δ_1^2 is a quadratic programming (QP) problem subject to suitable constraints.

The method of "least squares" can also be applied to this problem. By imposing the stationarity conditions,

$$\frac{\partial \Delta_i^2}{\partial a_{ij}} = \frac{\partial \Delta_i^2}{\partial b_{ij}} = 0, \quad j = 1, 2, ..., n, \qquad (11.2.24)$$

one obtains a set of linear equations in the place-dependent polynomial coefficients α_{ij}, β_{ij}.

Place-dependent grey level maps could also be used in the more general overlapping IFS case. The coefficients of the quadratic form in the a_{ij}, b_{ij} involve power moments. It is also quite straightforward to use place-dependent grey level maps in the "Local IFSM" formalism given in the previous section. In Figure 11.3 are shown some results for a PDLIFSM approach applied to the target $v(x) = \sin(\pi x)$. A comparison with Figures 11.1 and 11.2 shows that the PD method yields a better approximation even in the case where the domain block is $X = [0, 1]$.

Our numerical calculations in [91] confirmed the statements of some workers [187, 259] that there is little need for searching for optimal domain blocks when place-dependent grey level maps are used. We have found experimentally that for most domain-range block pairs, the minimum collage distances yielded by each of the eight possible affine IFS maps were equal to two, if not three, figures of accuracy. As well, we found that the minimal collage distances yielded by various domain blocks do not differ by much. Of course, the reduction of searching represents an enormous saving in computer time.

11.2.5 Application of IFSM Methods to Images

We summarize this section with a brief comparison of the various IFSM results outlined above, using the target image "Lena" in Figure 11.4 (512 × 512 pixel array with each pixel assuming one of 256 grey level values). In Figure 11.5 is shown the result of the nonoverlapping Local IFSM method of Jacquin using 16 × 16 pixel domain blocks and 8 × 8 pixel range blocks. No searching for optimal domain blocks was done - for each range block, only the domain block containing was used. All 8 possible IFS maps, however, were tested. In Figure 11.6 is shown the approximation from a place-dependent Local IFSM method - the "Bath Fractal Transform" using the same domain and range blocks as in Figure 11.5. The grey level maps were affine in both grey level value as well as spatial variable, i.e. $n_1 = 0$ and $n_2 = 1$ in Eq. (11.2.22):

$$\phi_i(t,s) = \alpha_{i,0}t + \beta_{i,1}s + \beta_{i,0}, \tag{11.2.25}$$

It is found that some improvement is made if a quadratic function for the offset term, i.e. $n_2 = 2$ in Eq. (11.2.22), is assumed. However, there is little, if any, improvement if higher order polynomials in the $\alpha_i(s)$ term, i.e. $n_1 > 1$ are assumed.

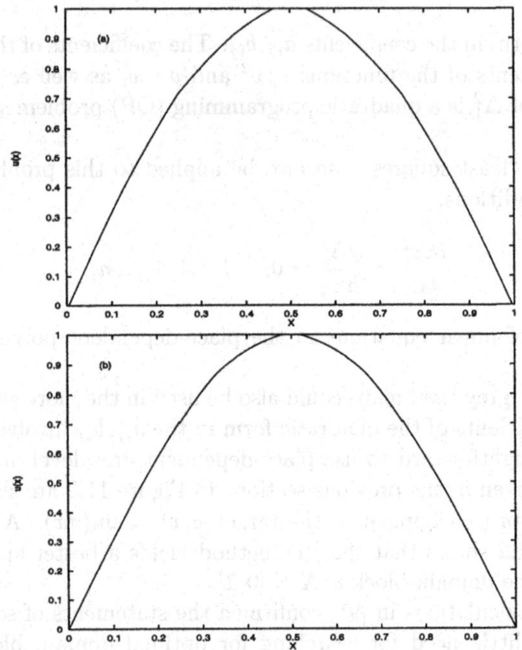

Figure 11.3: Approximations to the target set $v(x) = \sin(\pi x)$ on $X = [0,1]$ yielded by the place-dependent nonoverlapping Local IFSM method of Section 11.2.4 using N_I parent intervals and N_J child intervals. (a) $(N_I, N_J) = (1,2)$. Even with such low resolution, there is already a marked improvement in the approximation as compared to the and "local" IFSM methods of Figure 11.2. (b) $(N_I, N_J) = (2,4)$.

11.3 Approximation of Measures Using IFSP

The approximation of measures must employ indirect methods involving either moments or transforms. Here we outline the important steps behind a solution to the inverse problem for IFS approximation of measures using moments. An inverse problem may also be formulated in terms of Fourier transforms. A Collage Theorem for Fourier transforms of measures in $\mathcal{M}(X)$ is presented in the Appendix.

Figure 11.4: The target image "Lena", a 512 × 512 pixel array, 8 bits (256 grey-level values) per pixel.

Figure 11.5: Approximation to target image "Lena" using nonoverlapping Local IFSM method of Jacquin with $N_I = 32^2$ domain blocks (16 × 16 pixel blocks) and $N_J = 64^2$ range blocks (8 × 8 pixel blocks). For each child block, only the parent block containing it was used. All 8 possible maps were tested. $\mathcal{L}^1$ error $\| v - \overline{u} \|_1 = 0.029$, Relative $\mathcal{L}^1$ error = 0.068. (Unoptimized) coding time = 34 sec.

Figure 11.6: Approximation to target image "Lena" using place-dependent Local IFSM method with $N_I = 32^2$ domain blocks (16×16 pixel blocks) and $N_J = 64^2$ range blocks (8×8 pixel blocks). No search for optimal parent blocks. For each child, only the parent containing it was used. Only the map producing no rotation or inversion was used. $\| v - \overline{u} \|_1 = 0.022$, Relative $\mathcal{L}^1$ error $= 0.05$. (Unoptimized) coding time = 27 sec.

11.3.1 Approximation by "Moment Matching"

Much of the early work on the approximation of measures using IFSP was based on a knowledge of the moments of a target measure. Some form of "moment matching" was applied, in the following spirit: Given a target measure $\nu \in \mathcal{M}(X)$ ($X \subset \mathbf{R}$ for simplicity of notation), with moments $g_n = \int x^n d\nu$, $n = 0, 1, 2, \ldots$, find an IFS invariant measure $\bar{\mu}$ whose respective moments $\bar{g}_n \int x^n d\bar{\mu}$ are "close" to the g_n. In practical applications, moment matching is performed on a finite sequence of moments. In the case of an IFS with affine maps, the moments $\bar{g}_n$ of its invariant measure $\bar{\mu}$ may be computed recursively from the coefficients of the IFS maps as well as the associated probabilities. In [92], we provided a formal solution to the inverse problem for measure approximation by IFSP as well as an algorithm to approximate measures to arbitrary accuracy. Our method differs from previous efforts in two principal aspects:

1. We work with a *fixed*, infinite set of affine IFS maps which satisfy a density condition quite analogous to the μ-dense and nonoverlapping property of IFSM. Thus, only an optimization over the probabilities p_i is required.

2. The moment matching is accomplished by means of a *Collage Theorem for Moments.* The minimization of the squared collage distance in moment space is a quadratic programming (QP) problem in the p_i with a linear constraint. This problem can be solved numerically in a finite number of steps. The advantages of QP were outlined in Section 11.2.1 above.

Moment matching for the approximation of measures on $[0,1]^n$ can be justified by the fact that the convergence of moments is equivalent to the weak convergence of measures. Since we are working on compact spaces, the latter convergence is equivalent to convergence in Hutchinson metric d_H. This is summarized in the following theorem.

Theorem 11.3 [40] *For* $X = [0,1]$ *let* $\mu, \mu^{(n)} \in \mathcal{M}(X)$, $n = 1,2,3,\ldots$, *with power moments defined by*

$$g_k = \int_X x^k d\mu, \quad g_k^{(n)} = \int_X x^k d\mu^{(n)}, \quad k = 0,1,2,\ldots . \tag{11.3.1}$$

Then the following statements are equivalent:
(i) $g_k^{(n)} \to g_k$ *as* $n \to \infty$, $k = 0,1,2,\ldots$,
(ii) the sequence of measures $\mu^{(n)}$ *converges weak* to* μ, *i.e. for any* $f \in C(X)$, $\int f d\mu^{(n)} \to \int f d\mu$ *as* $n \to \infty$,
(iii) $d_H(\mu^{(n)}, \mu) \to 0$ *as* $n \to \infty$.

The results of this theorem apply to $[0,1]^n$.

11.3.2 Collage Theorem for Moments

Let $X = [0,1]$. (The extension to $[0,1]^n$ is straightforward.) As well, we consider only affine maps having the form

$$w_i(x) = s_i x + a_i, \qquad c_i = |s_i| < 1, \qquad 1 \le i \le N. \tag{11.3.2}$$

The use of affine maps leads to rather simple relations involving the moments of probability measures. That it will be sufficient to consider only affine maps is a result of the following theorem [40].

Theorem 11.4 *Let* (X,d) *denote a compact metric space and* $\mathcal{M}_{AIFS}(X) \subset \mathcal{M}(X)$ *the subset of invariant measures of affine IFS on X. Then* $\mathcal{M}_{AIFS}$ *is dense in* $(\mathcal{M}(X), d_H)$.

The above theorem is, in turn, a consequence of the following result.

Theorem 11.5 [208] *Let* (X,d) *denote a compact metric space and let* $\mathcal{M}_f(X) \subset \mathcal{M}(X)$ *denote the set of all measures with finite support. The* $\mathcal{M}_f(X)$ *is dense in* $\mathcal{M}(X)$.

We now provide the mathematical setting for the inverse problem applied to moments using IFSP. First we introduce the space $D(X)$ of all (infinite) moment vectors for probability measures in $\mathcal{M}(X)$:

$$D(X) = \{\mathbf{g} = (g_0, g_1, g_2, \ldots) \mid g_n = \int_X x^n d\mu, \; n = 0,1,2,\ldots, \text{ for some } \mu \in \mathcal{M}(X)\}. \tag{11.3.3}$$

(Note that $g_0 = 1$.) Then define the following metric on $D(X)$: For $\mathbf{u}, \mathbf{v} \in D(X)$,

$$\bar{d}_2(\mathbf{u}, \mathbf{v}) = \sum_{k=1}^{\infty} \frac{1}{k^2}(u_k - v_k)^2. \tag{11.3.4}$$

It was proved in [92] that $(D(X), \bar{d}_2)$ is a complete metric space.

Let $(\mathbf{w}, \mathbf{p})$ be an N-map affine IFS with Markov operator $M : \mathcal{M}(X) \to \mathcal{M}(X)$. Furthermore, let $\mu, \nu \in \mathcal{M}(X)$, with $\nu = M\mu$. Note that

$$\begin{aligned} \int_X f(x)d\nu(x) &= \int_X f(x)d(M\mu)(x) \\ &= \sum_{i=1}^{N} p_i \int_X (f \circ w_i)(x)d\mu(x). \end{aligned} \tag{11.3.5}$$

The moment vectors of μ and ν will be denoted by $\mathbf{g}, \mathbf{h} \in D(X)$, respectively. From the above equation, setting $f(x) = x^n$, $n = 1, 2, \ldots$, we have

$$h_n = \sum_{k=0}^{n} \binom{n}{k} \left[\sum_{i=1}^{N} p_i s_i^k a_i^{n-k} \right] g_k, \quad n = 1, 2, \ldots. \tag{11.3.6}$$

Thus to each Markov operator $M : \mathcal{M}(X) \to \mathcal{M}(X)$, there corresponds a linear operator $A : D(X) \to D(X)$ so that $\mathbf{h} = A\mathbf{g}$. In the basis $\{\mathbf{e}_i = (0, 0, \ldots, 0, 1, 0, \ldots)\}_{i=0}^{\infty}$, the (infinite) matrix representation of A is lower triangular. The diagonal elements of this matrix are $a_{00} = 1$ and

$$a_{nn} = \sum_{i=1}^{N} p_i s_i^n, \quad n \geq 1. \tag{11.3.7}$$

Since $|a_{nn}| \leq c^n < 1$ for $n \geq 1$, we have the following results.

Proposition 11.3 *The linear operator A is contractive in $(D(X), \bar{d}_2)$.*

Corollary 11.1 *The operator A has a unique and attractive fixed point $\bar{\mathbf{g}} \in D(X)$.*

The components $\bar{g}_n$ of $\bar{\mathbf{g}}$ are the moments of $\bar{\mu} = M\bar{\mu}$, the invariant measure of the IFSP $(\mathbf{w}, \mathbf{p})$. (These moments may be computed recursively by a slight rearrangement of Eq. (11.3.6) above.)

Corollary 11.2 *(Collage Theorem for Moments): Assume that (X, d) is a compact metric space and $\mu \in \mathcal{M}(X)$ with moment vector $\mathbf{g} \in D(X)$. Let $(\mathbf{w}, \mathbf{p})$ be an N-map IFSP with contractivity factor $c \in [0, 1)$ such that $\bar{d}_2(\mathbf{g}, \mathbf{h}) < \epsilon$, where $\mathbf{h} = A\mathbf{g}$ is the moment vector corresponding to $\nu = M\mu$. Then*

$$\bar{d}_2(\mathbf{g}, \bar{\mathbf{g}}) < \frac{\epsilon}{1 - c}, \tag{11.3.8}$$

where $\bar{\mathbf{g}} \in D(X)$ is the moment vector corresponding to $\bar{\mu}$, the invariant measure of the IFSP $(\mathbf{w}, \mathbf{p})$.

An inverse problem for the approximation of measures in $\mathcal{M}(X)$ may now be posed as follows: Let $\nu \in \mathcal{M}(X)$ be a target measure with moment vector $\mathbf{g} \in D(X)$. Given a $\delta > 0$, find an IFSP $(\mathbf{w}, \mathbf{p})$ with Markov operator $M : \mathcal{M}(X) \to \mathcal{M}(X)$ and associated linear operator $A : D(X) \to D(X)$ such that $d_2(\mathbf{g}, A\mathbf{g}) < \delta$. As for the IFSM case, we consider N-map affine IFSP $(\mathbf{w}, \mathbf{p})$ for which the IFS maps w_i are fixed. The problem then reduces to the determination of probabilities p_i which minimize the moment collage distance $\bar{d}_2(\mathbf{g}, \mathbf{h}) < \delta$, where $\mathbf{h} = A\mathbf{g}$.

We denote the squared collage distance between moment vectors in $(D(X), \bar{d}_2)$ as

$$S(\mathbf{p}) = \sum_{n=1}^{\infty} \frac{1}{n^2}(h_n(\mathbf{p}) - g_n)^2. \tag{11.3.9}$$

From Eq. (11.3.6) the moments h_n are given by

$$h_n = \sum_{i=1}^{N} A_{ni} p_i, \quad n = 1, 2, 3, \ldots, \tag{11.3.10}$$

where

$$A_{ni} = \sum_{k=0}^{n} \binom{n}{k} s_i^k a_i^{n-k} g_k. \tag{11.3.11}$$

Thus the function $S(\mathbf{p})$ may be written in the form

$$S(\mathbf{p}) = \mathbf{p}^T \mathbf{Q} \mathbf{p} + \mathbf{b}^T \mathbf{p} + s_0, \quad \mathbf{p} \in \Pi^N, \tag{11.3.12}$$

where

$$\Pi^N = \{\mathbf{p} = (p_1, p_2, \ldots, p_N) \mid \sum_{i=1} p_i = 1, \ p_i \geq 0\}. \tag{11.3.13}$$

The elements of the symmetric matrix $\mathbf{Q}$ are given by

$$q_{ij} = \sum_{n=1}^{\infty} \frac{1}{n^2} A_{ni} A_{nj}, \quad 1 \leq i, j \leq N. \tag{11.3.14}$$

The elements of $\mathbf{b}$ are

$$b_i = -2 \sum_{n=1}^{\infty} \frac{1}{n^2} A_{ni} g_n, \quad 1 \leq i \leq N \tag{11.3.15}$$

and

$$s_0 = \sum_{n=1}^{\infty} \frac{g_n^2}{n^2}. \tag{11.3.16}$$

The minimization of the squared moment collage distance $S(\mathbf{p})$ is a quadratic programming problem with linear constraints,

$$\text{minimize} \quad S(\mathbf{p}), \quad \sum_{i=1}^{N} p_i = 1, \quad p_i \geq 0. \tag{11.3.17}$$

11.3.3 Formal Solution to Inverse Problem

It will now be necessary to guarantee that the collage distance $S(\mathbf{p})$ can be made arbitrarily small. As in the IFSM case, we construct sequences of N-map IFSP, denoted as $(\mathbf{w}^N, \mathbf{p}^N)$, where the IFS maps in $\mathbf{w}^N$ are chosen from a fixed, infinite set $\mathcal{W}$ of contraction maps on X. A condition must be placed on this set, according to the following definition.

Definition: An infinite set of contraction maps $\mathcal{W} = \{w_1, w_2, \ldots\}$, $w_i \in Con(X)$ is said to satisfy an ϵ-contractivity condition on X if for each $x \in X$ and any $\epsilon > 0$, there exists an $i^* \in \{1, 2, \ldots\}$ such that $w_{i^*}(X) \subset N_\epsilon(x)$, where $N_\epsilon(x) = \{y \in X \mid d(x,y) < \epsilon\}$ denotes the ϵ-neighbourhood of x.

If $\mathcal{W}$ satisfies the ϵ-contractivity condition on X, then $\inf_{1\le i\le\infty} c_i = 0$. The set $\mathcal{W}$ provides N-map IFS with arbitrarily small degrees of refinement on (X,d). The "wavelet-type" basis functions of Eq. (11.2.13) conveniently satisfy such an ϵ-contractivity condition.

Now let $\mathcal{W} = \{w_1, w_2, \cdots\}$ be an infinite set of affine contraction maps on $X = [0,1]$ which satisfies the ϵ-contractivity condition. Let

$$\mathbf{w}^N = \{w_1, w_2, \ldots w_N\}, \quad N = 1, 2, \ldots, \tag{11.3.18}$$

denote N-map truncations of $\mathcal{W}$. As well, let

$$\Pi^N = \{\mathbf{p}^N = (p_1, p_2, \cdots, p_n) \mid p_i \ge 0, \ \sum_{i=1}^{N} p_i = 1\} \tag{11.3.19}$$

denote the set of all probability N-vectors for $\mathbf{w}^N$. Note that $\Pi^N \subset \mathbf{R}^N$ is compact in the natural topology on $\mathbf{R}^N$. Now let $\mu \in \mathcal{M}(X)$ be a target measure with moment vector $\mathbf{g} \in D(X)$. For a $\mathbf{p}^N \in \Pi^N$, let M^N be the Markov operator corresponding to the N-map IFS $(\mathbf{w}^N, \mathbf{p}^N)$. Also let $\nu_N = M^N \mu$, with associated moment vector $\mathbf{h}_N \in D(X)$. The collage distance between the moment vectors of μ and ν_N will be denoted as

$$\Delta^N(\mathbf{p}^N) \equiv \|\mathbf{g} - \mathbf{h}_N\|_{l^2} \, . \tag{11.3.20}$$

Since $\Delta^N : \Pi^N \to \mathbf{R}^+$ is continuous, it attains an absolute minimum value, to be denoted as $\Delta^N_{\min}$, on Π^N. The following result establishes that the above procedure provides a solution to the inverse problem for measure approximation.

Theorem 11.6 $\Delta^N_{\min} \to 0$ *as* $N \to \infty$.

Remarks:

1. Theorem 11.6 is a density result establishing that the set of invariant measures for all N-map IFS $(\mathbf{w}^N, \mathbf{p}^N)$ where $\mathbf{p}^N \in \Pi^N$, $N = 1, 2, \ldots$, is dense in $(\mathcal{M}(X), d_H)$. This result can be extended to $[0,1]^q, q \ge 2$.

2. Although not explicitly stated in the proof, the collage distances Δ^N and, in particular, the sequence $\Delta^N_{\min}$, are also dependent on the ordering of the w_i maps in the infinite set $\mathcal{W}$. However, at this point, we are not interested in any questions about the "optimal" ordering of the maps in $\mathcal{W}$ nor how N-map subsets $\mathbf{w}^N$ should be chosen.

11.3.4 Some Numerical Results

We present some results of the above approximation method to show its important features. The target measure μ considered here is the Lebesgue-Stieltjes measure generated by the following distribution $F(x)$ on $[0,1]$:

$$F(x) = \begin{cases} x, & x \in [0, \frac{1}{3}), \\ \frac{1}{3}, & x \in [\frac{1}{3}, \frac{2}{3}), \\ x, & x \in [\frac{2}{3}, 1], \end{cases} \tag{11.3.21}$$

i.e. $\mu(a,b] = F(b) - F(a)$, $a < b$. The moments of this measure are given by

$$\begin{aligned} g_n &= \int_0^{\frac{1}{3}} x^n dx + \frac{1}{3}\left(\frac{2}{3}\right)^n + \int_{\frac{2}{3}}^1 x^n dx \\ &= \frac{1}{n+1}\left[1 + \left(\frac{1}{3}\right)^{n+1} - \left(\frac{2}{3}\right)^{n+1}\right] + \frac{1}{3}\left(\frac{2}{3}\right)^n, \quad n \geq 0. \end{aligned} \tag{11.3.22}$$

The wavelet-type maps of Eq. (11.2.13) were used here. The truncated IFS map vectors $\mathbf{w}^N$ were formed by arranging the w_{ij} maps in the same manner as in Eq. (11.2.16). In practical calculations, only a finite number of moments can be matched. As such, the following function,

$$S_M^N(\mathbf{p}^N) = \sum_{n=1}^{M} \frac{1}{n^2}\left[\sum_{i=1}^{N} A_{ni} p_i - g_n\right]^2, \tag{11.3.23}$$

was minimized, subject to the constraints on the probabilities. In calculations reported here, $M = 30$ moments were matched. The minimization of the function S_M^N was performed with a quadratic programming algorithm developed by Best and Ritter [30].

Figure 11.7 shows approximations $\bar{F}_N(x)$ to $F(x)$ yielded by the optimal IFSP for values of $N = 2, 6, 14$, respectively. The $\bar{F}_N(x)$ functions were approximated by generating discrete approximations of the invariant measures of the $(\mathbf{w}^N, \mathbf{p}^N)$ on a lattice of 2000 points on [0,1]. There are two important features of these calculations:

1. As N increases, $\bar{F}_N(x)$ is seen to converge to $F(x)$, with better approximations to the jump at $x = \frac{2}{3}$.

2. For each $N \geq 3$, the minimum of $S_M^N(\mathbf{p}^N)$ located by the QP algorithm occurred on a boundary of the feasible region Π^N, implying that there were some zero probabilities. The actual number of nonzero probabilities which existed for the cases $N = 2, 6, 14$ in Figure 11.7 were $\bar{N} = 2, 4, 8$, respectively. As such, the QP algorithm has performed a data compression, eliminating the unnecessary IFS maps. (This would not necessarily have been done if gradient methods were used to locate the minima of the objective function.)

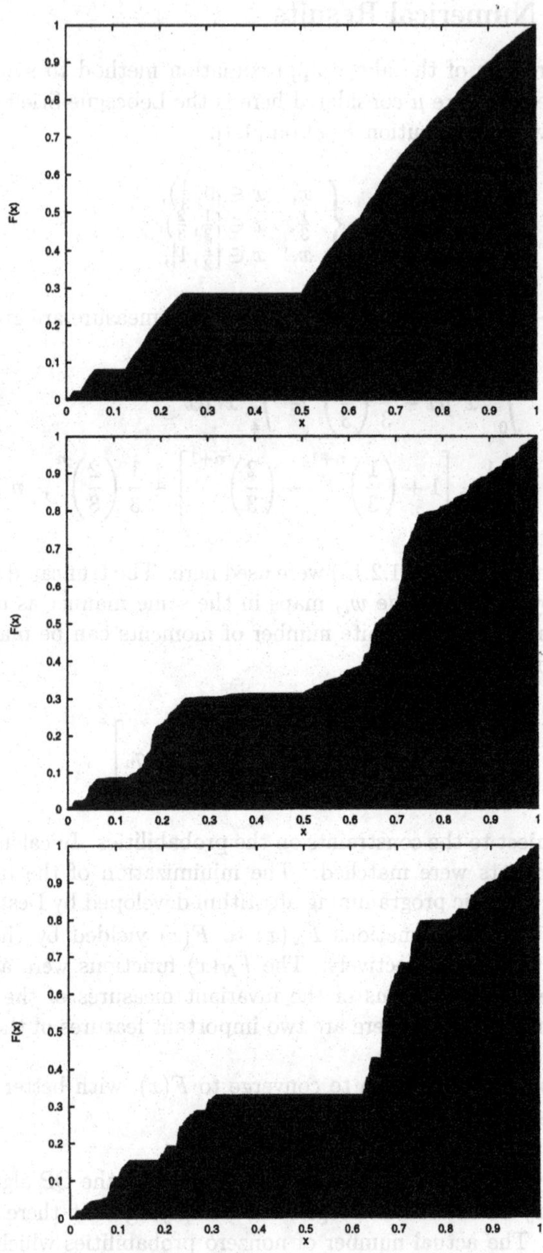

Figure 11.7: Approximations to the distribution $F(x)$ of Eq. (11.3.21) obtained by moment matching, using the wavelet-type basis of Eq. (11.3.3.13), with $N = 2, 6$ and 14 maps, respectively.

11.4 IFSM and Fractal Wavelet Compression

Let $(\mathbf{w}, \Phi)$ be an N-map affine IFSM on $X = [0,1]$ and suppose that its associated operator T is contractive on $(\mathcal{L}^p(X,m)$ for some (or all) $p \geq 1$. Then the fixed point $\bar{u}$ satisfies the equation (cf. Eq. (10.3.36) in Chapter 10)

$$\bar{u}(x) = \sum_{k=1}^{N} [\alpha_k \bar{u}(w_k^{-1}(x)) + \beta_k I_{X_k}(x)] \tag{11.4.1}$$

As noted in Chapter 10, $\bar{u}$ is expressed as a linear combination of dilated and translated copies of itself along with piecewise constant functions. This is somewhat reminiscent of using wavelet functions which are obtained by dilatations and translations of a "mother wavelet" function. The IFSM method may be viewed as an *adaptive* encoding since the "mother" function is the target itself. However, the copies $\psi_k(x) = \bar{u}(w_k^{-1}(x))$ are generally not orthogonal to each other. In fact, the above expansion may be considered as highly redundant - the piecewise constant functions are sufficient to create a linearly independent basis. Nevertheless, some workers have been investigating the idea of constructing orthonormal basis functions from the scaled copies $\psi_k(x)$ of the target function $v(x)$ [249, 196].

If the affine grey level maps are now assumed to be place-dependent, having the form

$$\phi_k(t,s) = \alpha_k t + \gamma_k(s), \quad 1 \leq k \leq N, \tag{11.4.2}$$

(with the offset term β_k being absorbed by the function $\gamma_k(s)$), the fixed point equation for $\bar{u}$ then becomes

$$\bar{u}(x) = \sum_{k=1}^{N} [\alpha_k \bar{u}(w_k^{-1}(x)) + \gamma_k(w_k^{-1}(x))] \tag{11.4.3}$$

If $\alpha_k = 0$ for $1 \leq k \leq N$ (implying that the contraction factor of T is $C_p = 0$), then the above equation reduces to an expansion of $\bar{u}$ in terms of functions $\gamma_k : X \to \mathbf{R}$. In most practical applications to date (e.g. the Bath Fractal Transform [187, 188, 259]) these functions have been assumed to be polynomial. However, one could equally assume that the $\gamma_k(s)$ were linear combinations of orthonormal functions, e.g. $\{\cos(n\pi x)\}$, etc..

In the place-dependent local IFSM formalism (PDIFSM). the functions $\gamma_k(s)$ now map *domain blocks* D_k to the grey level range $\mathbf{R}$. Moreover, the compositions $(\gamma_k \circ w_{j(k),k}^{-1})(x)$ are now translations and dilatations of the $\gamma_k(s)$ functions. If the $\gamma_k(s)$ are chosen to be *wavelet* functions, then the orthogonality is guaranteed. In the special case that all grey-level scaling parameters α_k are zero, the place-dependent local IFSM method can be made to coincide with wavelet expansions. What remains is to examine wavelet expansions of functions/images from an *indirect* approach as was done for measures, i.e. through their moments.

For simplicity we restrict our attention to the one-dimensional case, i.e. $X = [0,1]$. Let $\{q_n\}_{n=0}^{\infty}$, with $q_0(x) = 1$, denote a complete set of orthonormal basis functions on

X. Then for a given $u \in \mathcal{L}^2(X,m)$,

$$u(x) = \sum_{k=0}^{\infty} c_k q_k(x), \tag{11.4.4}$$

where

$$c_k = < u, q_k > = \int_X u(x) q_k(x) dx. \tag{11.4.5}$$

Now let $(\mathbf{w}, \Phi)$ denote an N-map affine IFSM on X with associated operator T and let $v = Tu$. Then

$$v(x) = \sum_{k=0}^{\infty} d_k q_k(x), \tag{11.4.6}$$

where

$$\begin{aligned} d_k &= < q_k, v > \\ &= < q_k, Tu > \\ &= \sum_{i=1}^{N} \alpha_i < q_k, u \circ w_i^{-1} > + \sum_{i=1}^{N} \beta_i < q_k, I_{w_i(X)} > . \end{aligned} \tag{11.4.7}$$

Using Eq. (11.4.4) to expand $u(w_i^{-1}(x))$ yields

$$d_k = \sum_{l=1}^{\infty} a_{kl} c_l + e_k, \tag{11.4.8}$$

where

$$a_{kl} = \sum_{i=1}^{N} \alpha_i < q_k, q_l \circ w_i^{-1} >, \qquad e_k = \sum_{i=1}^{N} \beta_i < q_k, I_{w_i(X)} > . \tag{11.4.9}$$

This is the affine mapping $A : \mathbf{c} \to \mathbf{d}$ associated with the IFSM operator T. In this general case, there is a simple, yet important result:

Proposition 11.4 *Given an orthonormal basis $\{q_n\}$ on X as above. Suppose that the N-map IFSM operator T is contractive in $\mathcal{L}^2(X)$ with fixed point $\bar{u}$. Then the mapping $A : l^2(\mathbf{N}) \to l^2(\mathbf{N})$ in Eq. (11.4.9) is contractive in the $l^2(\mathbf{N})$ metric. The fixed point $\bar{\mathbf{c}} = A\bar{\mathbf{c}}$ is the vector of Fourier coefficients of $\bar{u}$ in the q_i basis.*

Proof: Let $u, v \in \mathcal{L}^2(X,m)$ with Fourier coefficients $\mathbf{c}, \mathbf{d} \in l^2(\mathbf{N})$, respectively, in the q_i basis. Since T is contractive, there exists a $C \in [0,1)$ such that

$$\| Tu - Tv \|_{\mathcal{L}^2} \leq C \| u - v \|_{\mathcal{L}^2} . \tag{11.4.10}$$

From Parseval's relation, i.e. $\| u \|_{\mathcal{L}^2} = \| \mathbf{c} \|_{l^2}$, etc.,

$$\| Tu - Tv \|_{\mathcal{L}^2} = \| A\mathbf{c} - A\mathbf{d} \|_{l^2} \tag{11.4.11}$$

so that

$$\| A\mathbf{c} - A\mathbf{d} \|_{l^2} \leq C \| \mathbf{c} - \mathbf{d} \|_{l^2} . \tag{11.4.12}$$

Therefore A is contractive in $l^2(\mathbf{N})$. Since $l^2(\mathbf{N})$ is complete, there exists an element $\bar{\mathbf{c}} \in l^2(\mathbf{N})$ such that $A\bar{\mathbf{c}} = \bar{\mathbf{c}}$. The vector $\bar{\mathbf{c}}$ defines a unique function $\bar{u} \in \mathcal{L}^2(X)$ through the expansion in Eq. (11.4.4). From Eqs. (11.4.7)-(11.4.9), it follows that $\bar{\mathbf{c}} = A\bar{\mathbf{c}}$ implies $T\bar{u} = \bar{u}$. □

In general, the matrix elements $< q_k, q_l \circ w_i^{-1} >$ in Eq. (11.4.9) do not vanish, leading to a rather full (i.e. not sparse) matrix representation of A. This would occur, for example, in the case of the Discrete Cosine Transform. However, in the special case that the q_k are localized in space, e.g. *wavelets*, many of these matrix elements vanish. The relations between the d_k and the c_l simplify even further when the IFS maps w_i are *local* and map domain blocks which support wavelets of lower resolution/frequency to range blocks which support wavelets of higher resolution/frequency. This is the basis of what has been referred to in one form or another as "wavelet-based fractal compression" [56, 146, 233, 247, 245]. In what follows, we consider only the one-dimensional case for simplicity. The extension to functions/images in $\mathbf{R}^2$ is straightforward but more tedious.

The following standard dyadic multiresolution approximation of $\mathcal{L}^2(\mathbf{R})$ is assumed [165]:

1. A sequence of nested subspaces $V_j \in \mathcal{L}^2(\mathbf{R})$, $j \in \mathbf{Z}$, where $V_j \subset V_{j+1}$. V_j contains the set of all approximations of functions $f \in \mathcal{L}^2(\mathbf{R})$ at resolution 2^j. Moreover $\lim_{n\to\infty} V_n = \mathcal{L}^2(\mathbf{R})$.

2. The sequence of orthogonal complements $W_j \perp V_j$ such that $W_j \oplus V_j = V_{j+1}$, $j \in \mathbf{Z}$. This implies that for any $k \in \mathbf{Z}$ and $n > 0$,

$$V_k \oplus W_k \oplus W_{k+1} \oplus \ldots \oplus W_{k+n} = V_{k+n+1}. \tag{11.4.13}$$

3. A "scaling function" $\phi \in \mathcal{L}^2(\mathbf{R})$ such that the functions

$$\phi_{ij}(x) = 2^{i/2}\phi(2^i x - j), \qquad j \in \mathbf{Z}, \tag{11.4.14}$$

form an orthonormal basis for V_i.

4. The "orthogonal wavelet" $\psi \in \mathcal{L}^2(\mathbf{R})$ such that the functions

$$\psi_{ij}(x) = 2^{i/2}\psi(2^i x - j), \qquad j \in \mathbf{Z}, \tag{11.4.15}$$

form an orthonormal basis for W_i. It follows that the set $\{\psi_{ij}\}$, $i, j \in \mathbf{Z}$, forms a complete orthonormal basis for $\mathcal{L}^2(\mathbf{R})$.

In the case of the Haar wavelet system,

$$\phi_{00}(x) = 1, \;\; x \in [0,1), \qquad \psi_{00}(x) = \begin{cases} 1, & x \in [0, \frac{1}{2}), \\ -1, & x \in [\frac{1}{2}, 1). \end{cases} \tag{11.4.16}$$

Then $u \in \mathcal{L}^2(\mathbf{R})$ may be expanded as

$$u(x) = \sum_{j=-\infty}^{\infty} < u, \phi_{kj} > \phi_{kj}(x) + \sum_{l=0}^{\infty} \sum_{j=-\infty}^{\infty} < u, \psi_{k+l,j} > \psi_{k+l,j}(x). \qquad (11.4.17)$$

For functions on $[0,1]$, all contributions from V_j, $j < 0$ vanish. (We assume some kind of periodic extension of the functions on [0,1] to **R**.) Then $k = 0$ in Eq. (11.4.17) so that

$$u(x) = b_{00}\phi_{00}(x) + c_{00}\psi_{00}(x) + \sum_{i=1}^{\infty} \sum_{j=0}^{2^i-1} c_{ij}\psi_{ij}(x), \qquad (11.4.18)$$

where, of course, $b_{00} =< u, \phi_{00} >$ and $c_{ij} =< u, \psi_{ij} >$. In the case of the Haar wavelets, the supports of the functions ψ_{ij} are the dyadic intervals

$$I_{ij} = \left[\frac{j}{2^i}, \frac{j+1}{2^i}\right], \quad 0 \leq j < 2^i. \qquad (11.4.19)$$

For more generalized wavelets, the supports of the ψ_{ij} will be larger than these intervals. Nevertheless the orthogonality of the ψ_{ij} is preserved. In this case, we shall consider the wavelets ψ_{ij} to be centered on the intervals I_{ij}. The coefficients b_{00} and c_{ij} may be conveniently arranged in a table such as the one shown in Figure 11.8 which reflects the degree of resolution as well as localization in space.

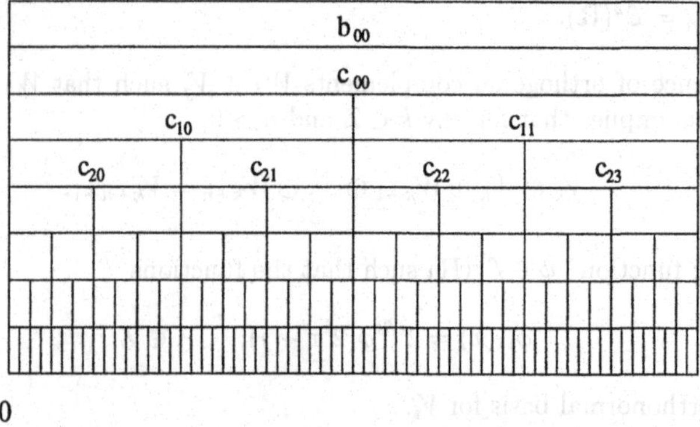

Figure 11.8: The table of wavelet coefficients b_{00} and $c_{i,j}$ associated with the expansion in Eq. (11.4.18). The location of each coefficient reveals the resolution as well as the spatial localization of the wavelet ψ_{ij}.

For simplicity, we consider the following "nonoverlapping" local IFSM. (Our analysis may be extended to cover more general cases where "quadtree partitioning" has been used.)

1. Domain blocks given by the intervals $I_{i^*,j}$, $0 \leq j \leq 2^{i^*} - 1$,

2. Range blocks given by the intervals $I_{k^*,l}$, $0 \leq l \leq 2^{k^*} - 1$, where $k^* > i^*$.

Suppose that for each range block $I_{k^*,l}$, $l \in \{0,1,\ldots,2^{k^*}-1\}$, we choose a domain block $I_{i^*,j(l)}$. The local IFSM map for this pair, denoted as $w_l : I_{i^*,j(l)} \to I_{k^*,l}$, will have the contraction factor $2^{i^*-k^*}$. There will be an associated affine grey level map $\phi_l(t) = \alpha_l t + \beta_l$ (not to be confused with the scaling function $\phi_{ij}(x)$). From Eq. (11.4.8) and the orthogonality property of the ψ_{ij}, we have

$$d_{k^*,l} = \alpha_l c_{i^*,j(l)} < \psi_{kl}, \psi_{ij} \circ w_l^{-1} > . \tag{11.4.20}$$

Since $(\psi_{ij} \circ w_l^{-1})(x) = 2^{(i-k)/2}\psi_{kl}(x)$, it follows that

$$d_{k^*,l} = \alpha_l 2^{(i^*-k^*)/2} c_{i^*,j(l)}. \tag{11.4.21}$$

As well, all coefficients lying below $d_{k^*,l}$ in the table will be scaled copies of the corresponding entries below $c_{i^*,j}$, that is,

$$d_{k^*+k',2^{k'}l+l'} = \alpha_l 2^{(i^*-k^*)/2} c_{i^*+k',2^{k'}j(l)+l'}, \quad k' \geq 0, \quad 0 \leq l' \leq 2^{k'} - 1. \tag{11.4.22}$$

This equation represents the transformation on the wavelet coefficients c_{ij} for $i \geq k^*$ induced by the local IFSM. The domain and range regions in the wavelet coefficient table are schematically illustrated in Figure 11.9. Such a transformation has been derived independently in the literature in the context of wavelet representations. Writing this transformation as an IFS-type method on wavelet coefficients has been referred to as "fractal wavelet compression". The purpose of this Section has been to show how IFSM/local IFSM induces an IFS-type transform on the wavelet coefficients.

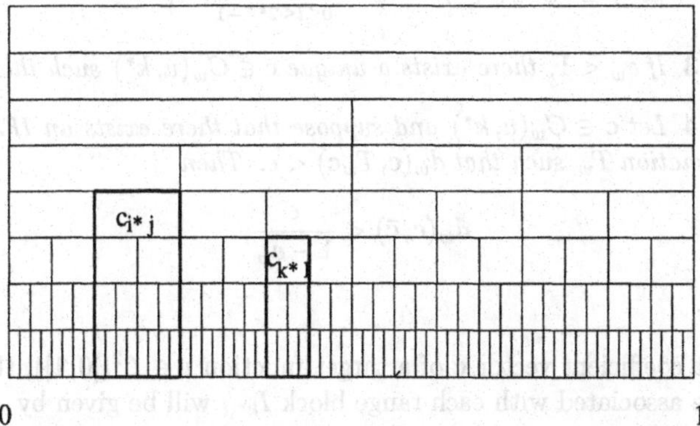

Figure 11.9: The domain and range blocks associated with the nonoverlapping IFSM in the text, cf. Eqs. (11.4.21) and (11.4.22).

Note that no "offset" terms involving the β_i appear in Eqs. (11.4.21) and (11.4.22). Such terms would appear only if the resolution level of the domain blocks would be $i^* = 0$, i.e. "traditional IFS", where the domain block is X itself. Again by orthogonality, the offset terms involving the β_i would contribute only to the coefficient b_{00}. (See note later in this section.)

The coefficients b_{00} and c_{ij}, $0 \leq i \leq k^* - 1$, $0 \leq j \leq 2^i - 1$ remain fixed. Therefore, once computed using Eq. (11.4.5), these coefficients are stored and then used to generate the higher resolution coefficients. In what follows, we provide the framework for an IFS-type inverse problem on the wavelet coefficients.

For a given function $u \in \mathcal{L}^2([0,1])$ and the above local IFSM, define the following wavelet coefficient space:

$$\begin{aligned} C_w(u,k^*) &= \{b_{00}, c_{kl},\ k \geq 0,\ 0 \leq l \leq 2^k - 1 \mid \sum_{k,l}^{\infty} c_{kl} < \infty, \text{ where} \\ & \quad c_{ij} = < u, \psi_{ij} >,\ 0 \leq i \leq k^* - 1,\ 0 \leq j \leq 2^i - 1\}. \end{aligned} \tag{11.4.23}$$

Let $T_w : C_w(u,k^*) \to C_w(u,k^*)$ denote the transformation induced by the LIFSM on the wavelet coefficients. We consider the following metric on $C_w(u,k^*)$: For $\mathbf{c}, \mathbf{d} \in C_w(u,k^*)$, define

$$d_w(\mathbf{c},\mathbf{d}) = \max_{0 \leq l \leq 2^{k^*}-1} \Delta_l^2, \tag{11.4.24}$$

where

$$\Delta_l^2 = \sum_{k'=0}^{\infty} \sum_{l'=0}^{2^{k'}-1} (c_{k^*+k', 2^{k'} l + l'} - d_{k^*+k', 2^{k'} l + l'})^2. \tag{11.4.25}$$

Proposition 11.5 *The metric space $(C_w(u,k^*), d_w)$ is complete.*

Proposition 11.6 *For $\mathbf{c}, \mathbf{d} \in C_w(u,k^*)$,*

$$d_w(T_w\mathbf{c}, T_w\mathbf{d}) \leq c_w d_w(\mathbf{c},\mathbf{d}), \quad c_w = \max_{0 \leq l \leq 2^{k^*}-1} |\alpha_l| 2^{(i^*-k^*)/2}. \tag{11.4.26}$$

Corollary 11.3 *If $c_w < 1$, there exists a unique $\bar{\mathbf{c}} \in C_w(u,k^*)$ such that $T_w\bar{\mathbf{c}} = \bar{\mathbf{c}}$.*

Corollary 11.4 *Let $\mathbf{c} \in C_w(u,k^*)$ and suppose that there exists an IFSM with associated transformation T_w such that $d_w(\mathbf{c}, T_w\mathbf{c}) < \epsilon$. Then*

$$d_w(\mathbf{c},\bar{\mathbf{c}}) < \frac{\epsilon}{1 - c_w}, \tag{11.4.27}$$

where $\bar{\mathbf{c}} = T_w\bar{\mathbf{c}}$.

For the wavelet coefficient vector $\mathbf{c}$ of a target function $v \in \mathcal{L}^2([0,1])$, the squared $\mathcal{L}^2$ collage distance associated with each range block $I_{k^*,l}$ will be given by

$$\Delta_l^2 = \sum_{k'=0}^{\infty} \sum_{l'=0}^{2^{k'}-1} (c_{k^*+k', 2^{k'} l + l'} - \alpha_l 2^{(i^*-k^*)/2} c_{i^*+k', 2^{k'} j(l) + l'})^2. \tag{11.4.28}$$

The absence of terms involving β_l greatly simplifies the minimization of this distance. The optimal scaling factor is given by

$$\bar{\alpha}_l = 2^{(k^*-i^*)/2} \frac{S_{k^*,l,i^*,j}}{S_{i^*,j,i^*j}}, \tag{11.4.29}$$

where

$$S_{\alpha,\beta,\gamma,\delta} = \sum_{k'=0}^{\infty} \sum_{l'=0}^{2^{k'}-1} c_{\alpha+k',2^{k'}\beta+l'} c_{\gamma+k',2^{k'}\delta+l'}. \tag{11.4.30}$$

The minimum collage distance is

$$\Delta_{l,\min} = [S_{k^*,l,k^*,l} - \bar{\alpha}_l 2^{(i^*-k^*)/2} S_{k^*,l,i^*,j}]^{1/2}. \tag{11.4.31}$$

One may now proceed in a fashion similar to that of the usual LIFSM method: For a given range block $I_{k^*,l}$, find the optimal domain block $I_{k^*,j(l)}$, i.e. the block yielding the lowest minimum collage distance $\Delta_{l,\min}$. When this has been done for all range blocks, an operator T_w has then been defined. One may then generate the fixed point $\bar{\mathbf{c}}$ of T_w by the iteration process $T_w^n \mathbf{c}_0$, where $\mathbf{c}_0 \in C_w(u, k^*)$. The corresponding approximation $\bar{v}(x)$ to $v(x)$ may then be constructed by summing the resulting wavelet series in the coefficients $\bar{c}_{ij}$.

One further note regarding offset terms β_k and their role in the wavelet coefficient transformation: As stated earlier, offset terms would appear in this transformation only if $i^* = 0$, i.e. the domain blocks $D_k = X$. However, note that our local IFSM method in Chapter 10 involves offset terms in the grey level maps. The explanation is that the wavelet expansion implicitly assumed by the LIFSM method is not Eq. (11.4.17) but rather the following:

$$\begin{aligned} u(x) = {} & \sum_{j=0}^{2^{i^*}-1} b_{i^*,j}\phi_{i^*,j}(x) + \sum_{j=0}^{2^{i^*}-1} c_{i^*,j}\psi_{i^*,j}(x) \\ & + \sum_{i=1}^{\infty} \sum_{j=0}^{2^{i^*+i}-1} c_{i^*+i,j}\psi_{i^*+i,j}(x). \end{aligned} \tag{11.4.32}$$

In other words, the minimum resolution k in Eq. (11.4.17) is i^*. Each domain block D_k has been expanded separately in a wavelet series expansion. The structure of the corresponding wavelet coefficient table is sketched in Figure 11.10.

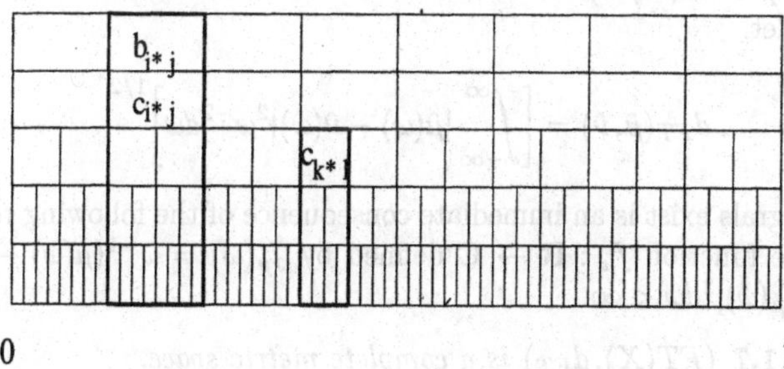

Figure 11.10: The domain and range blocks associated with the nonoverlapping IFSM in the text.

When the wavelet system used above is the Haar system, then the above fractal transform method is identical to the usual nonoverlapping LIFSM method. However, the above method also applies to more generalized wavelets ψ_{ij} which are supported on range blocks R_k which overlap with each other. Even though these blocks overlap, the wavelets remain orthogonal to each other. (Of course, the form of the local IFSM maps will have to be altered accordingly.) The overlapping of these wavelets has been shown to be beneficial as it can reduce the usual "block" effects exhibited by normal "nonoverlapping" fractal transforms [247, 245]. An example is shown in Figure 11.11. Both images use 16×16 pixel domain blocks and 8×8 pixel range blocks. However, Figure 11.11(a) was produced using the Haar wavelet basis and Figure 11.11(b) was produced using the "Coifman 12" wavelet basis [256]. Wavelet functions of the latter type which are centered on neighbouring blocks overlap with each other while remaining orthogonal to each other. The "blockiness" of the Haar expansion has been reduced somewhat by the Coifman 12 expansion. We thank Mr. A. Van de Walle for providing these images as computed from his fractal wavelet compression routine [247, 245].

11.5 Collage Theorem for Fourier Transforms

In Section 11.3, it was shown that an N-map IFSP $(\mathbf{w}, \mathbf{p})$ with associated Markov operator $M : \mathcal{M}(X) \to \mathcal{M}(X)$ induces a linear operator A which maps $D(X)$, the space of moment vectors for measures in $\mathcal{M}(X)$, into itself. In this section, we derive the mapping which is induced on the space of Fourier transforms of measures in $\mathcal{M}(X)$. The treatment may easily be modified to treat the discrete cosine transforms (DCT) of measures. The structure of our discussion will closely follow that of Section 11.3.2 on moments. It is again assumed that $X = [0, 1]$.

Given a measure $\mu \in \mathcal{M}(X)$, we define its Fourier transform (FT) $\hat{\mu} : \mathbf{R} \to \mathbf{C}$ as

$$\hat{\mu}(\omega) = \int_X e^{-i\omega x} d\mu(x), \qquad \omega \in \mathbf{R}. \tag{11.5.1}$$

Note that $\hat{\mu}(0) = 1$ and that $|\hat{\mu}(\omega)| \leq 1$, $\forall\, \omega \in \mathbf{R}$. Now let $FT(X)$ denote the set of FT's for all measures in $\mathcal{M}(X)$. As is well known, $\hat{\mu}(\omega) \in FT(X)$ does not necessarily imply that $\hat{\mu}(\omega) \in \mathcal{L}^2(\mathbf{R}, m)$. We thus define the following metric on $FT(X)$: For $\hat{\mu}, \hat{\nu} \in FT(X)$, let

$$d_{FT}(\hat{\mu}, \hat{\nu}) = \left[\int_{-\infty}^{\infty} |\hat{\mu}(\omega) - \hat{\nu}(\omega)|^2 \omega^{-2} d\omega \right]^{1/2}. \tag{11.5.2}$$

That these integrals exist is an immediate consequence of the following result: For any $\mu \in \mathcal{M}(X)$, the function $f_\mu : \mathbf{R} \to \mathbf{C}$ defined by $f_\mu(\omega) = \omega^{-1}(\hat{\mu}(\omega) - 1)$ for $\omega \neq 0$ satisfies $\int_{-\infty}^{\infty} |f_\mu(\omega)|^2 d\omega < \infty$.

Proposition 11.7 $(FT(X), d_{FT})$ *is a complete metric space.*

Proposition 11.8 *Let* $\mu, \nu^{(n)} \in \mathcal{M}(X)$, $n = 1, 2, 3, \ldots$, *with FT's* $\hat{\mu}, \hat{\nu}^{(n)} \in FT(X)$. *Then* $d_{FT}(F, F^{(n)}) \to 0$ *as* $n \to \infty$ *iff* $d_H(\mu, \nu^{(n)}) \to 0$ *as* $n \to \infty$.

Now, as in Section 11.3.1, let $(\mathbf{w}, \mathbf{p})$ be an N-map affine IFS with associated Markov operator $M : \mathcal{M}(X) \to \mathcal{M}(X)$. Let $\mu, \nu \in \mathcal{M}(X)$, with $\nu = M\mu$ and FT's $\hat{\mu}, \hat{\nu} \in$

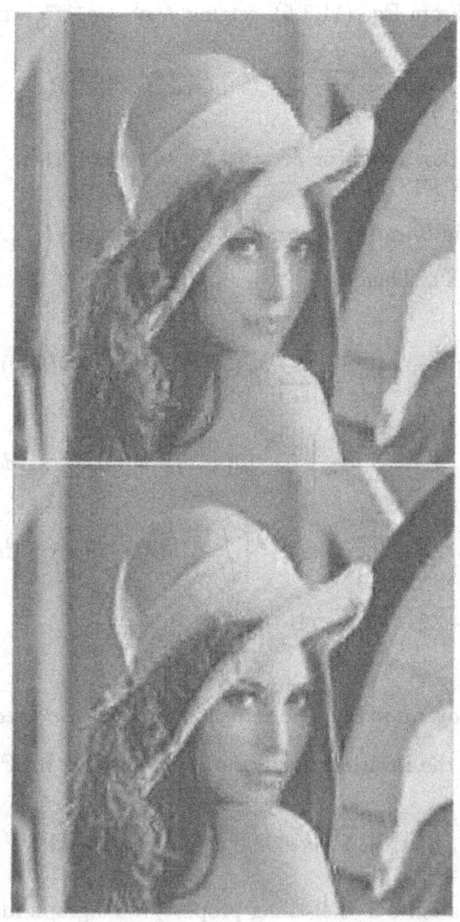

Figure 11.11: Approximations to target image "Lena" using discrete wavelet fractal transform method of Section 11.5. Domain blocks are 16 × 16 pixel arrays, range blocks are 8 × 8 pixel arrays. (a) Haar wavelet basis. (b) "Coifman 12" wavelet basis. (Courtesy of A. Van de Walle.)

$FT(X)$, respectively. From Eq. (11.3.5), setting $f(x) = e^{-i\omega x}$, we have

$$\hat{\nu}(\omega) = \sum_{k=1}^{N} p_k e^{-i\omega a_k} \hat{\mu}(s_k \omega), \quad \omega \in \mathbf{R}. \tag{11.5.3}$$

Therefore, to each Markov operator $M : \mathcal{M}(X) \to \mathcal{M}(X)$, there corresponds a linear operator $B : FT(X) \to FT(X)$.

Proposition 11.9 *The linear operator B is contractive in $(FT(X), d_{FT})$.*

Proof: Let $\hat{\mu}, \hat{\nu} \in FT(X)$. Then

$$\begin{aligned} d_{FT}(B(\hat{\mu}), B(\hat{\nu})) &= \left[\int_{-\infty}^{\infty} \left| \sum_{k=1}^{N} p_k e^{-i\omega a_k} [\hat{\mu}(s_k\omega) - \hat{\nu}(s_k\omega)] \right|^2 \omega^{-2} d\omega \right]^{1/2} \\ &\le \sum_{k=1}^{N} p_k \left[\int_{-\infty}^{\infty} |\hat{\mu}(s_k\omega) - \hat{\nu}(s_k\omega)|^2 \omega^{-2} d\omega \right]^{1/2} \\ &= \sum_{k=1}^{N} p_k |s_k|^{1/2} \left[\int_{-\infty}^{\infty} |\hat{\mu}(\tau) - \hat{\nu}(\tau)|^2 \tau^{-2} d\tau \right]^{1/2} \\ &\le c^{1/2} d_{FT}(\hat{\mu}, \hat{\nu}). \end{aligned} \tag{11.5.4}$$

□

Corollary 11.5 *The operator B has a unique and attractive fixed point $\hat{\bar{\mu}} \in D(X)$.*

Note: $\hat{\bar{\mu}}$ is the FT of the invariant measure $\bar{\mu}$ of the affine IFSP $(\mathbf{w}, \mathbf{p})$ and satisfies the relation

$$\hat{\bar{\mu}}(\omega) = \sum_{k=1}^{N} p_k e^{-i\omega a_k} \hat{\bar{\mu}}(s_k \omega), \quad \omega \in \mathbf{R}. \tag{11.5.5}$$

In other words, $\hat{\bar{\mu}}(\omega)$ satisfies a "self-tiling property": its graph may be expressed as a linear combination of dilated copies of itself. However, the graphs $\hat{\bar{\mu}}(s_k\omega)$ are copies of $\hat{\bar{\mu}}(\omega)$ which are *stretched* along the ω-axis, unlike the case for IFSM functions. (This is perfectly in accord with the "duality" of space and frequency variables in Fourier transforms.)

Corollary 11.6 *(Collage Theorem for Fourier Transforms): Let (X, d) be a compact metric space and $\mu \in \mathcal{M}(X)$ with FT $\hat{\mu} \in FT(X)$. Let $(\mathbf{w}, \mathbf{p})$ be an N-map IFSP with contractivity factor $c \in [0, 1)$ such that $d_{FT}(\hat{\mu}, \hat{\nu}) < \epsilon$, where $\hat{\nu} = B(\hat{\mu})$ is the FT corresponding to $\nu = M\mu$. Then*

$$d_{FT}(\hat{\mu}, \hat{\bar{\mu}}) < \frac{\epsilon}{1 - c^{1/2}}, \tag{11.5.6}$$

where $\hat{\bar{\mu}} \in FT(X)$ is the FT corresponding to $\bar{\mu} = M\bar{\mu}$, the invariant measure of the IFSP $(\mathbf{w}, \mathbf{p})$.

The above treatment now allows us to formulate an inverse problem for Fourier transforms of measures as a minimization of the collage distance $d_{FT}(\hat{\mu}, B\hat{\mu})$.

Acknowledgements

We wish to thank Mr. J. Kominek and Mr. A. Van de Walle for many discussions on the inverse problem and once again thank the latter for providing the images in Figure 11.11. This research was supported in part by the Natural Sciences and Engineering Council of Canada (NSERC) in the form of individual Operating Grants as well as a Collaborative Projects Grant (with C. Tricot and J. Lévy Véhel), all of which are gratefully acknowledged.

Chapter 12

Fractal Compression of ECG Signals

Geir E. Øien and Geir Nårstad

Most of the theory of fractal compression is independent of the type of signal under consideration, or can easily be adapted to hold for almost any kind of signal we may encounter in practice. Despite this, there have been very few attempts to design and test fractal coders for other classes of signals than grey tone or color images, – for which there actually exists a number of more established methods (e.g. subband coding) that perform as good as – or better than – existing fractal coders.

In this chapter we describe the application of fractal coding techniques to a class of signals which to the authors' knowledge has never before been attempted compressed with such methods: *electrocardiogram* (ECG) signals. Initially, we describe how these medical signals are generated, why they are important, and why there is a need for compressing them. Thereafter, we will briefly describe existing coding techniques for ECG signals. The main part of the paper is dedicated to the development of a fractal compression system for ECG signals. Theory and empirical tests are used to adapt the parameters of the basic method commonly used for images to the ECG signal characteristics, and test results are shown on a range of typical sample signals. The tests show that fractal compression is well suited to the compression of ECG signals and can compete with many existing methods over a range of compression factors.

12.1 Introduction to ECG Signals

In this section we give an introduction to electrocardiogram (ECG) signals – their generation, medical importance, and physical characteristics. We discuss the need for compressing these signals and conclude with an brief overview of existing methods for doing so.

12.1.1 ECG Measurement and Characteristics

Simply stated, ECG signals are curves displaying *heartbeat rhythms*. They are most usually obtained by measuring the potential difference between electrodes which are placed on the surface of the patients' skin. Electrical fields resulting from the patient's heart beating are thus detected and the field variations transferred to a voltage signal. An example of such a signal with typical features for one beat of a normal heart is shown in Figure 12.1.

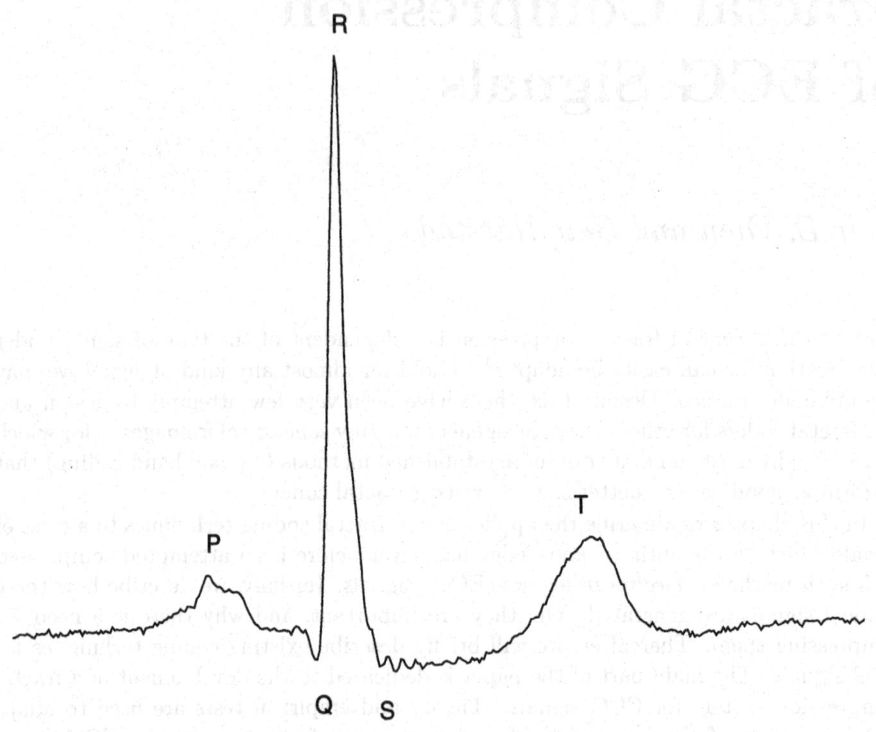

Figure 12.1: Typical features of a normal ECG signal.

The voltage signal can be amplified, displayed, recorded, digitized, and digitally processed/analyzed in various ways depending on the application at hand. The signal may be either *single-channel* or *multi-channel*: the last situation occurs when several pairs of electrodes are used, at various parts of the body at the same time. Each pair of electrodes then gives rise to one channel, each channel representing a different "view" of the heart rhythm.

In this paper, only single-channel ECG signals are considered. However, the need for effective compression is of course even greater in the case of a multi-channel signal, – the amount of information being directly proportional to the number of channels.

ECG signals are commonly used in medical care, for monitoring, diagnosis, and treatment of patients suffering from (possible) heart diseases. Experienced cardiologists can use even minute features of such signals to obtain important knowledge about the heart function of their patients. Figure 12.2 shows sample curves from four different important classes of ECG signals: first a normal heart (sinus) rhythm again, and then signals featuring three kinds of anomalies: ventricular tachycardia, ventricular flutter, and ventricular fibrillation.

A common feature to all of the curves is their *quasiperiodic* nature, akin to that of the vowel parts of a speech signal (although the variations are much slower). This feature of course stems from the main heartbeat rhythm. The peaked area, commonly called the *QRS complex* (again, see Figure 12.1) is, together with its neighbouring parts (the *P*- and *T-waves*; see Figure 12.1), the portion of the signal thought to contain most of the diagnostically important information. It is therefore of the utmost importance that these parts of the signal can be rendered with good precision after any attempt at compression.

12.1.2 Digitization and Storage of ECG Signals

Heartbeat rhythms typically consist of 50 – 200 beats per minute, which define the "pitch", or periodicity, of the ECG signal. In addition to this, there of course exist overharmonic frequencies, as well as other frequency components arising from the nonperfect periodicity. A sampling frequency somewhere between 125 and 500 Hz has mostly been used when digitizing ECG signals. In our study we have used test signals from one of the standard databases of ECG research, the *MIT-BIH Arrhythmia Database CD-ROM (2nd edition)*. Here, a sampling frequency of 360 Hz is used, and each sample is quantized using a 12 bit uniform quantizer[1]. This adds up to 16.2 kBytes per minute per channel, or about 1 Mbyte per hour per channel. This may not sound too dramatic compared to the numbers we are used to hearing in conjunction with audio, image, or video storage – but there are several points to consider:

- Hospitalized patients are often monitored over periods of many hours or days at a time, and there may be a need to analyze the resulting signals at some other time than that of their recording.
- There is a need for building large data bases of ECG signals for medical research purposes.
- There may be a need for transferring ECG signals via the telecommunications network for telemedicine applications.
- One important and growing application of ECG monitoring and storage is in conjunction with *portable, semi-automatic defibrillators*, which are used in the field on patients possibly suffering from cardiac arrest. A defibrillator is a device used for giving electrical shocks in order to get the heart beating again if it has stopped. However, such shocks may be fatal if given to persons *not* requiring them. The defibrillator must therefore be able to measure and display ECG signals for on-the-spot diagnosis, and to store the signals (and possibly also spoken

[1] 8 to 12 bits has been used – with 12 bits being most common.

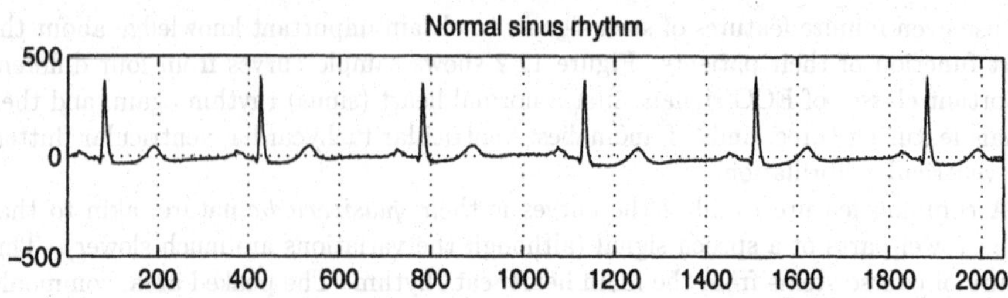

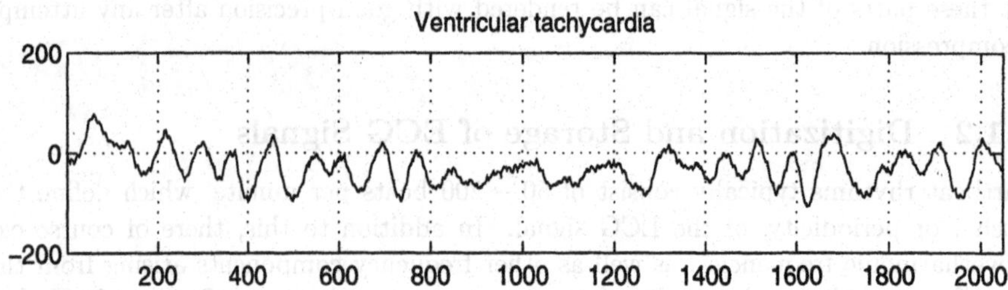

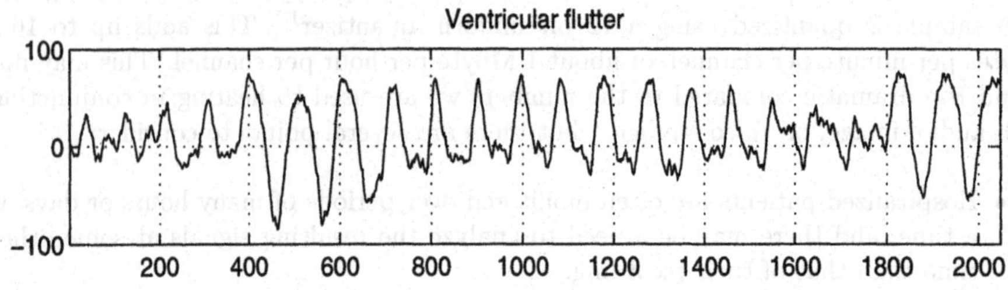

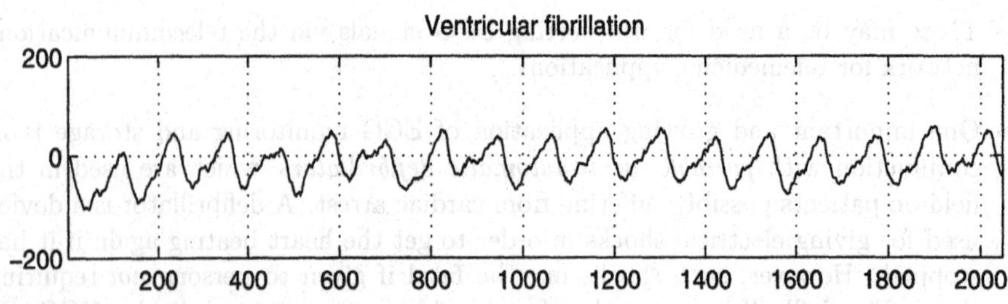

Figure 12.2: Four different classes of ECG signals. Horizontal axis: Sample number. Vertical axis: Voltage amplitude in mV.

comments concerning the patient's condition) for later analysis [25]. The portability of such a device gives it limited storage capacity, however, which makes the need for effective compression more imminent.

12.1.3 History of ECG Compression – a Brief Review

Over the last three decades, a wide variety of ECG compression techniques have been proposed. Some of the most commonly used traditional ones are summed up and discussed in [135]. They include the *AZTEC* (Amplitude Zone Time Epoch Coding) method, the *Fan/SAPA* (Fan/Scan Along Polygonal Approximation) method, the *TP* (Turning Point) technique, and the *CORTES* (Coordinate Reduction Time Encoding System) algorithm. All of these methods, as well as the *SAIES* scheme [134], and the *SLOPE* and *AZDIS* schemes described by Tai [238, 237], are based on redundancy removal by simple and "direct" sample analysis, such as linear polygonal approximation, adaptive subsampling, and so on.

This is in contrast to more recent methods built on *transform domain* or *frequency plane* analysis/signal decomposition. Coding schemes employing such principles (also called *waveform coders*), well known from speech and image coding, are in the process of becoming more common also in ECG compression, and there exists a number of newer publications discussing the use of Karhunen-Loève transforms [59], cosine transforms [5], wavelets [51, 239], and filterbanks (subband coders) [11, 119, 109, 120, 1], in ECG compression. Vector quantization [168] and artificial neural networks [104] have also been applied.

Most of these recent methods may in general be said to offer better compression ratios – at the expense of the need of more computations – than the more traditional ones. Due to the relatively low sampling frequency of the ECG signal, the computational need is not necessarily a problem, however, and future research will probably continue along the lines of waveform coding.

12.2 Fractal Compression and ECG Signals

In this section we initially discuss the general concepts of redundancy and irrelevance in signals, and then provide an empirically based argument for why the fractal method might be a viable technique for ECG compression. We then develop a complete fractal ECG coding system, with an encoder consisting of a modelling part and a quantization part, and a decoder retrieving the attractor from the quantized code.

12.2.1 Redundancy and Irrelevance

The reason why signal compression is possible at all, is that all information signals encountered in practice exhibit *redundancy* and possibly *irrelevance* [137, 97]. A direct sample-by-sample description is in general not strictly necessary as there exists *correlation* – or some other kind of dependence – between signal samples or blocks of samples. By describing and removing this dependence – or redundancy – a more compact signal description may be found.

The concept of *irrelevance* is linked to the fact that much of the importance of many signals – such as speech, audio, images, and also ECG signals – lies in the *human perception* (for the most part, either visual or auditory) of their appearance. Two signals may be *perceptually* identical even though they are *not numerically* identical – which implies that there is no need to render every sample value in such signals with infinite precision. Even a slight perceptual degradation is often acceptable as long as the "important" features of a signal can be distinguished by the observer. This makes *lossy* compression viable – and thereby much higher compression ratios than those experienced in *lossless* coding.

12.2.2 Why Should Fractal Compression Work?

In fractal compression, the property to be exploited is a signal's *self-similarity*. By this we mean what might, in more traditional signal processing language, be termed *generalized cross-correlation*: We consider parts of the signal not necessarily adjacent to each other in space or time, we allow for a certain kind of *mathematical transformation* of one of the signal parts before measuring the correlation, and we make our comparisons between versions of the signal at *different resolutions*. We measure the distance (most often the l_2-distance) between every *range* block in the signal itself, and a number of *transformed domain blocks* in a *decimated* version of the signal. The transform and domain block yielding the smallest distance for each range block is kept. The description of this best transform and the address of the corresponding domain block makes up the range block code, and the sum total of all the range block codes acts as a description of a *contractive mapping* on the signal space, having an *attractor* resembling the signal.

The question is: Do ECG signals exhibit this kind of redundancy? To get a feeling for the answer to this question, consider Figure 12.3. Here we have chosen a sample signal and a range block r of length 16 samples from it, and then searched the signal for the best domain block d of length 32 samples[2]. The range block and the optimal decimated domain block are displayed below the signal itself, and at the bottom the resulting *collage approximation* is given.

As can be seen, the degree of visual similarity between the range block, the domain block, and the collage block is quite striking. From this experiment (and others; not shown here, but similar) we conclude that there seems to exist quite a lot of self-similarity in typical ECG signals. We therefore proceed to apply the whole theory of fractal compression to such signals.

12.2.3 The Fractal Signal Model

We shall conform to the restrictions which have been used in most practical fractal coding research, which is to say we basically apply the fractal signal model introduced by Jacquin in 1989 [123] and later extended and/or modified by, among others, Øien and Lepsøy [198, 149, 197, 152, 202, 200, 199, 201], Fisher [84], and Baharav *et.al.* [14]. I.e. we assume that a signal $\mathbf{x}$ consisting of M real-valued samples can be well

[2] By "best" we mean closest in l_2-distance after decimation and an optimal affine transformation of the structure discussed in the next subsection.

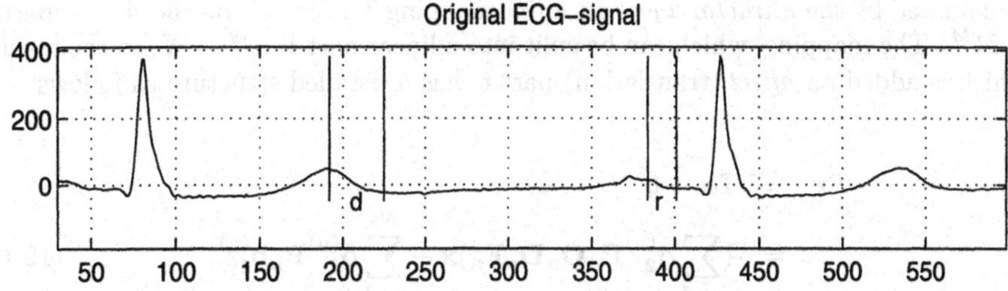

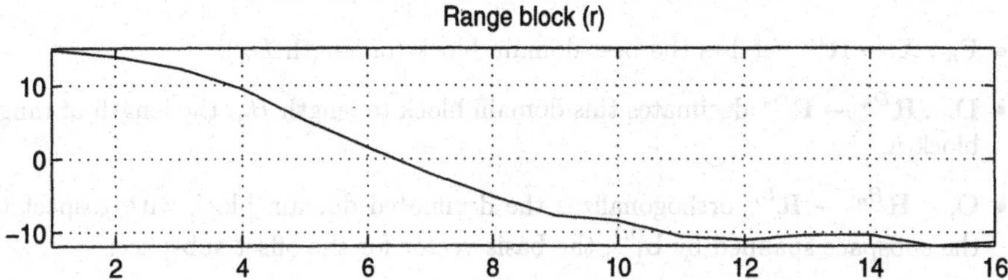

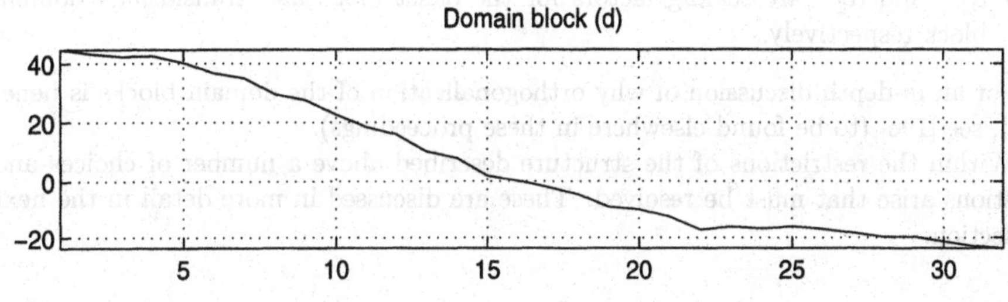

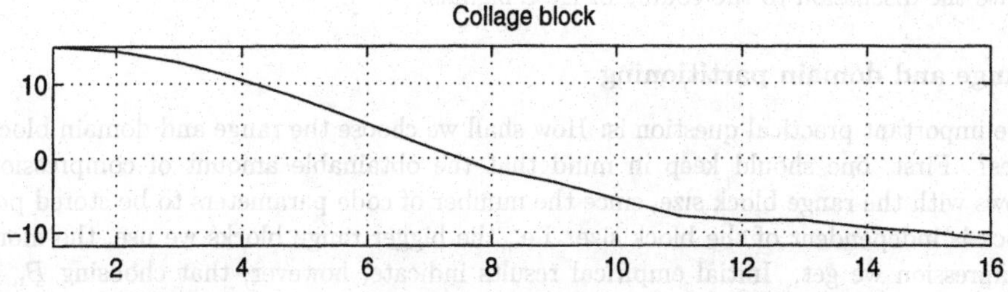

Figure 12.3: Illustration of self-similar parts of an ECG signal.

approximated by the *attractor* $\mathbf{x}_T$ of an *affine* mapping $T : X \to X$ on the signal space, $X = \mathbf{R}^M$. The mapping, which can be split into a linear part $\mathbf{L} : X \to X$, to the result of which is added an *offset* (translation) part $\mathbf{t}$, has a detailed structure as follows:

$$\begin{aligned} T\mathbf{x} &= \mathbf{L}\mathbf{x} + \mathbf{t} \\ &= (\sum_{n=1}^{N_r} \alpha_2^{(n)} \mathbf{P}_n \mathbf{O}_n \mathbf{D}_n \mathbf{F}_n)\mathbf{x} + \sum_{n=1}^{N_r} \alpha_1^{(n)} \mathbf{P}_n \mathbf{b}_1^{(n)}, \end{aligned} \tag{12.1}$$

Here,

- N_r is the number of range blocks we divide the signal into;
- $\mathbf{F}_n : X \to \mathbf{R}^{D_n}$ fetches the best domain block (of length D_n);
- $\mathbf{D}_n : \mathbf{R}^{D_n} \to \mathbf{R}^{B_n}$ decimates this domain block to length B_n, the length of range block n;
- $\mathbf{O}_n : \mathbf{R}^{B_n} \to \mathbf{R}^{B_n}$ orthogonalizes the decimated domain block with respect to the subspace spanned by $\mathbf{b}_1^{(n)}$, the basis vector for the offset subspace;
- $\mathbf{P}_n : \mathbf{R}^{B_n} \to X$ places the sum of the offset block and the transformed domain block in the position of range block n;
- $\alpha_1^{(n)}$ and $\alpha_2^{(n)}$ are scaling factors for the offset block and transformed domain block respectively.

For an in-depth discussion of why orthogonalization of the domain blocks is beneficial, see [196] (to be found elsewhere in these proceedings).

Within the restrictions of the structure described above a number of choices and questions arise that must be resolved. These are discussed in more detail in the next subsection.

12.2.4 Parameter Choices

We now discuss some important practical aspects of a fractal compression system, and relate the discussion to the coding of ECG signals.

Range and domain partitioning

One important practical question is: How shall we choose the range and domain block sizes? First, one should keep in mind that the obtainable amount of compression grows with the range block size, since the number of code parameters to be stored per block is independent of the block size. I.e., the bigger range blocks we use, the more compression we get. Initial empirical results indicate, however, that choosing B_n ¿ 16 in practice gives too coarse collage (and hence attractor) approximations for most ECG signals. We therefore constrain the maximum range block size to be $B_{max} = 16$ samples.

In fractal *image* coding, it is furthermore known that a constant range block size generally gives too poor a fit in areas containing a lot of detail/high frequencies. This turns out to be the case also for ECG signals: If we are to keep the distortion per sample below an acceptable level in all blocks, we must allow for adaptivity in the range block size. Typically, the QRS complex is the region requiring the smallest block sizes. We have used a *binary tree range partition* to account for this: Starting with the biggest possible range blocks, we recursively split each block in two, and model its parts as two separate blocks, until either an acceptable distortion limit or a maximum tree depth has been reached. This construction is basically a 1-D version of the *quadtree* partition known from fractal image coding [87], and has the benefits of compact description and regularity of computation. As a compromise between attainable quality and compression ratio we have used a tree depth of 4, such that the possible range sizes are 2, 4, 8, and 16.

As to the domain block size, it is known that the domain blocks have to be bigger than the range blocks in order for the attractor to be able to fall outside the offset subspace [163, 198]. For images, typically the domain blocks are mostly chosen as 2 – 4 times the range block size in each spatial direction [123, 84, 198, 149, 87]. From empirical studies we have performed, it seems like $D_n/B_n = 2$ or 4 yields the best result for ECG signals. We have chosen the former in our experiments, which gives domain block sizes in the range 4 to 32 samples.

The attainable quality is also a function of the number N_d of domain blocks used. In fact N_d plays the same role as the *codebook size* in an ordinary vector quantizer [97, 149]. However, the more domain blocks used, the more bits must be used to specify their address ($\log_2(N_d)$); thus the codebook size must also be constrained. We have found a codebook size of $N_d = 256$ domain blocks per range block to be favourable; the address overhead per block is then only $\log_2(256) = 8$ bits, and larger N_d do not seem to provide us with much improvement in the reconstruction quality. This might have to do with the fact that there will be a lot of similarity between many of the domain blocks in a large domain codebook. This has been shown to be true for images by Lepsøy [150], and experiments by Nårstad seem to confirm this tendency also for ECG signals [192]. In a large domain codebook, there are many redundant blocks, – as opposed to what is the case in a large VQ codebook, where the blocks have been optimized with respect to the signal statistics.

The decimation operator

Since we only decimate by a factor of 2, the choices of decimation operator open to us are basically *subsampling* and *averaging*. Based on the experience from images (for which subsampling typically yields a 1 dB original-attractor PSNR loss compared to averaging [84]), as well as theoretical issues discussed in [198], we have chosen to use averaging in our coder.

The offset subspace basis

As shown in [198], if the signals to be coded are samples from a stationary random process, the optimal basis block (as concerns coding gain) for a 1-D offset subspace is the eigenvector corresponding to the largest eigenvalue of the autocorrelation matrix of the process. We have estimated such offset subspace basis blocks (one for each range block size) in our system. When constructing the data set used for this estimation procedure for a given range size, we have (by means of manual segmentation based on experience) excluded the parts of the signal which would most probably be coded by range blocks of some *other* size. This is done to ensure the best possible correspondence between the data used to optimize the offset block, and the data for which it is to be used. It also makes the stationarity assumption somewhat more reasonable.

For the smallest range block sizes, the resulting offset block is, for all practical purposes, a constant value (DC) block. Even for range blocks the size of 8 or 16 the results do not differ much from the DC block most commonly used in fractal image coding. The resulting basis block for $B_n = 16$ is shown in Figure 12.4.

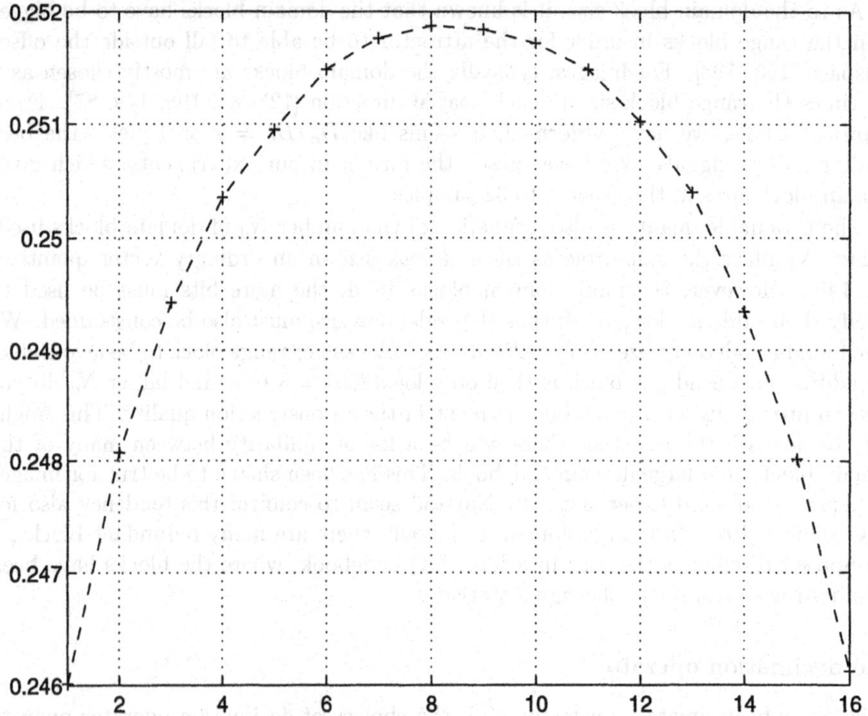

Figure 12.4: Estimated optimal offset subspace basis block, range block size = 16.

12.2.5 Parameter Optimization

The parameter values are found using the l_2-optimization scheme originally suggested in conjunction with fractal compression by Lundheim and later adapted by Øien and Lepsøy [163, 198, 149]. Due to the use of orthogonalization we may perform unconstrained optimization without having to fear a noncontractive mapping as a result[3]. The expressions of the scaling constants for range block n are therefore those given in [196], to be found elsewhere in these proceedings.

12.2.6 The Coefficient Quantization Scheme

As shown in [198, 196], $\alpha_1^{(n)}$ and $\alpha_2^{(n)}$ are decorrelated when orthogonalization is used and the signals are samples from a stationary random process. However, due to the quasiperiodicity of the ECG signals, stationarity is not really a good assumption. Still, experiments have shown that the crosscorrelation between these two coefficients is negligible after all [192]. We may therefore apply *scalar* quantization to each of the coefficients without losing too much coding gain relative to what we would have if optimal *vector* quantization of the $[\alpha_1^{(n)}, \alpha_2^{(n)}]^T$ vectors were used.

Coefficient distributions

The probability density functions of the two coefficients have been estimated by histograms, using 74394 sample values of each taken from a variety of signals covering both normal rhythms and various kinds of arhythmia. The results are displayed in Figure 12.5 and Figure 12.6.

As can be seen, both distributions are highly nonuniform, tending towards zero mean Gaussian or Laplacian pdfs[4]. For α_1 this is in contrast to what has been experienced with images, where it is more or less uniformly distributed [198].

Using this knowledge, we may opt for *pdf-optimized (Lloyd-Max) scalar quantization* of each coefficient. The iterative Lloyd-Max algorithm [137] is used to design nonuniform quantizer characteristics for given target bit rates. The empirically optimal allocation of bits among the coefficients at various overall target rates is discussed in Subsection 12.3.2, where the resulting quantizer characteristics are also displayed.

Temporal coefficient correlations

It is also of interest to see whether *temporal correlation* (i.e. correlation between adjacent values of each coefficient viewed as a time series) exists in the coefficients, in order to decide if more sophisticated quantization techniques than ordinary sample-by-sample quantization can be of use. Due to our choice of offset basis vector, α_1 can be easily interpreted as representing the *lowpass* information of the signal. It should therefore exhibit quite a bit of temporal correlation between samples. These facts are readily confirmed from the upper two curves in Figure 12.7, which displays a sample signal together with α_1 and α_2 as functions of time (range block number) for the same signal.

The autocorrelation functions and of α_1 and α_2 have also been explicitly estimated and are displayed in Figure 12.8.

[3]Strictly speaking, this has only been formally shown for an exact DC offset block. However, all experiments, as well as our intuition, tell us that it in practice holds also with our signal-optimized offset subspace.

[4]With the exception of the lack of α_2 values close to zero.

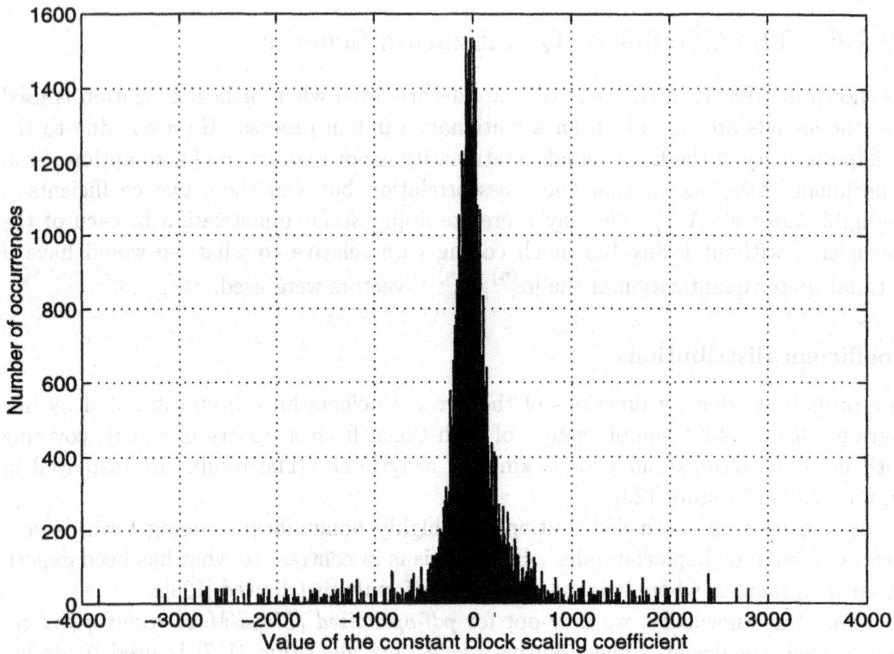

Figure 12.5: Histogram showing the distribution of the coefficient α_1.

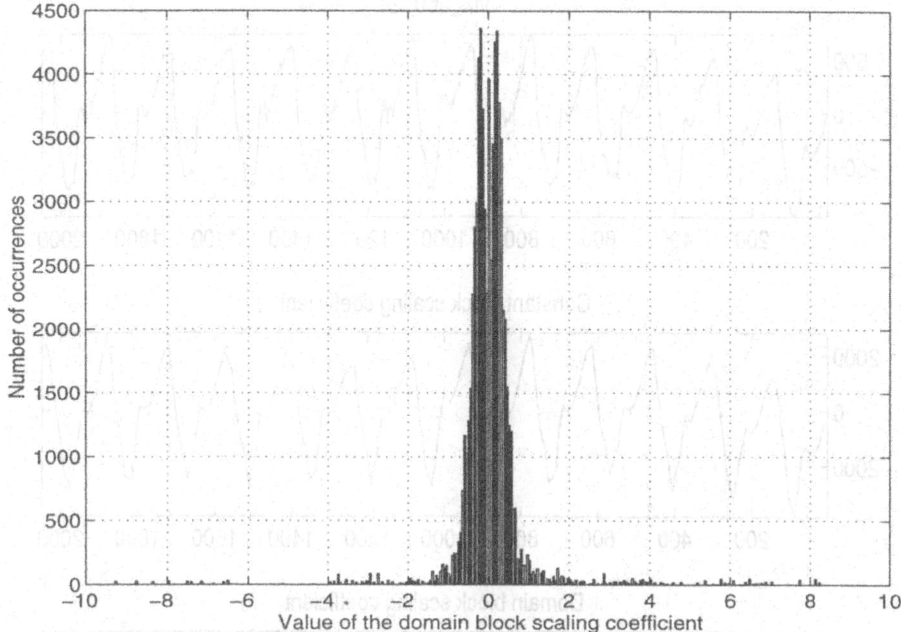

Figure 12.6: Histogram showing the distribution of the coefficient α_2.

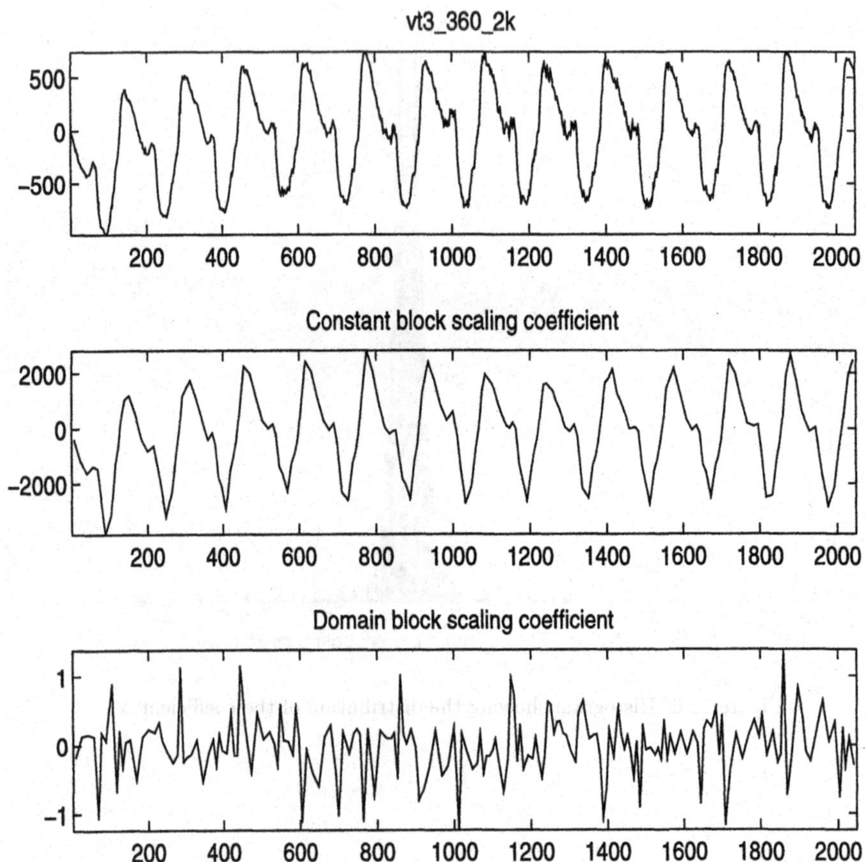

Figure 12.7: A sample ECG signal and its α_1 and α_2 coefficients.

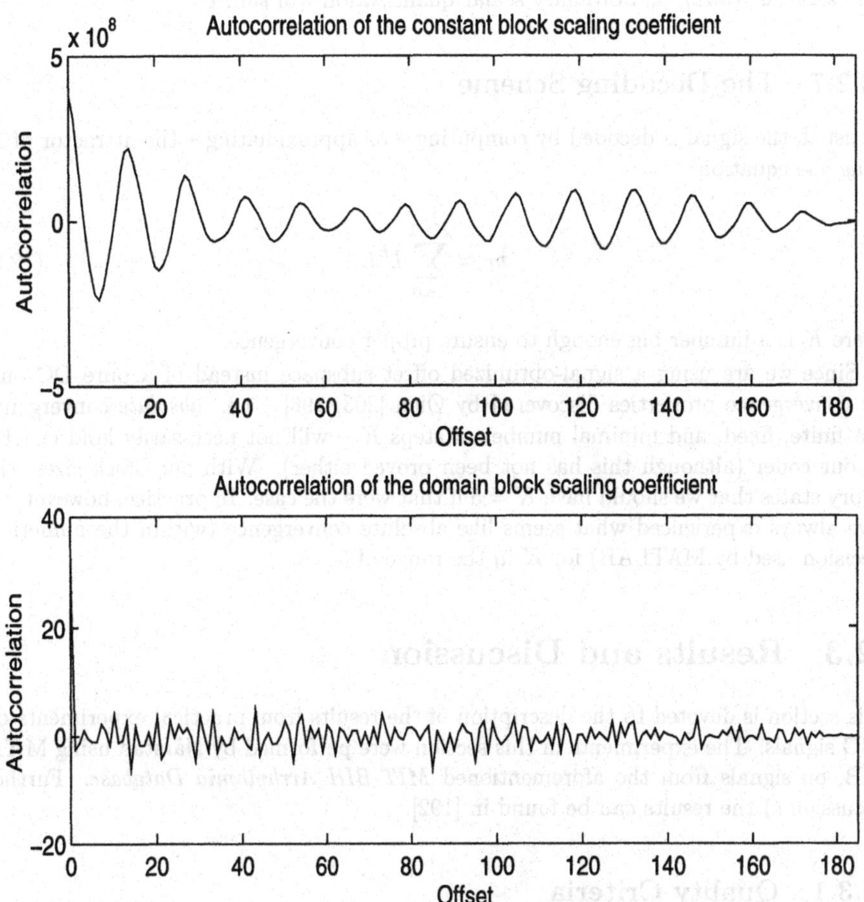

Figure 12.8: Temporal correlation in α_1 and α_2.

The degree of temporal correlation in α_1 is such that *differential/predictive coding* [137] is sure to yield a coding gain over direct quantization. This has yet to be implemented, however. In the case of α_2, there is no discernible temporal correlation (the time series is *white*), and ordinary scalar quantization will suffice.

12.2.7 The Decoding Scheme

As usual, the signal is decoded by computing – or approximating – the attractor of T, using the equation

$$x_T = \sum_{k=0}^{K-1} L^k t \tag{12.2}$$

where K is a number big enough to ensure proper convergence.

Since we are using a signal-optimized offset subspace instead of a pure DC one, the convergence properties discovered by Øien [203, 196] – i.e. *absolute* convergence in a finite, fixed, and minimal number of steps K – will not necessarily hold exactly for our coder (although this has not been proved either). With our block sizes, the theory states that we should have $K = 5$ if that were the case. In practice, however, we have always experienced what seems like absolute convergence (within the numerical precision used by MATLAB) for K in the range of 5 – 8.

12.3 Results and Discussion

This section is devoted to the description of the results from practical experiments on ECG signals. The experiments in this section were performed by Nårstad using MATLAB, on signals from the aforementioned *MIT-BIH Arrhythmia Database*. Further discussion of the results can be found in [192].

12.3.1 Quality Criteria

The objective criterion most commonly used for judging the quality of coded ECG signals, and also the one used here, is the *Percentile RMS Difference* or *PRD*, defined as

$$PRD = 100 \cdot \sqrt{\frac{\sum_{i=1}^{M}(x_i - x_{T,i})^2}{\sum_{i=1}^{M} x_i^2}} \tag{12.3}$$

where x_i are the samples of the original signal (of length M), and $x_{T,i}$ those of the decoded signal. The PRD is smaller the better the coding is.

In addition to the objective quality measure given by the PRD, we shall use subjective evaluation in our discussion of the coding quality.

12.3.2 Bit Allocation Experiments

As was previously mentioned, given a total of $B = B_1 + B_2$ bits to use for the quantization of α_1 (by B_1 bits) and α_2 (by B_2 bits), there is no theory prescribing exactly how these bits are to be distributed. However, it is known that the low frequency content of an ECG signal is of great importance and hence must be represented with high accuracy. Tentatively, we are therefore prepared to spend more bits on the quantization on α_1[5].

This suspicion has been confirmed by our experiments on the bit allocation problem. For various values of B we have tried out (on the four ECG signals displayed in Figure 12.2, at a constant target bit rate of 2 bits per sample) all possible combinations of B_1 and B_2 such that $B_1 + B_2 = B$. The empirically optimal values of B_1 and B_2 (in terms of lowest possible PRD) resulting from this are given in Table 12.1.

B	B_1	B_2
8	4	4
9	5	4
10	6	4
11	7	4
12	8	4
13	8	5
14	8	6
15	8	7
16	8	8

Table 12.1: Optimal bit allocation.

We see that for the most part, α_1 requires more bits than α_2. The bit allocation results are therefore in accordance with intuition/theory.

12.3.3 Quantizer Characteristics

The quantizer characteristics resulting from Lloyd-Max-optimization as previously described are displayed in the figures 12.9 and 12.10. In order to exploit the training set more effectively, an assumption of symmetry in the distributions has been applied when running the optimization.

It can be seen that the resulting quantizers are highly nonuniform for all the values of B_1, B_2 tried out.

12.3.4 Coding Experiments

In the figures 12.11 – 12.14 we have displayed an original sample signal together with a coded version at 2.0 bits per sample, for the four main types of ECG signals. The number of bits used per coefficient were 8 bits for α_1 and 5 bits for α_2. A rate of 2.0

[5] Alternatively, a more sophisticated coding scheme.

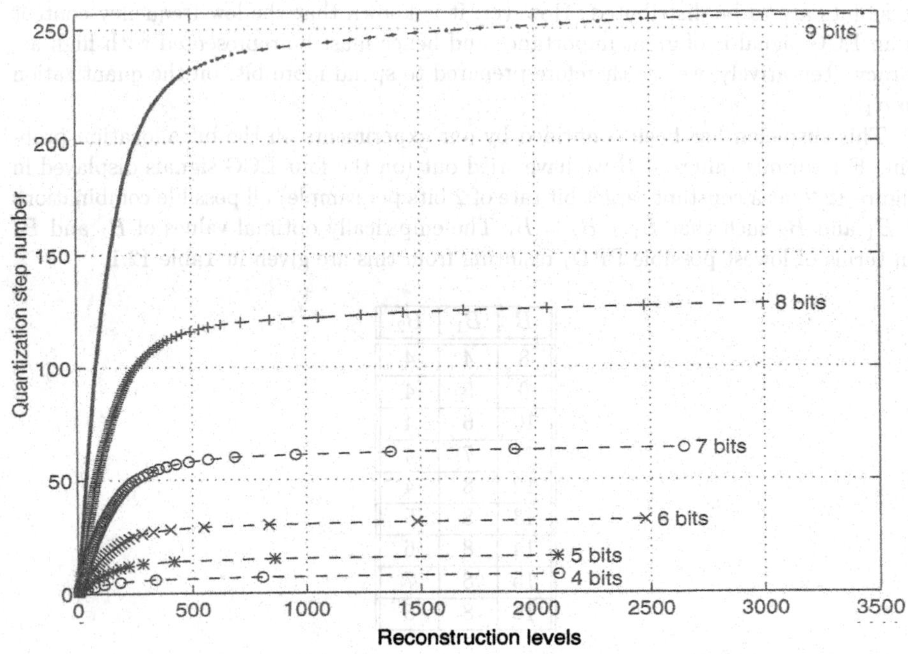

Figure 12.9: Quantizer characteristics for α_1.

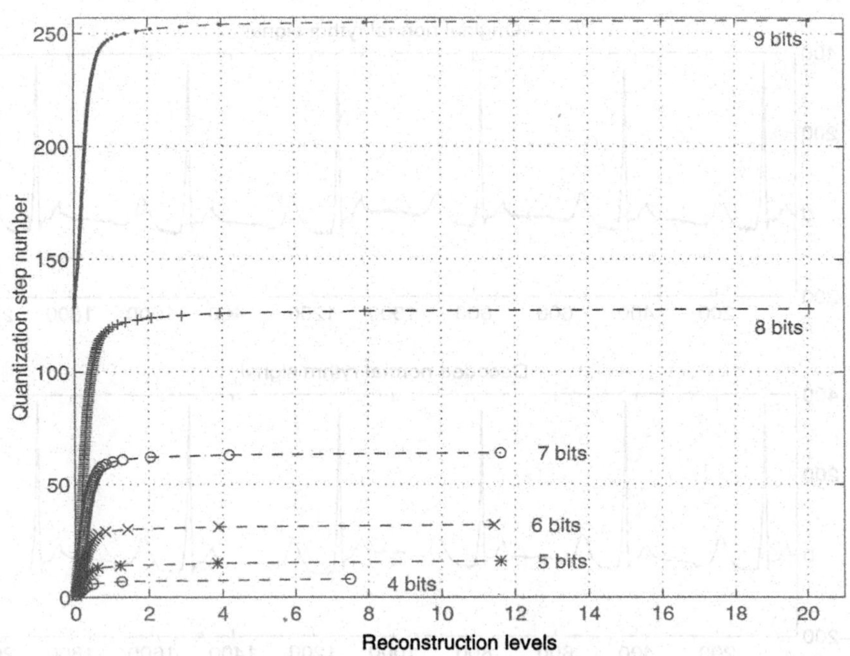

Figure 12.10: Quantizer characteristics for α_2.

bps means that we have achieved a compression factor of 6.0 compared to the original signal.

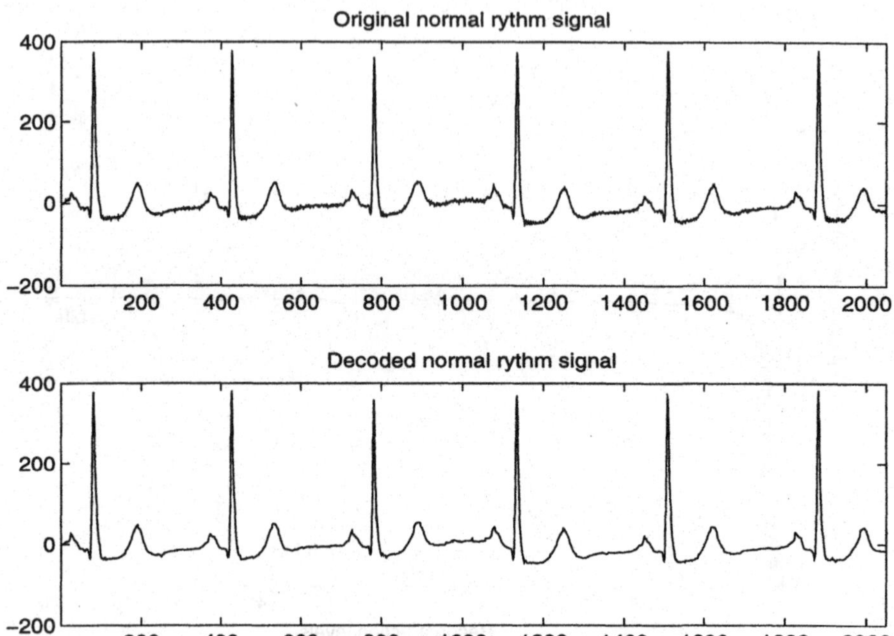

Figure 12.11: Original and coded "normal" ECG signal at 2.0 bits per sample.

As can be seen, the perceptual quality/fidelity is very good; all of the main distinguishing features are in place. The only thing lost seems to be the very high frequency "dithering" — which is most probably noise and carries no diagnostical information anyway. Neither is there any discernable "blockiness" – a feature which can be disturbing in fractal *image* coding.

We have also run experiments where the above four sample signals have been coded at various bit rates. In each case the PRD was measured. Some results from these experiments are collected in Figure 12.15. Here, 'no1_360_2k' is the normal signal, 'vt1_360_2k' is ventricular tachycardia, 'vl1_360_2k' is ventricular flutter, and 'vf1_360_2k' is ventricular fibrillation.

As can be seen, the curves tend to become quite steep for bit rates below 2 bits per sample. This should imply that fractal coding of ECG signals is best suited at medium bit rates. We should also note that the normal heart rhythm seems to be the hardest to code. For diagnosis purposes this may not be such a bad thing, as it should imply that the features occuring in conjunction with heart diseases — i.e. those that are most important to detect — are the features easiest to compress.

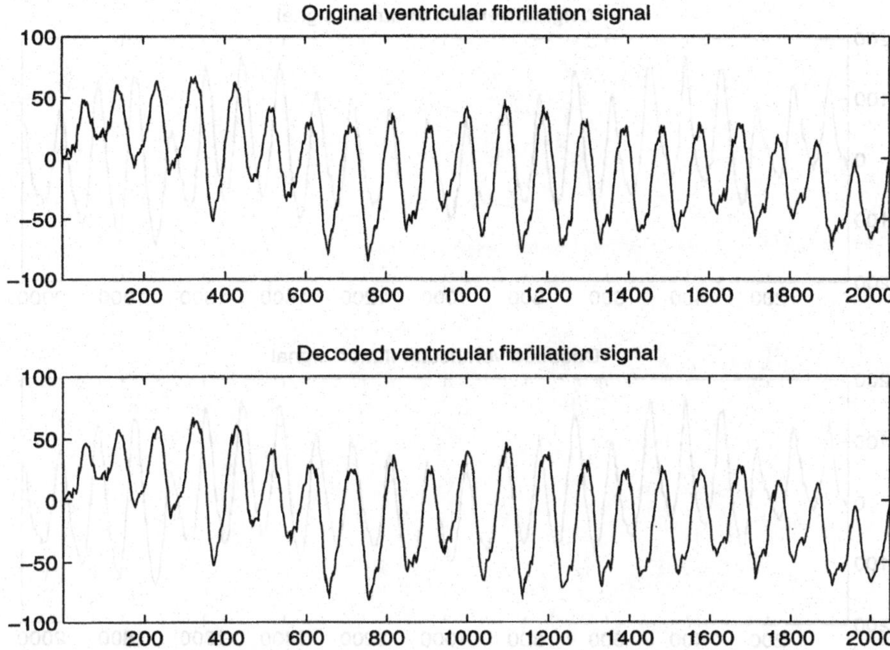

Figure 12.12: Original and coded ventricular fibrillation ECG signal at 2.0 bits per sample.

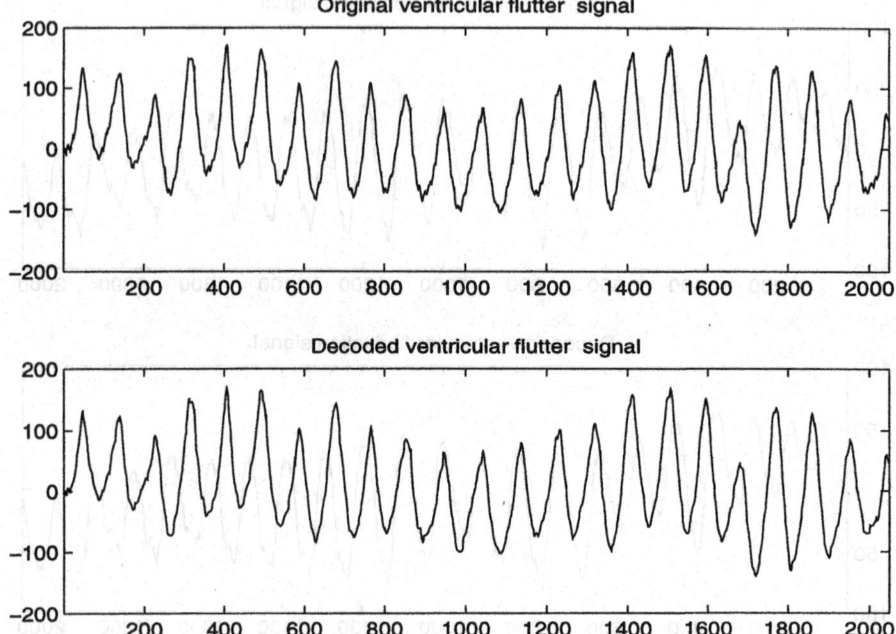

Figure 12.13: Original and coded ventricular flutter ECG signal at 2.0 bits per sample.

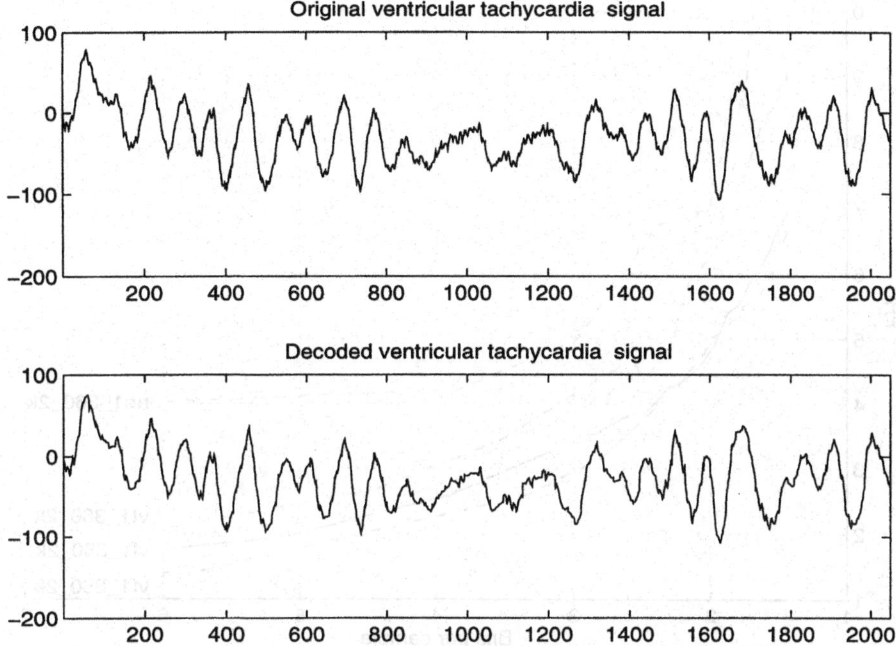

Figure 12.14: Original and coded ventricular tachycardia ECG signal at 2.0 bits per sample.

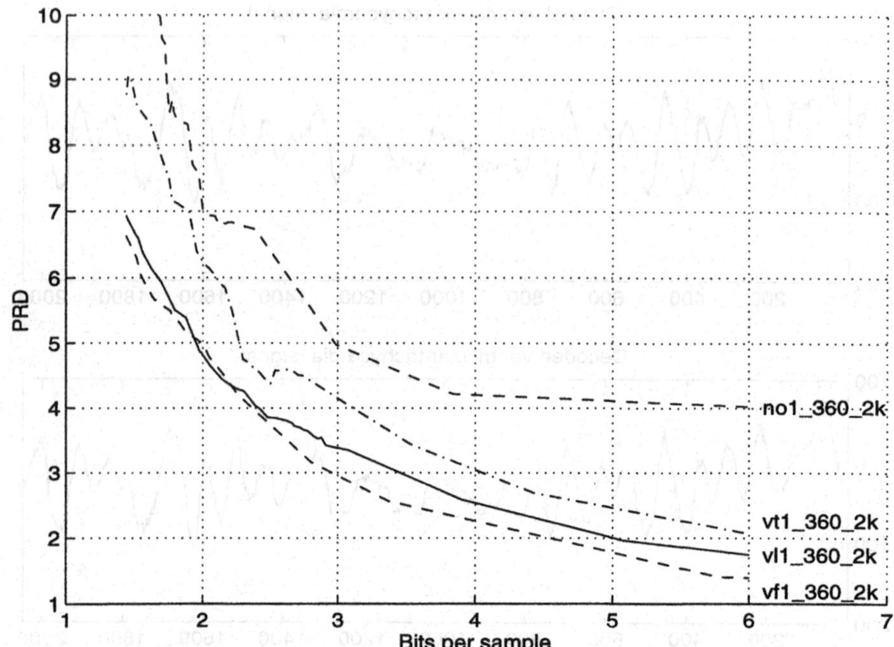

Figure 12.15: PRD versus bit rate for the four signals displayed in the figures 12.11 – 12.14.

12.3.5 Comparisons to Other Techniques

As previous ECG compression research has been carried out on a bewildering multitude of data bases, sampling frequencies, and bits per original sample, it is quite hard to make qualified judgements about the preference of one method over the other. However, Table 12.2 (most of the results in which are adopted from [135, 134]) is an attempt to sum up some important results from previous work and relate them to some of our own results (PRD averaged over all coding experiments at 2.0 bits per sample). The subband coding result is taken from [108], and is the result most easily comparable to ours, as it is obtained on similar signals from the same data base.

Method	*CR*	*SF* [Hz]	Precision [bits]	*PRD* [%]
AZTEC	10.0	500	12	28.0
TP	2.0	200	12	5.3
CORTES	4.8	200	12	7.0
Fan/SAPA	3.0	250	—	4.0
SAIES	5.9	166	10	16.3
Peak-Picking with Entropy Encoding	10.0	500	8	14.0
DPCM-Linear Pred., Interp. and Entropy Coding	7.8	500	8	3.5
Fourier Descriptors	7.4	250	12	7.0
Subband coding	6.5	360	12	5.1
Fractal-based compression	6.0	360	12	5.8

Table 12.2: Comparison of ECG data compression schemes.

It seems that we may conclude that fractal compression works at least as good as, and often much better than, the more traditional methods — although the results are certainly not directly comparable. However, the most modern waveform coders still seems to have a slight edge over fractal methods — which is the same situation as with images. However, the results from subband and fractal coding are definitely in the same quality range.

12.4 Conclusion

We have demonstrated the use of fractal compression on a new class of signals, namely electrocardiogram signals. A complete fractal coding system has been designed, with parameters optimized as far as possible with respect to this particular signal class. A coder structure using l_2-optimal affine mappings with adaptive binary range and domain partitions and basis orthogonalization has been used.

The results show that fractal coding outperforms the methods traditionally used for ECG compression, and also competes in the same league as more modern waveform coders based on transforms and filter banks. The fractal method has the benefit of a fast decoding algorithm with the possibility of a multiresolution implementation, but the encoder is at present rather slow, a recurring problem with fractal coding methods. However, taking into account various complexity reduction methods previously used in conjunction with digital images [198, 149, 151, 223, 68] one might expect a significant complexity reduction without any serious loss of quality. This may still enable implementations fast enough to be used in portable equipment. Along with more effective coding of α_1, encoder complexity reduction would seem to be the most fruitful area of future research into fractal coding of ECG signals.

Part II

Fractal Image Analysis

Chapter 13

Dimensions of Fractals and Multifractals

Kenneth Falconer

Abstract: The concept of dimension is fundamental in the analysis of fractals and multifractals. We survey some definitions and properties of dimensions, and discuss some of the situations when calculation of dimension has been possible, for example for self-similar sets, self-affine sets, graph-directed sets, etc. We also consider the related topic of finding the dimension spectra of multifractal measures.

13.1 Dimensions of Fractals

13.1.1 Hausdorff and Packing Dimensions

Hausdorff and packing dimensions are defined in terms of corresponding measures.

Let E be a subset of n-dimensional Euclidean space $\mathcal{R}^n$, and let $0 \leq s \leq n$. For each $\delta > 0$ we define

$$H^s_\delta(E) = \inf\left\{\sum_{i=1}^{\infty}|U_i|^s : E \subseteq \bigcup_{i=1}^{\infty} U_i \text{ with } 0 < |U_i| \leq \delta\right\}.$$

Here $|U| = \sup\{|x-y| : x, y \in U\}$ is the diameter of the set U; thus the infimium is taken over all coverings of E by a countable collection of sets with diameters of most δ. Decreasing δ reduces the allowable coverings of E for the infimium, and so $H^s_\delta(E)$ increases as δ decreases. Therefore, the limit

$$H^s(E) = \lim_{\delta \to 0} H^s_\delta(E)$$

exists (though it may be 0 or ∞) and we term this number the *s-dimensional Hausdorff measure* of E. It may be shown that H^s is a Borel measure so that is particular

$$H^s\left(\bigcup_{i=1}^{\infty} E_i\right) = \sum_{i=1}^{\infty} H^s(E_i)$$

for every countable collection of disjoint Borel sets $\{E_i\}_{i=1}^{\infty}$. Moreover, if E is a smooth curve then $H^1(E)$ gives the length of E, and if E is a smooth surface then $H^2(E)$ is the area of E to within a constant factor. Thus Hausdorff measures generalize the familiar notions of length, area, etc.

We use Hausdorff measure to define Hausdorff dimensions. Suppose that $t > s$ and $E \subseteq \bigcup_{i=1}^{\infty} U_i$ with $|U_i| \leq \delta$ for all i. Then

$$\sum_i |U_i|^t = \sum_i |U_i|^s \, |U_i|^{t-s} \leq \delta^{t-s} \sum_i |U_i|^s,$$

so on taking infima

$$H_\delta^t(E) \leq \delta^{t-s} H_\delta^s(E).$$

Letting $\delta \to 0$, it follows that if $t > s$ and $H^s(E) = \lim_{\delta\to 0} H_\delta^s(E) < \infty$ then $H^t(E) = \lim_{\delta\to 0} H_\delta^t(E) = 0$. Thus there is a number at which $H^s(E)$ 'jumps' from ∞ to 0 as s increases. This is the *Hausdorff dimension* $\dim_H E$ of E. Thus

$$\dim_H E = \sup\{s : H^s(E) = \infty\} = \inf\{s : H^s(E) = 0\}.$$

It is easy to see that if E is a smooth (or rectifiable) curve then $\dim_H E = 1$, if E is a smooth surface then $\dim_H E = 2$, etc. However, there are many examples of 'fractal sets' for which $\dim_H E$ is non-integral.

Packing measures and dimensions were introduced relatively recently [242]; in some ways these may be though of as 'dual' to Hausdorff measures and dimensions. For $E \subseteq \mathcal{R}^n$ and $0 \leq s \leq n$ and $\delta > 0$ we define

$$P_\delta^s(E) = \sup\left\{\sum_{i=1}^{\infty} r_i^s : E \subseteq \bigcup_{i=1}^{\infty} B_{r_i}(x_i)\right.$$
$$\left.\text{with } x_i \in E, r_i \leq \delta \text{ and this union disjoint}\right\}.$$

(We write $B_r(x)$ for the ball of center x and radius r.) Then the limit

$$P_0^s(E) = \lim_{\delta\to 0} P_\delta^s(E)$$

exists, but is not a measure. However,

$$P^s(E) = \inf\left\{\sum_{i=1}^{\infty} P_0^s(E_i) : E \subseteq \bigcup_{i=1}^{\infty} E_i\right\}$$

does define a Borel measure on $\mathcal{R}^n$ called *packing measure* of E.

In just the same way as for Hausdorff dimension we may define the *packing dimension* $\dim_p E$ of E by

$$\dim_p E = \sup\{s : P^s(E) = \infty\} = \inf\{s : H^s(E) = 0\}.$$

Again $\dim_p E = 1$ if E is a smooth curve and $\dim_p E = 2$ if E is a smooth surface. In general, $\dim_H E \leq \dim_p E$ but there are many examples for which equality does not hold.

A number of nice properties ensure that Hausdorff and packing dimensions provide useful generalizations of the 'classical' ideas of (integral) dimensions. The following properties (which may easily be verified from the definitions) hold when 'dim' is either Hausdorff or packing dimension.

1. *Open sets* If $E \subseteq \mathcal{R}^n$ contains a (non-empty) open region then $\dim E = n$.
2. *Smooth sets* If E is a smooth m-dimensional submanifold of $\mathcal{R}^n$ then $\dim E = m$.
3. *Monotonicity* If $E \subseteq F$ then $\dim E \leq \dim F$.
4. *Countable stability* $\dim \bigcup_{i=1}^{\infty} E_i = \sup_{1 \leq i < \infty} \dim E_i$.
5. *Countable sets* If E is countable then $\dim E = 0$.
6. *Lipschitz mappings* If $E \subseteq \mathcal{R}^n$ and $f : E \to \mathcal{R}^m$ is Lipschitz (i.e. $|f(x) - f(y)| \leq c|x - y|$ for all $x, y \in E$) then $\dim f(E) \leq E$.
7. *Bi-Lipschitz mappings* If $f : E \to f(E)$ satisfies

$$c_1 |x - y| \leq |f(x) - f(y)| \leq c_2 |x - y| \; (x, y \in E)$$

$(0 < c_1 \leq c_2)$ then $\dim E = \dim f(E)$. In particular, this is true when f is a congruence, similarity or affine transformation.

8. *Projections* Let $\text{proj} : \mathcal{R}^n \to P$ denote onthogonal projection onto an m-dimensional subspace P. Then $\dim \text{proj}\, E \leq \min\{m, \dim E\}$ for $E \subseteq \mathcal{R}^n$.

In fact $\dim_H \text{proj} E = \min\{m, \dim_H E\}$ for 'almost all' subspaces V; but this is not true for $\dim_p$.

For further properties of dimensions see [74].

13.1.2 Calculation of Dimensions

These are a variety of ways of calulating dimensions of given sets. Many methods involve supporting a measure on the set (often in a 'natural' way) and examining its properties. The following criteria, involving the local behavior of such a measure, will be convenient for our applications.

Let $E \subseteq \mathcal{R}^n$, let $s > 0$ and let μ be a (Borel) measure (or 'mass distribution') with $0 < \mu(E) < \infty$. If for all $x \in E$ we have

(i) $\limsup_{r \to 0} \mu(B_r(x))/r^s < \infty$ then $\dim_H E \geq s$

(ii) $\limsup_{r \to 0} \mu(B_r(x))/r^s > 0$ then $\dim_H E \leq s$

(iii) $\liminf_{r \to 0} \mu(B_r(x))/r^s < \infty$ then $\dim_p E \geq s$

(iv) $\liminf_{r \to 0} \mu(B_r(x))/r^s > 0$ then $\dim_p E \leq s$.

(Of course, if for some μ we have that $0 < a \le \mu(B_r(x))/r^s \le b < \infty$ for all $x \in E$ and all sufficiently small r,then all four conclusions hold simultaneously.) Note that the behaviour as $r \to 0$ of $\mu(B_r(x))/r^s$ and of $\log \mu(B_r(x))/\log r$ are very closely related, for example

$$\liminf_{r\to 0} \log \mu(B_r(x))/\log r \ge s \text{ if and only if}$$
$$\limsup \mu(B_r(x))/r^{s-\varepsilon} = 0 \text{ for all } \varepsilon > 0.$$

To give the 'feel' for these criteria we give a sample proof of part (i). Suppose that $\limsup_{r\to 0} \mu(B_r(x))/r^s < \infty$ for all $x \in E$ where $\mu(E) > 0$. Then we may find $E_0 \subseteq E$ with $\mu(E_0) > 0$ where this happens uniformly, that is with

$$\mu(B_r(x)) \le cr^s \text{ for all } r < \delta \text{ and } x \in E_0$$

for some $\delta > 0$ and $0 < c < \infty$. Suppose $E_0 \subseteq \bigcup_{i=1}^{\infty} U_i$ where $|U_i| \le \delta$. Take $x_i \in E_0 \cap U_i$ (we may assume this intersection to be non-empty); then $U_i \subseteq B_{r_i}(x_i)$ where $r_i = |U_i| \le \delta$. Then $E_0 \subseteq \bigcup_{i=1}^{\infty} B_{r_i}(x_i)$ so

$$\mu(E_0) \le \sum_{i=1}^{\infty} \mu(B_{r_i}(x_i)) \le c \sum_{i=1}^{\infty} r_i^s = c \sum_{i=1}^{\infty} |U_i|^s.$$

Hence $H^s_\delta(E_0) \ge \mu(E_0)/c$, so $H^s(E_0) \ge \mu(E_0)/c > 0$ giving $\dim_H E \ge \dim_H E_0 \ge s$.

The proof of (iv) is rather similar; (ii) and (iii) require a covering lemma.

13.1.3 Iterated Function Systems

The most basic examples to which these criteria may be applied are to the attractors of certain iterated function schemes.

Let $D \subseteq \mathcal{R}^n$ be a closed domain (often $D = \mathcal{R}^n$). Let $S_1, \ldots, S_m : D \to D$ be contractions, that is $|S_i(x) - S_i(y)| \le c_i\, |x - y|$ for all $x, y \in D$ where $c_i < 1$. A basic result of [121] is that there exists a unique non-empty compact set $E \subseteq D$ satisfying

$$E = \bigcup_{i=1}^{m} S_i(E)\ ;$$

such a set E is called *invariant* or an *attractor* of the iterated function system $\{S_1, ..., S_m\}$. Moreover, writing $S(X) = \bigcup_{i=1}^{m} S_i(X)$ for a set X, we have for any non-empty compact set X satisfying $\bigcup_{i=1}^{m} S_i(X) \subseteq X$ that

$$E = \bigcap_{k=0}^{\infty} S^k(X)$$

where S^k is the k-th iterate of S.

To see all this, we note that C, the class of all non-empty compact subsets of D, is complete under the Hausdorff metric $d(A,B) = \sup_{a \in A}\{\inf_{b \in B} |b-a|\} + \sup_{b \in B}\{\inf_{a \in A} |b-a|\}$. Then $S : C \to C$ is a contraction with respect to this metric, that is $d(S(A), S(B)) \leq c\, d(A,B)$ where $c = \max_{1 \leq i \leq m} c_i < 1$, so by Banach's contraction mapping theorem S has a unique fixed point, i.e. there is a unique set $E \in C$ with $E = S(E) = \bigcup_{j=1}^{m} S_j(E)$. Moreover, $S^k(X) \to E$ for any $X \in C$.

The attractors defined in this way are usually fractals, and it is natural to try to find the dimensions of such sets.

13.1.4 Self-similar Sets

For the simplest situation, we assume that the $S_i : \mathcal{R}^n \to \mathcal{R}^n$ are contracting similarities with ratios r_i; thus $|S_i(x) - S_i(y)| = r_i|x-y|$ for all $x, y \in \mathcal{R}^n$. The attractors E thus obtained are called *self-similar sets*; this class of set includes the middle-third Cantor set, the von Koch curve, the Sierpinski gasket and many other standard examples. In this situation $\dim_H E = \dim_p E = s$ where s is the non-negative solution of

$$\sum_{i=1}^{m} r_i^s = 1,$$

provided that the $S_i(E)$ are disjoint, or more generally the *open set condition* is satisfied, that is $O \supseteq \bigcup_{i=1}^{m} S_i(O)$ with this union disjoint for some non- empty bounded open set O.

We may verify this using criteria of 1.1.2. We assume the disjoint case. We may define a measure μ on E by repeated subdivision in the ratio $r_1^s : r_2^s : \cdots : r_m^s$. Let X be a set (which could be E itself) such that $S_i(X) \subseteq X$ for all i and $S_i(X) \cap S_j(X) = \emptyset$ if $i \neq j$. We let $\mathbf{i}$ denote the sequence
$(i_1, \cdots, i_k)$ where $i_j \in \{1, 2, \ldots, m\}$ for all j and write

$$X_{\mathbf{i}} = X_{i_1,\ldots,i_k} = S_{i_1} \circ \cdots \circ S_{i_k}(X)\,.$$

We define

$$\mu(X_{\mathbf{i}}) = r_{i_1}^s r_{i_2}^s \cdots r_{i_k}^s$$

and extend this to a measure on the Borel sets in the usual way. Then

$$|X_{\mathbf{i}}| = r_{i_1}\, r_{i_2} \cdots r_{i_k}\, |X|$$

so

$$\mu(X_{\mathbf{i}}) = |X|^{-s} |X_{\mathbf{i}}|^s.$$

If $x \in E$ and $r < |X|$, by suitable choice of k we may find $X_{i_1,\ldots,i_k} \ni x$ with $X_{i_1,\ldots,i_k} \subseteq B_r(x)$ and $r \leq |X_{i_1,\ldots i_k}|$, so $\mu(B_r(x)) \geq \mu(X_{i_1}, \cdots_{i_k}) \geq |X|^{-s} r^s$. Similarily, we may find $X_{i_1,\ldots,i_k} \ni x$ with $B_r(x) \cap E \subseteq X_{i_1,\ldots,i_k}$ and $|X_{i_1,\ldots,i_k}| \leq ar$ where a is independent of x (using the separation condition). Thus $\mu(B_r(x)) \leq \mu(X_{i_1,\ldots,i_k}) \leq |X|^{-s} a^s\, r^s$. Combining these estimates

$$c_1 \leq \mu(B_r(x))/r^s \leq c_2$$

for small r, where $0 < c_1 \leq c_2$. Letting $r \to 0$ and applying (i) - (iv) of 1.1.2 it follows that $\dim_H E = \dim_p E = s$. The same idea using a slightly more involved upper estimate for $\mu(B_r(x))$ deals with the open set condition case.

13.1.5 Self-affine Sets

It is natural to try to extend these dimension calculations for self-similar sets to self-affine sets. A set E is termed *self-affine* if it is the attractor of an iterated function system $\{s_1, \cdots, s_m\}$ of affine transformations, that is with

$$s_i(x) = T_i(x) + a_i$$

where T_i is a linear contraction and $a_i \in \mathcal{R}^n$ effects a translation. Self-affine sets are much harder to analyse than self-similar sets: the problem is that in an iterated construction the sets $X_{\mathbf{i}} = X_{i_1,\dots,i_k}$ can become 'very long and thin' for large k. This leads to two problems:

(i) In calculating Hausdorff measures, it is natural to try and use the sets $X_{\mathbf{i}}$ as covers of E. However, in minimizing $\sum |U_i|^s$, if $X_{\mathbf{i}}$ is long and thin it might be better to cut $X_{\mathbf{i}}$ into a large number of roughly cube like pieces. For example, in two dimensions, if $X_{\mathbf{i}}$ is roughly a rectangle with sides $\alpha_1 \geq \alpha_2$ we have a choice between covering $X_{\mathbf{i}}$ with a single set, giving a contribution to $\sum |U_i|^s$ of about α_1^s, or dividing $X_{\mathbf{i}}$ into about α_1/α_2 pieces of diameter roughly α_2^s, giving a contribution of roughly $\alpha_1\, \alpha_2^{-1}\, \alpha_2^s = \alpha_1\, \alpha_2^{1-s}$. If $0 \leq s \leq 1$ the former option is best, if $1 < s \leq 2$ the latter becomes the more effective cover when it comes to calculating dimensions.

(ii) It may be that two of the component sets $X_{\mathbf{i}}$ at $X_{\mathbf{i}'}$ lie alongside each other, in which case it may be more efficient to use a covering by set U_i that cover parts of both $X_{\mathbf{i}}$ and $X_{\mathbf{i}'}$ rather than covering the sets individually. In a sense this happens in exceptional cases; unfortunately it leads to the phenomena that $\dim_H E$ need not vary continuously as the s_i are varied. For an example of this consider the system

$$S_1(x,y) = (\frac{1}{2}x + \lambda, \frac{1}{3}y); \quad S_2(x,y) = (\frac{1}{2}x, \frac{1}{3}y + \frac{2}{3})$$

where $0 \leq \lambda < \frac{1}{2}$ is a parameter. By considering projections onto the x-axis, it is easy to see that if $\lambda > 0$ then the attractor has Hausdorff dimension 1. However, if $\lambda = 0$ the attractor is the middle-third Cantor set (in the y-axis) of dimension $\log 2/\log 3 < 1$. Thus the dimension is discontinuous at $\lambda = 0$.

Nevertheless, we can obtain some formula for dimensions of self-affine sets. For a linear contraction T on $\mathcal{R}^n$ we define the *singular values* $1 > \alpha_1 \geq \alpha_2 \geq \cdots \geq \alpha_m > 0$ of T to be the positive square roots of the eigenvalues of T^*T where T^* is the adjoint of T, or equivalently the lengths of the semi-axes of the ellipsoid $T(B)$ where B is the unit ball in $\mathcal{R}^n$. For $0 \leq s \leq n$ we define the *singular value function* φ^s by

$$\varphi^s(T) = \alpha_1\, \alpha_2 \cdots \alpha_{q-1}\, \alpha_q^{s-q+1}$$

where q is the integer such that $q-1 < s \leq q$. If we assume (without loss of generality) that X is contained in a ball of unit radius, then $X_{\mathbf{i}} = X_{i_1,\dots,i_k} = S_{i_1} \circ \cdots \circ S_{i_k}(X)$ is contained in an ellipsoid with semi-axes given by the singular values of $T_{i_1} \circ \cdots \circ T_{i_k} = T_{\mathbf{i}}$, so that by dividing $X_{\mathbf{i}}$ into pieces depending on the values of s, we get that $X_{\mathbf{i}}$ may be covered by at most $c\frac{\alpha_1}{\alpha_q} \cdots \frac{\alpha_{q-1}}{\alpha_q}$ pieces of diameter α_q. Writing U_i for the sets of this covering of $X_{\mathbf{i}}$ we have $\sum_i |U_i|^s \leq c\, \alpha_1 \cdots \alpha_{q-1}\, \alpha_q^{s-q+1} = c\, \varphi^s(T_{\mathbf{i}})$. Thus, given

$\delta > 0$ if we choose k large enough so that $|X_{\mathbf{i}}| \leq \delta$ for all $|\mathbf{i}| = \mathbf{k}$ when $|\mathbf{i}| = \mathbf{k}$ is the number of terms in the sequence $\mathbf{i} = \mathbf{i_1}, \ldots, \mathbf{i_k}$, we have

$$H^s_\delta(E) \leq c \sum_{|\mathbf{i}|=\mathbf{k}} \varphi^s(T_{\mathbf{i}}).$$

It may be shown (using the submultiplicativity of these singular value sums) that $\Phi(s) \equiv \lim_{k\to\infty}(\sum_{|\mathbf{i}|=\mathbf{k}} \varphi^s(T_{\mathbf{i}}))^{1/k}$ exists for each $s \geq 0$, and it follows that $\dim_H E \leq s$ where $\Phi(s) = 1$.

In fact

$$\dim_p E = \dim_H E = s \text{ where } \Phi(s) = 1$$

'very often.' In particular, if $S_i(x) = T_i(x) + a_i$ with T_i fixed, then (13.1.5) holds for almost all $(a_1, \cdots, a_m) \in \mathcal{R}^{mn}$ (in the sense of mn-dimensional Lebesgue measure); see [73] for full details.

Moreover, [111] have shown that (13.1.5) holds for $S_i(x) = T_i(x) + a_i$ for *all* a_i for certain T_i with positive matrix entries, provided certain distortion and separation conditions are satisfied. This is the case, for example, if

$$T_1 = \frac{1}{120}\begin{pmatrix} 4 & 1 \\ 4 & 2 \end{pmatrix}, \; T_2 = \frac{1}{120}\begin{pmatrix} 4 & 3 \\ 4 & 4 \end{pmatrix}, \; T_3 = \frac{1}{120}\begin{pmatrix} 3 & 4 \\ 2 & 4 \end{pmatrix}.$$

Suppose now that $T_1 = T_2 = \cdots = T_m = T$ and the eigenvectors of T are mutually perpendicular. Then $\varphi^s(T_{i_1} \circ \cdots \circ T_{i_k}) = \varphi^s(T^k) = (\varphi^s(T))^k$, so for almost all $(a_1, \ldots, a_k)$ we have $\dim_H E = s$ where $m\varphi^s(T) = 1$.

This should be compared with the special and exceptional case analyzed in [184] and [26] where the unit square is divided into a p by q array of rectangles with sides p^{-1} and q^{-1} (where $p < q$). A subcollection of these rectangles is selected, with N_j rectangles selected from the j-th column. For the selected rectangles we define S_i to be the affine mapping that maps the unit square onto that rectangle (with no change in orientation). If E is the attractor of this iterated function scheme,

$$\dim_H E = \log(\sum_{j=1}^{p} N_j^{\log p/\log q})/\log p.$$

Because of the way that the rectangles 'line up' here, this dimension is in general smaller than the value given by (13.1.5).

13.1.6 Related Constructions

We mention briefly some related constructions of sets whose dimensions can be estimated in a similar manner.

(i) Graph directed constructions

Let $S_{i,j}$ $(1 \leq i, j \leq m)$ be contracting similarities on $\mathcal{R}^n$ of ratios $r_{i,j}$. A generalization of the method of Section 13.1.3 shows that there exists a unique family of non-empty compact sets $E_1, \ldots, E_m$ such that

$$E_i = \bigcup_{j=1}^{m} S_{i,j}(E_j)$$

for $i = 1, 2, \ldots, m$. Such sets were studied in [26] and [183] where it was shown that $\dim_H E_i = \dim_p E_i = s$ for all i, where s is the non-negative number such that

$$1 = \text{largest eigenvalue of} \begin{pmatrix} r_{1,1}^s & \cdots & r_{1,m}^s \\ \vdots & & \vdots \\ r_{m,1}^s & \cdots & r_{m,m}^s \end{pmatrix}.$$

(ii) Statistically self-similar sets

Let $S_i : \mathcal{R}^n \to \mathcal{R}^n$ be *random* similarities with random ratios R_i for $i = 1, 2, \ldots, m$. We assume that there is a compact set $X \subseteq \mathcal{R}^n$ such that with probability one $S_i(X) \subseteq X$ for all i, that for all $i \neq j$ the random sets $S_i(X)$ and $S_j(X)$ are separated by distance at least $\varepsilon > 0$, and that $a \leq R_i \leq b$ where $0 < a \leq b < 1$. For all sequences $i_1, \ldots, i_k$ $(1 \leq i_j \leq m)$ we let $T_{i_1,\ldots,i_k}$ be independent random similiarities with the same distribution as S_{i_1}. We define similarities $S_{i_1,\ldots,i_k} = T_{i_1} \circ T_{i_1,i_2} \circ \cdots \circ T_{i_1,i_2,\ldots,i_k}$ and define a decreasing sequence of random sets $E_k = \bigcup_{i_1,\cdots,i_k} S_{i_1,\cdots,i_k}(x)$. Then the random set $E = \bigcap_{k=1}^{\infty} E_k$ is called a ***statistically self-similiar set*** which is generally a fractal. It may be shown that with probability 1 we have that $\dim_H E = \dim_p E = s$ where s is the non-negative number satisfying the expectation equation $1 = \mathrm{E}(\sum_{i=1}^m R_i^s)$, where E denotes expectation. For further details, see [72, 99, 100].

13.2 Fractal Measures

13.2.1 Local Dimensions and Multifractal Spectra

A measure of μ on $\mathcal{R}^n$ (with $0 < \mu(\mathcal{R}^n) < \infty$) can give rise to a hierarchy of fractal sets. For $\alpha \geq 0$ we define sets

$$F_\alpha = \left\{ x : \lim_{r \to 0} \log(B_r(x)) / \log r = \alpha \right\}.$$

We think of $\lim_{r\to 0} \log \mu(B_r(x))/\log r$ as the *local dimension of* μ at x; thus F_α is the set of points at which μ has local dimensions α. Under certain circumstances, F_α may be a fractal for a range of α. (Of course, for many measures $\lim_{r\to 0} \log(B_r(x))/\log r$ may not in general exist, in which case we need to work with upper and lower limits. However, for many natural examples the limit exists for many x leading to an adequate theory.)

It is natural to ask 'how big' the sets F_α are for various α. There are two approaches to this. We may consider $\mu(F_\alpha)$ as α varies: this approach was adopted in [218] and more recently by [52] and leads to a 'decomposition' of the measure μ into components that are α-dimensional for a range of α.

The other approach is to find the (Hausdorff or packing) dimension of F_α: although for many α we may have $\mu(F_\alpha) = 0$, the dimension $\dim_H F_\alpha$ may nevertheless be significant. The function $f(\alpha) = \dim_H F_\alpha$ is termed the *multifractal spectrum* of μ, and it is this we concentrate on here.

13.2.2 Invariant Measures

One way of defining measures with multifractal properties is using a variant of iterated function systems. Let $S_1, \ldots, S_m$ be contractions on a region $D \subseteq \mathcal{R}^n$ as before, and let $p_1, \ldots, p_m$ be 'probabilities', so that $0 < p_i < 1$ and $\sum_{i=1}^m p_i = 1$. Then there exists a unique Borel measure on D satisfying

$$\mu(E) = \sum_{i=1}^{m} p_i \mu(s_i^{-1}(E))$$

for all Borel sets E; we term μ an *invariant measure* for this system. This implies that $\operatorname{spt}\mu = \bigcup_{i=1}^{m} S_i(\operatorname{spt}\mu)$ where $\operatorname{spt}\mu$ denotes the support of μ (by definition $x \in \operatorname{spt}\mu$ if and only if $\mu(B_r(x)) > 0$ for every $r > 0$). Since $\operatorname{spt}\mu$ is compact and non-empty, $\operatorname{spt}\mu$ is the invariant set of the iterated function scheme $\{S_1, \ldots, S_m\}$.

The existence of a unique invariant measure follows from a contraction mapping argument analogous to that used to demonstrate the existence of invariant sets for an IFS. The measure μ may be constructed directly. As before, we choose X to be a compact non-empty set with $S_i(x) \subseteq X$ for all i, and write $X_{\mathbf{i}} = X_{i_1,\ldots,i_k} = S_{i_1} \circ \cdots \circ S_{i_k}(X)$, and we assume that the $S_i(X)$ are disjoint. Then $\mu(X_{\mathbf{i}}) = p_{i_1} p_{i_2} \cdots p_{i_k}$ and this extends to a Borel measure in the usual way.

13.2.3 The Multifractal Spectrum of Self-similar Measures

If $S_1, \ldots, S_m$ are now similarities on $\mathcal{R}^n$ of ratios $r_1, \ldots, r_m$ then the measure μ defined by (13.2.2) is called a *self-similar measure*; from Section 13.1.4 $\operatorname{spt}\mu$ is a self-similar set with dimension given by $\sum_{i=1}^m r_i^s$. We assume a strong separation condition, that there is a non-empty compact set X such that $S_i(X) \subseteq X$ for all i with $S_i(X) \cap S_j(X) = \emptyset$ for $i \neq j$, and for convenience we assume $|X| = 1$. In this section we calculate the multifractal spectrum of μ; that is we find $\dim_H F_\alpha$ where

$$F_\alpha = \{x \in \mathcal{R}^n : \log \mu(B_r(x)) / \log r \to \alpha\}.$$

The idea is to define a suitable measure on F_α so that the criterion of 1.1.2 gives the dimension of F_α. This method is the prototype for many further examples.

Let

$$\Phi(q, \beta) = \sum_{i=1}^{m} p_i^q r_i^\beta$$

for real numbers q and β, and define $\beta = \beta(q)$ by

$$\Phi(q, \beta(q)) = 1.$$

It is easy to see that $\beta : \mathcal{R} \to \mathcal{R}$ is analytic, decreasing and convex with $\lim_{q \to -\infty} \beta(q) = \infty$ and $\lim_{q \to \infty} \beta(q) = -\infty$. Define

$$\alpha = -\frac{d\beta}{dq};$$

since $1 = \sum_{i=1}^m p_i^q r_i^{\beta(q)}$ we have

$$\alpha = \frac{\sum_{i=1}^m p_i^q r_i^\beta \log p_i}{\sum_{i=1}^m p_i^q r_i^\beta \log r_i}.$$

Define

$$f = q\alpha + \beta = -q\frac{d\beta}{dq} + \beta.$$

Then f is the Legendre transform of $\beta(q)$, that is to say the tangent to the curve $\beta = \beta(q)$ of slope α cuts the β axis at $f(\alpha)$. Assuming (as we do from now on) that we do not have the special situation where $p_i = cr_i^\gamma$ for all i for some c and γ, then $\alpha \mapsto f(\alpha)$ is defined and is convex over a proper interval $[\alpha_{\min}, \alpha_{\max}]$.

We note that if $q = 0$ then $f = \beta = \dim_H(\operatorname{spt}\mu)$, and also $\frac{df}{dq} = -q\frac{d^2\beta}{dq^2} = 0$, so $\dim_H(\operatorname{spt}\mu)$ equals the maximum of $f(\alpha)$.

We shall show that

$$\dim_H F_\alpha = \dim_p F_\alpha = f(\alpha)$$

for $\alpha \in [\alpha_{\min}, \alpha_{\max}]$ (with $F_\alpha = \emptyset$ if $\alpha \notin [\alpha_{\min}, \alpha_{\max}]$).

We require the following property of Φ:

Given $\varepsilon > 0$ we have

$$\Phi(q+\delta, \beta(q) + (-\alpha+\varepsilon)\delta) < 1$$

and

$$\Phi(q-\delta, \beta(q) + (\alpha+\varepsilon)\delta) < 1$$

for all sufficiently small $\delta > 0$.

To see this, recall that $\frac{d\beta}{dq} = -\alpha$, so $\beta(q+\delta) = \beta(q) - \alpha\delta + 0(\delta^2) < \beta(q) + (-\alpha+\varepsilon)\delta$ if δ is small enough. Since $\Phi(q+\delta, \beta(q+\delta)) = 1$ and Φ is strictly decreasing in the second argument, the first inequality follows. The second inequality is similar.

The crux of the calculation of $\dim_H F_\alpha$ is to concentrate a measure on F_α and to apply the criteria of 1.1.2. For a given q we set $\beta = \beta(q)$ and define ν on $\operatorname{spt}\mu$ by

$$\nu(X_{i_1,\dots,i_k}) = (p_{i_1}p_{i_2}\cdots p_{i_k})^q (r_{i_1}r_{i_2}\cdots r_{i_k})^\beta$$

or

$$\nu(X_{\mathbf{i}}) = p_{\mathbf{i}}^q r_{\mathbf{i}}^\beta$$

where $p_{\mathbf{i}} = p_{i_1}p_{i_2}\cdots p_{i_k}$ and $r_{\mathbf{i}} = r_{i_1}r_{i_2}\cdots r_{i_k}$ for $\mathbf{i} = i_1, \dots, i_k$. As usual, ν may be extended to all Borel sets to give a Borel measure on $\operatorname{spt}\mu$. Recall that we also have

$$|X_{\mathbf{i}}| = r_{\mathbf{i}}$$

and

$$\mu(X_{\mathbf{i}}) = p_{\mathbf{i}}.$$

With α and f defined in terms of q above we claim that:
(i) $\nu(F_\alpha) = 1$
(ii) For all $x \in F_\alpha$ we have $\log \nu(B_r(x))/\log r \to f$ as $r \to 0$.

To prove (i): For all $x \in \operatorname{spt}\mu$ and $k \in Z^+$ we write $X_k(x) = X_{i_1,\dots,i_k}$ where $x \in X_{i_1,\dots,i_k}$. Then given $\varepsilon > 0$ we have for all $\delta > 0$

$$\begin{aligned}
\nu\left\{x : \mu(X_k(x)) \le |X_k(x)|^{\alpha+\varepsilon}\right\} & \\
&= \nu\left\{x : 1 \le \mu(X_k(x))^{-\delta}|X_k(x)|^{(\alpha+\varepsilon)\delta}\right\} \\
&\le \sum_{|\mathbf{i}|=k} \nu(X_{\mathbf{i}})\mu(X_{\mathbf{i}})^{-\delta}|X_{\mathbf{i}}|^{(\alpha+\varepsilon)\delta} \\
&\le \sum_{|\mathbf{i}|=k} p_{\mathbf{i}}^{q-\delta} r_{\mathbf{i}}^{\beta+(\alpha+\varepsilon)\delta} \\
&= \left(\sum_{i=1}^{m} p_i^{q-\delta} r_i^{\beta+(\alpha+\varepsilon)\delta}\right)^k \\
&= \gamma^k
\end{aligned}$$

where $\gamma < 1$ if δ is small enough, by the property of Φ . Thus for all $\varepsilon > 0$, we have

$$|X_k(x)|^{\alpha+\varepsilon} < \mu(X_k(x)) < |X_k(x)|^{\alpha-\varepsilon}$$

ν-almost all x using the above and a symmetrical argument. Taking a sequence of $\varepsilon \searrow 0$, it follows that for ν-almost all x we have $\log \mu(X_k(x))/\log |X_k(x)| \to \alpha$.

Just as when we were looking at the dimension of self-similar sets, we may compare $\mu(X_k(x))$ with $\mu(B_r(x))$ where the radius r is comparable with $|X_k(x)|$ to get that $\log \mu(B_r(x))/\log r \to \alpha$ for ν-almost all x, thus $\nu(F_\alpha) = 1$.

To prove (ii) we note that if $x \in F_\alpha$, and $x \in X_k(x)$ then

$$\begin{aligned}
\frac{\log \nu(X_k(x))}{\log |X_k(x)|} &= \frac{q \log \mu(X_k(x))}{\log |X_k(x)|} + \frac{\beta \log |X_k(x)|}{\log |X_k(x)|} \\
&\to q\alpha + \beta = f
\end{aligned}$$

since $x \in F_\alpha$, again using that $\mu(X_k(x))$ is comparable to $\mu(B_r(x))$ when $|X_k(x)|$ is comparable to r. Thus for $x \in F_\alpha$ we have $\log \nu(B_r(x))/\log r \to f$ as $r \to 0$.

Applying the criteria 1.1.2 to (i) and (ii) here it is immediate that $\dim_H F_\alpha = \dim_p F_\alpha = f(\alpha)$, with $f(\alpha)$ as above, if $\alpha \in (\alpha_{\min}, \alpha_{\max})$. We remark that $F_\alpha = 0$ if $\alpha \notin [\alpha_{\min}, \alpha_{\max}]$, since $\alpha_{\min} \le \log p_{\mathbf{i}}/\log r_{\mathbf{i}} \le \alpha_{\max}$ for all $\mathbf{i}$.

This dimension calculation is essentially due to [34], see [46] for a contemporary account.

13.2.4 Other Examples

Methods akin to the above have been used to find the multifractal spectra of a variety of other measures. Some of these calculations become quite technical, and here we merely give some references.

(i) Self-similar measures with the open set condition: see [8].

(ii) Self-affine measures: see [227] for 'almost sure' results, and [140] and [204] for the case with rectangular alignment.

(iii) Graph-directed self-similar measures: see [69].

(iv) Random self-similar measures: see [170, 169, 8, 75, 76, 205].

(v) Self-conformal measures: see, e.g. [213, 160].

(vi) Vector-valued self-similar measures: see [77].

(vii) Self-similar measures for infinite IFS : see [217].

13.2.5 Other Approaches to Multifractal Theory

(i) *Coarse theory.* For many purposes, one wishes to analyse the local intensity of measures over a finite range of scales rather than using sets defined in terms of limiting behavior. A 'coarse theory' which in many ways parallels the 'fine theory' described above has been developed for this situation.

Let μ be a finite measure on $\mathcal{R}^n$. We term the lattice of cubes (or squares, etc.) of side r the set of *r-mesh cubes.* We define

$$N_r(\alpha) = \#\{r\text{-mesh cubes } B \text{ with } \mu(B) \geq r^\alpha\}$$

and

$$S_r(q) = \sum_{r\text{-mesh cubes } B_i} \mu(B_i)^q$$

(some care is required with the domain of the sum for $q < 0$.) Assume that

$$N_r(\alpha + \delta\alpha) - N_r(\alpha) \sim r^{-f(\alpha)}\delta\alpha$$

for small r so that $f(\alpha)$ is a sort of 'dimension' of $\{x : \mu(B_r(x)) \sim r^\alpha\}$.

Then for given q, very roughly:

$$S_r(q) \sim \int_{\alpha=0}^{\infty} r^{q\alpha - f(\alpha)}\, d\alpha \sim r^{-\beta(\alpha)}$$

where $\beta(q) = \sup_{\alpha\geq 0}(f(\alpha) - q\alpha)$. At the supremum, $\frac{df}{d\alpha} - q = 0$, so $q = \frac{df}{d\alpha}(\alpha(q))$, giving

$$\beta(q) = f(\alpha(q)) - q\alpha(q)$$

and

$$\frac{d\beta}{dq} = \frac{df}{d\alpha}\frac{d\alpha}{dq} - \alpha - q\frac{d\alpha}{dq} = -\alpha.$$

Thus, as in the 'fine theory' $\beta(q)$ and $f(\alpha)$ form a Legendre transform pair.

This can be justified in many instances, see [71, 74, 216] for various accounts of this. Moreover, for many 'regular' measures μ, such as self-similar measures, the $\beta(q)$ and $f(\alpha)$ of the coarse theory and fine theory coincide.

(ii) *Generalized Hausdorff measures*

We may generalize Hausdorff measures (and also packing measures) in the following way. For real numbers q and β set

$$H^{q,\beta}_{\mu,\delta}(E) = \inf\left\{\sum_i \mu(B_{r_i}(x_i))^q r_i^\beta : E \subseteq \bigcup_i B_{r_i}(x_i) \text{ with } x_i \in E, r_i \leq d\right\}$$

$$H^{q,\beta}_{\mu,0}(E) = \lim_{\delta\to 0} H^{q,\beta}_{\mu,\delta}(E)$$

$$H^{q,\beta}_{\mu}(E) = \sup_{F\subseteq E} H^{q,\beta}_{\mu,0}(F)$$

(the last step is a tedious technical device to ensure motonicity). Then $H^{q,\beta}_{\mu}$ is a Borel measure. For each q we define $\beta(q)$ to be the number such that $H^{q,\beta}_{\mu}(E) = \infty$ if $\beta < \beta(q)$ and $H^{q,\beta}_{\mu}(E) = 0$ if $\beta > \beta(q)$.

Given $\beta(q)$ we can use Legendre transformation to define a function $f(\alpha)$, and it turns out that, for a large class of measures, $f(\alpha)$ equals the dimension of the set of x at which $\log\mu(B_r(x))/\log r \to \alpha$. These generalized Hausdorff measures introduced in [205, 207] provide a natural and very powerful tool for studying multifractal properties of very general measures. Moreover, using these measures, it is possible to obtain geometrical results, for example relating the multifractal spectrum of a measure to its sections by hyperplanes.

Chapter 14

Velocity, Length, Dimension

Claude Tricot

14.1 Introduction

The length of a curve may be calculated as the integral of the *instantaneous velocity*. This can be done only for a special class of regular curves. For other curves, even with finite length, such a relationship between length and velocity is not possible (see the devil's staircase, Section 14.2). And what can we say about curves of infinite length?

The fractal approach consists of defining a length *at the precision* ϵ, for all $\epsilon > 0$. In some sense, the curve is approximated, more and more closely, by a sequence of finite-length curves. But there are several ways to create such sequences. Some are purely geometrical, others rely upon analysis. In the latter case, the curve must be parametrized, and an ϵ-length is defined with help of sums or integrals; a notion of ϵ-velocity may be introduced at that point. Three methods are investigated in this paper. Amongst several consequences, this study may lead to a better understanding of the fractal dimension, new algorithms to evaluate it, and, more importantly, to the characterization of new classes of curves which generalize the classical self-similar or self-affine curves.

Section 2 considers finite-length curves, with the two approaches, geometric and analytic. Section 3 introduces a geometric notion of ϵ-length, for a curve of infinite length. Section 4 contains technical lemmas about coverings by intervals, which are useful in the following sections. Being given a parametrization of the curve, and a covering of the definition interval, a general notion of length is given and sudied in Sections 5 and 6. Two applications are given, one with arcs of equal diameter (compass method, Section 7), and one with arcs of equal breadth (Section 8). Section 9 shows the relationships between the different ϵ-lengths. Section 10 deals with the fractal dimension, and 11 presents some examples and typical classes of curves.

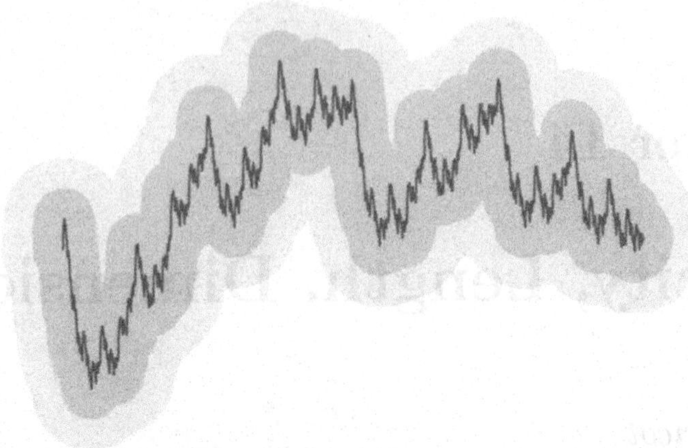

Figure 14.1: Minkowski sausages around a curve.

14.2 Finite Length

A *curve* is the image of a continuous application $\gamma : [a,b] \longrightarrow \Re^2$. All curves in this paper are supposed to be bounded. We will use the notation

$$\Gamma = \gamma([a,b]) .$$

The notion of *length of Borchardt-Minkowski* (a terminology used by Lebesgue [147]) goes as follows:

Definition 14.1 *Let $\mathcal{A}(E)$ denote the area (two-dimensionnal Lebesgue measure in the plane) of a set E, and Γ be a bounded curve. For all $\epsilon > 0$, let $\Gamma(\epsilon)$ denote the set of all points at distance $\leq \epsilon$ from Γ. Then the length of Γ is*

$$L(\Gamma) = \lim_{\epsilon \to 0} \frac{\mathcal{A}(\Gamma(\epsilon))}{2\,\epsilon} . \tag{14.1}$$

The set $\Gamma(\epsilon)$ is known as the ϵ-*Minkowski sausage* of Γ. This is a geometrical approach, independant of the parametrization. One can check that this length is the same as the length $l(\Gamma)$ obtained with sequences of inscribed polygonal curves. For a proof of this old result, see [243], and [261] for a short proof of the inequality

$$\mathcal{A}(\Gamma(\epsilon)) \leq 2\,\epsilon\, l(\Gamma) + \pi\,\epsilon^2 .$$

Anyway, this definition is useless for a precise evaluation of the length, since $\mathcal{A}(\Gamma(\epsilon))$ is hard to evaluate, and the convergence in Equation 1 is slow. For an analytical approach. let us use the parametrization of Γ. Let $\operatorname{dist}(x,y)$ denote the distance between two points x and y of $\Re^2$, and

$$\mathrm{S}(t_1,t_2) = \operatorname{dist}(\gamma(t_1),\gamma(t_2)) \text{ for all } a \leq t_1 \leq t_2 \leq b .$$

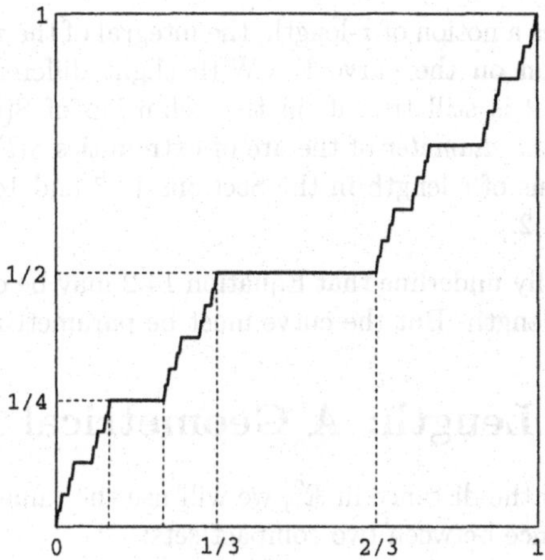

Figure 14.2: The devil's staircase is a curve of length 2.

The function $\mathrm{S}(t_1, t_2)$ is continuous. The *instantaneous velocity* at t may be defined as

$$v(t) = \lim_{\tau \to 0} \frac{\mathrm{S}(t, t+\tau)}{\tau}.$$

If the function v is defined, and continuous, over $[a, b[$, then the classical equality

$$L(\Gamma) = \int_a^b v(t)\, dt$$

holds. But this is not the case for all curves Γ.

Example 1 The devil's staircase is represented in Fig. 14.2. It is the graph of an increasing, continuous function, which is constant over all complementary intervals of the triadic Cantor set F. The velocity $v(t)$ is equal to 1, except on F where it is not defined. Consequently,

$$\int_{[0,1]-F} v(t)\, dt = 1.$$

The length of Γ is 2, and is not related to an integral of the velocity.

For any parametrized curve Γ, it is still possible to calculate $L(\Gamma)$ with the help of integrals, by using the following result:

Theorem 1 *Let*

$$L_\tau(\Gamma) = \frac{1}{\tau} \int_a^{b-\tau} \mathrm{S}(t, t+\tau)\, dt.$$

Then

$$L(\Gamma) = \lim_{\tau \to 0} L_\tau(\Gamma). \tag{14.2}$$

This theorem introduces a notion of τ-length, the integral of the τ-velocity $S(t, t+\tau)/\tau$, without any assumption on the curve Γ. With slight differences, it can be found in [243]. Equation 14.2 is still true if, in the definition of $S(t, t+\tau)$, one replaces $\text{dist}(\gamma(t), \gamma(t+\tau))$ by the *diameter* of the arc of extremities $\gamma(t)$,$\gamma(t+\tau)$ in Γ. Under this form, the definitions of ϵ-length in the Sections 14.7 and 14.8 may be considered as generalizations of 14.2.

Remark 1 Let us finally underline that Equation 14.2 may be quite useful for a practical evaluation of the length. But the curve must be parametrized.

14.3 Infinite Length: A Geometrical Approach

Since $\text{dist}(x, y)$ denotes the distance in $\Re^2$, we will use the same notation $\text{dist}(E_1, E_2)$ for the Hausdorff distance between two compact sets:

$$\text{dist}(E_1, E_2) = \inf\{\epsilon > 0 \text{ s.t. } E_1 \subset E_2(\epsilon) \text{ and } E_2 \subset E_1(\epsilon)\}.$$

For every curve Γ, we define an ϵ-length as follows:

$$L_\epsilon(\Gamma; \text{Minkowski}) = \inf\{L(C) \text{ s.t. } C \text{ is a curve, and } \text{dist}(\Gamma, C) \leq \epsilon\}. \tag{14.3}$$

The value $L_\epsilon(\Gamma; \text{Minkowski})$ is finite. It is defined for all curves, whether they have finite or infinite length. It verifies the following properties:

- $L_\epsilon(\Gamma; \text{Minkowski})$ is a decreasing, continuous function of ϵ;
- If $L(\Gamma) < \infty$, then $L(\Gamma) = \lim_{\epsilon \to 0} L_\epsilon(\Gamma; \text{Minkowski})$.

Notation 1 Let $f(\epsilon)$ and $g(\epsilon)$ be two functions defined in a neighbourhood of 0. Let us write

$$\begin{aligned} f &\simeq g \text{ if } 0 < \liminf_{\epsilon \to 0} \frac{f(\epsilon)}{g(\epsilon)} \leq \limsup_{\epsilon \to 0} \frac{f(\epsilon)}{g(\epsilon)} < \infty; \\ f &\preceq g \text{ if } \limsup_{\epsilon \to 0} \frac{f(\epsilon)}{g(\epsilon)} < \infty; \\ f &\succeq g \text{ if } 0 < \liminf_{\epsilon \to 0} \frac{f(\epsilon)}{g(\epsilon)}. \end{aligned}$$

Theorem 2 *For any bounded curve* Γ*:*

$$\mathcal{A}(\Gamma(\epsilon)) \simeq \epsilon\, L_\epsilon(\Gamma; \text{Minkowski}). \tag{14.4}$$

Recall that the Minkowski-Bouligand dimension (shortly called *fractal dimension*) of Γ is

$$\Delta(\Gamma) = \limsup_{\epsilon \to 0}\left(2 - \frac{\log \mathcal{A}(\Gamma(\epsilon))}{\log \epsilon}\right). \tag{14.5}$$

Equation 14.4 gives:

$$\Delta(\Gamma) = \limsup_{\epsilon \to 0}\left(1 + \frac{\log L_\epsilon(\Gamma; \text{Minkowski})}{|\log \epsilon|}\right). \tag{14.6}$$

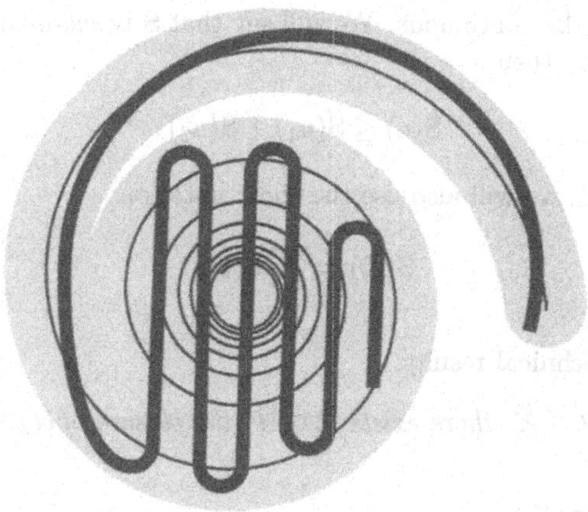

Figure 14.3: The curve C (in bold) can be considered as an approximation to the spiral–but only in the sense of the Hausdorff distance.

Remark 2 The length $L_\epsilon(\Gamma; \text{Minkowski})$ is related to the Minkowski sausage of Γ. Curves C such that $\text{dist}(C, \Gamma) \leq \epsilon$ may be considered as approximation curves. From a visual point of view, they are not good approximation in general. Fig. 14.3 shows a spiral Γ, and a curve C of finite length, whose length is close to $L_\epsilon(\Gamma; \text{Minkowski})$. This curve does not look like a spiral. In order to get better approximations, we need to construct finite length curves which "follow" closely the curve Γ. This may be done by using a parametrization of Γ.

14.4 The Metric Set of Intervals: Preliminary Results

Let $\mathcal{X}$ be the set of all closed intervals in $[a, b]$ (including the points). Then $\mathcal{X}$ is a metric space, with respect to the Hausdorff distance. The limit of a converging sequence of intervals is an interval, so that $\mathcal{X}$ is complete.

The length of an interval u is $|u|$. The interior of u is $\mathring{u}$. For any subfamily $\mathcal{R} \subset \mathcal{X}$ of intervals, $\cup\mathcal{R}$ denotes the union of all intervals in $\mathcal{R}$. The *index* of $\mathcal{R}$ is the integer

$$\omega(\mathcal{R}) = \max\{n \text{ s.t. there exists } u_1, \ldots, u_n \text{ in } \mathcal{R} \text{ with } \cap_{i=1}^{n} \mathring{u}_i \neq \emptyset\}.$$

Thus, $\omega(\mathcal{R}) = 1$ if, and only if, any two intervals of $\mathcal{R}$ have at most one extremity in common.

The minimum length in $\mathcal{R}$ is

$$\rho_{\min}(\mathcal{R}) = \inf\{|v| \text{ s.t. } v \in \mathcal{R}\}.$$

And the maximum length:

$$\rho_{\max}(\mathcal{R}) = \sup\{|v| \text{ s.t. } v \in \mathcal{R}\}.$$

Let $\mathbf{S} : \mathcal{X} \longrightarrow \Re^+$ be continuous. We will say that $\mathbf{S}$ is *sub-additive* if, for all $u \in \mathcal{X}$ such that $u = u_1 \cup u_2$, then

$$\mathbf{S}(u) \leq \mathbf{S}(u_1) + \mathbf{S}(u_2)\,.$$

For any $\mathcal{R} = \{u_i\}$, we will also use the short notation

$$\mathbf{S}(\mathcal{R}) = \sum_i \mathbf{S}(u_i)\,.$$

Here are a few technical results.

Lemma 1 *For all $\mathcal{U} \subset \mathcal{X}$, there exists $\mathcal{R} \subset \overline{\mathcal{U}}$ (the closure of $\mathcal{U}$) such that $\cup\mathcal{U} \subset \cup\mathcal{R}$, and $\omega(\mathcal{R}) \leq 2$.*

For a reference, see [241].

Lemma 2 *If $\omega(\mathcal{R}) < \infty$, there exists an integer $n \leq \omega(\mathcal{R})$, and n subfamilies $\mathcal{R}_1, \ldots, \mathcal{R}_n$, such that $\mathcal{R} = \cup_{i=1}^n \mathcal{R}_i$ and $\omega(\mathcal{R}_i) = 1$ for all i.*

Lemma 3 *For all $\mathcal{R}_1$, $\mathcal{R}_2$ included in $\mathcal{X}$:*

$$\omega(\mathcal{R}_1 \cup \mathcal{R}_2) \leq \omega(\mathcal{R}_1) + \omega(\mathcal{R}_2)\,.$$

Lemma 4 *Let $\mathcal{R} \subset \mathcal{X}$, and an interval $u \subset \cup\mathcal{R}$ (which does not necessarily belong to $\mathcal{R}$). Then, $\mathring{u}$ is covered by at most $(2 + (|u|/\rho_{\min}))\omega(\mathcal{R})$ intervals of $\mathcal{R}$.*

Lemma 5 *Let $\mathcal{R}_1$ and $\mathcal{R}_2$ be two subsets of $\mathcal{X}$, such that $\cup\mathcal{R}_1 = \cup\mathcal{R}_2$. If $\mathbf{S}$ is sub-additive,*

$$\frac{1}{\omega(\mathcal{R}_2)\left(2 + \frac{\rho_{\max}(\mathcal{R}_1)}{\rho_{\min}(\mathcal{R}_2)}\right)}\mathbf{S}(\mathcal{R}_2) \leq \mathbf{S}(\mathcal{R}_1) \leq \omega(\mathcal{R}_1)\left(2 + \frac{\rho_{\max}(\mathcal{R}_2)}{\rho_{\min}(\mathcal{R}_1)}\right)\mathbf{S}(\mathcal{R}_2)\,. \tag{14.7}$$

Now this is a result which relates a finite sum to an integral:

Lemma 6 *Let $\mathcal{R} \subset \mathcal{X}$, τ a real number such that $0 < \tau < b - a$, and $\mathbf{S}$ sub-additive. Then*

$$\frac{\tau}{2 + \frac{\rho_{\max}(\mathcal{R})}{\tau}}\mathbf{S}(\mathcal{R}) \leq \int_a^{b-\tau} \mathbf{S}([t, t+\tau])\, dt \leq \omega(\mathcal{R})\tau(2 + \frac{\tau}{\rho_{\min}(\mathcal{R})})\mathbf{S}(\mathcal{R})\,. \tag{14.8}$$

14.5 A Generalized Notion of Length

Let $\mathbf{S}$ be continuous, and sub-additive, as in Section 14.4.

Let $\mathcal{U} \subset \mathcal{X}$ be a covering of $[a, b]$, with the two following properties:

- No interval of $\mathcal{U}$ is included in another interval;
- $\mathcal{U} = \overline{\mathcal{U}}$.

Notation 2 Let

$$\mathcal{L}(\mathcal{U}) = \inf\{\mathbf{S}(\mathcal{R}) \text{ s.t. } [a,b] \subset \cup\mathcal{R}, \text{ and } \mathcal{R} \subset \mathcal{U}\}. \tag{14.9}$$

Remark 3 In the next sections, a parametrized curve $\Gamma = \gamma([a,b])$ is given, and $\mathbf{S}(u)$ is interpreted as the diameter of the subarc $\gamma(u)$. Therefore, we may think of $\mathcal{L}(\mathcal{U})$ as an approximation of the length of the curve Γ, even though the mathematical setting of $\mathcal{L}(\mathcal{U})$ lies only in $\mathcal{X}$.

Proposition 1 *We have:*

$$\frac{1}{2}\mathcal{L}(\mathcal{U}) \le \sup\{\mathbf{S}(\mathcal{R}) \text{ s.t. } \mathcal{R} \subset \mathcal{U}, \text{ and } \omega(\mathcal{R}) = 1\} \le 2\,\mathcal{L}(\mathcal{U})\,; \tag{14.10}$$

and

$$\frac{1}{2}\mathcal{L}(\mathcal{U}) \le \sup\{\mathbf{S}(\mathcal{R}) \text{ s.t. } \mathcal{R} \subset \mathcal{U}, \text{ and } \omega(\mathcal{R}) \le n\} \le 2\,n\,\mathcal{L}(\mathcal{U})\,. \tag{14.11}$$

This result shows that there are many ways to evaluate $\mathcal{L}(\mathcal{U})$, with help of discrete sums. Now we look for relationships between $\mathcal{L}(\mathcal{U})$ and an integral. First, let us introduce two notions of *velocity*, based upon the intervals of $\mathcal{U}$:

Notation 3 Let

$$v(t,\mathcal{U}) = \inf\{\frac{\mathbf{S}(u)}{|u|} \text{ for all } u \in \mathcal{U} \text{ s.t. } t \in u\}\,;$$

$$V(t,\mathcal{U}) = \sup\{\frac{\mathbf{S}(u)}{|u|} \text{ for all } u \in \mathcal{U} \text{ s.t. } t \in u\}.$$

Proposition 2 *We have*

$$\frac{1}{8}\int_a^b v(t,\mathcal{U})\,dt \le \mathcal{L}(\mathcal{U}) \le \int_a^b V(t,\mathcal{U})\,dt\,. \tag{14.12}$$

The first inequality in 14.12 may be replaced by the following:

Proposition 3 *Let*

$$J(t) = \overline{\cup\{u \in \mathcal{U} \text{ s.t. } t \in \overset{\circ}{u}\}}\,.$$

Then

$$\int_a^b \frac{\mathbf{S}(J(t))}{|J(t)|}\,dt \le 16\,\mathcal{L}(\mathcal{U})\,. \tag{14.13}$$

This is, in some sense, an improvement of 14.12, since $v(t,\mathcal{U})$ and $\mathbf{S}(J(t))/|J(t)|$ may not have the same order of magnitude. The inequality

$$v(t,\mathcal{U}) \le 2\frac{\mathbf{S}(J(t))}{|J(t)|}$$

is always true.

Finally, Lemma 6 implies the following:

Proposition 4 *Let $\mathcal{R}$ be any covering of $[a,b]$, such that $\omega(\mathcal{R}) \le 2$. Let $\tau > 0$, $c_1 = \min\{\tau, \rho_{\min}(\mathcal{R})\}$, and $c_2 = \max\{\tau, \rho_{\max}(\mathcal{R})\}$. Then*

$$\frac{1}{3}\frac{c_1^2}{c_2}\mathcal{L}(\mathcal{U}) \le \int_a^b \mathbf{S}([t,t+\tau])\,dt \le 24\frac{c_2^2}{c_1}\mathcal{L}(\mathcal{U})\,. \tag{14.14}$$

14.6 Infinite Length: An Analytical Approach

Let $\Gamma = \gamma([a,b])$ be a parametrized curve. If $a \le t_1 \le t_2 \le b$, $\gamma([t_1,t_2])$ is an arc of Γ. Let diam denote the diameter:

$$\mathrm{diam}(E) = \sup\{\mathrm{dist}(x,y) \text{ where } x \in E,\, y \in E\}.$$

For any interval $u \in \mathcal{X}$, let

$$\mathbf{S}(u) = \mathrm{diam}(\gamma(u)).$$

We will also use the convenient notation, for every $u = [t_1, t_2]$:

$$\mathbf{S}(u) = \mathrm{S}(t_1, t_2).$$

To define $\mathbf{S}$ or S, one can replace the diameter by any continuous *size* function, like the radius of the circumscribed circle, the length of the circumscribed rectangle with sides parallel to two given axes, etc... (see [243]). A size function must be continuous and subadditive, so that all results in Sections 14.4 and 14.5 are valid. Being given a covering $\mathcal{U}$ as in Section 14.5, the value $\mathcal{L}(\mathcal{U})$ may be interpreted as a length, evaluated with the diameter of arcs $\gamma(u)$, $u \in \mathcal{U}$.

Now let us introduce another continuous function $\mathbf{T} : \mathcal{X} \longrightarrow \Re^+$ (a *gauge* function), with the two following properties:

- $u_1 \subset u_2 \Longrightarrow \mathbf{T}(u_1) \le \mathbf{T}(u_2)$;
- If $x \in [a,b]$, $\mathbf{T}(\{x\}) = 0$.

Definition 14.2 *An interval* u *is* **T**-maximal *if*

$$u \subset v \Longrightarrow \mathbf{T}(u) < \mathbf{T}(v).$$

For any ϵ, such that $0 \le \epsilon \le \mathbf{T}([a,b])$, we will consider the following covering of $[a,b]$:

$$\mathcal{U}_\epsilon = \{u \text{ s.t. } u \text{ is maximal, and } \mathbf{T}(u) = \epsilon\}.$$

The corresponding length will be denoted

$$\mathcal{L}(\mathcal{U}_\epsilon) = L_\epsilon(\Gamma, \mathbf{T}).$$

As ϵ tends to 0, both $\rho_{\max}(\mathcal{U}_\epsilon)$ and $\sup\{\mathrm{S}(u) \text{ where } u \in \mathcal{U}_\epsilon\}$ tend to 0. In other words, the length $L_\epsilon(\Gamma, \mathbf{T})$ is calculated with smaller and smaller arcs.

Definition 14.3 *The function* **T** *is* uniform *over* Γ *if there exists a constant* c*, and for all* $\epsilon \le \mathbf{T}([a,b])$*, there exists a covering* $\mathcal{R}_\epsilon$ *of* $[a,b]$*, such that for all* u_1*,* u_2 *in* $\mathcal{R}_\epsilon$*:*

$$\mathbf{T}(u_1) = \mathbf{T}(u_2) = \epsilon, \text{ and } \frac{1}{c} \le \frac{|u_1|}{|u_2|} \le c.$$

Intervals of same image by **T**, have about the same length. The following is a direct consequence of Proposition 4:

Proposition 5 *Let* **T** *be uniform over* Γ*,* $\mathcal{R}_\epsilon$ *as in Definition 14.3, and* $\tau(\epsilon) = \rho_{\max}(\mathcal{R}_\epsilon)$*. Then*

$$L_\epsilon(\Gamma, \mathbf{T}) \simeq \frac{1}{\tau(\epsilon)} \int_a^{b-\tau(\epsilon)} \mathrm{S}(t, t+\tau(\epsilon))\, dt. \tag{14.15}$$

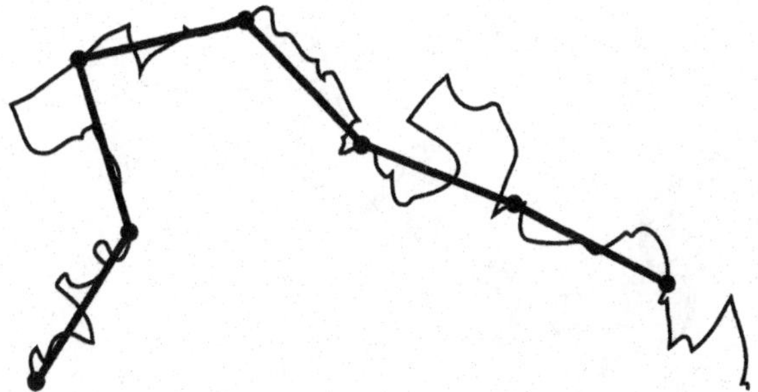

Figure 14.4: The compass method goes along the curve using steps of equal size.

14.7 The Compass Method

As a first application, we choose $\mathbf{T} = \mathbf{S}$. For every $t \in [a, b]$, such that $\text{dist}(\gamma(t), \gamma(b)) \geq \epsilon$, let

$$p_\epsilon(t) = \inf\{s > t \text{ s.t. } \mathrm{S}(t,s) > \epsilon\}.$$

The arc of extremities $\gamma(t)$, $\gamma(p_\epsilon(t))$ have a diameter equal to ϵ, the distance between the extremities.

One can construct a sequence $t_0 = a$, t_1, ..., $t_{N(\epsilon)}$ such that $t_{i+1} = p_\epsilon(t_i)$. The uniform continuity implies that $\inf\{p_\epsilon(t) - t\} > 0$, so that the sequence stops at a value $t_{N(\epsilon)}$ at distance $< \epsilon$ from b. The polygonal curve C_ϵ of vertices $x_0 = \gamma(a)$, x_1, ..., $x_{N(\epsilon)}$, where $x_i = \gamma(t_i)$, has length $\epsilon\, N(\epsilon)$. It is easy to verify that $\text{dist}(\Gamma, C_\epsilon) \leq \epsilon$. Let

$$\mathcal{U}_\epsilon = \{u \text{ s.t. } u \text{ is maximal, and } \mathbf{T}(u) = \epsilon\}.$$

The corresponding length is denoted

$$\mathcal{L}(\mathcal{U}_\epsilon) = L_\epsilon(\Gamma; \text{diameter}).$$

Then Proposition 1 implies that

$$L_\epsilon(\Gamma; \text{diameter}) \simeq \epsilon\, N(\epsilon). \tag{14.16}$$

We now apply Proposition 2: Let

$$\begin{aligned} \rho_{\min}(t) &= \inf\{|u| \text{ for all } u \in \mathcal{U}_\epsilon \text{ s.t. } t \in u\} \\ \rho_{\max}(t) &= \sup\{|u| \text{ for all } u \in \mathcal{U}_\epsilon \text{ s.t. } t \in u\}. \end{aligned}$$

The local velocities on Γ at t are

$$v(t, \mathcal{U}_\epsilon) = \frac{\epsilon}{\rho_{\max}(t)}, \text{ and } V(t, \mathcal{U}_\epsilon) = \frac{\epsilon}{\rho_{\min}(t)}.$$

Equation 14.12 gives:

$$\int_a^b \frac{dt}{\rho_{\max}(t)} \preceq N(\epsilon) \preceq \int_a^b \frac{dt}{\rho_{\min}(t)}. \tag{14.17}$$

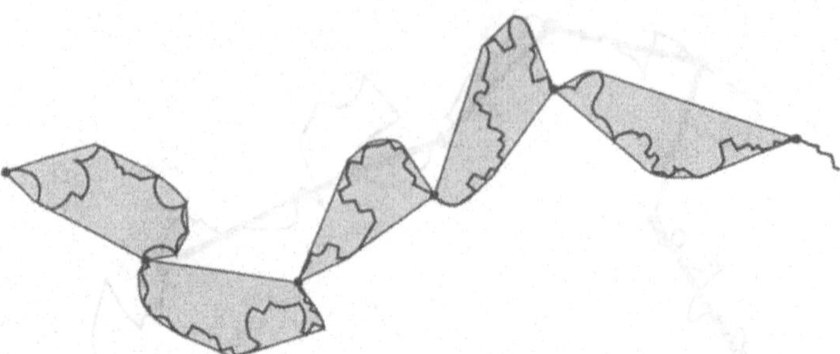

Figure 14.5: The method of constant breadth goes along the curve using steps of variable size, but same breadth.

If the diameter is *uniform* over Γ (see Definition 14.3), then Equation 14.15 gives

$$N(\epsilon) \simeq \frac{1}{\tau(\epsilon)},$$

a trivial result in this case.

14.8 The Method of Constant Breadth

As a second application of Section 14.6, we choose for $\mathbf{T}(u)$, the *breadth* of the arc $\gamma(u)$. If $u = [t_1, t_2]$, we will use the following notations:

$$\mathbf{T}(u) = \text{breadth}(\gamma(u)) = \mathrm{B}(t_1, t_2).$$

Recall that the *convex hull* $\mathcal{K}(u)$ of u is the smallest convex set containing u; and that the breadth of $\mathcal{K}(u)$ (or the breadth of u) is the smallest distance between two parallel lines on both sides of $\mathcal{K}(u)$. Alike the diameter, which may be replaced by the more general notion of *size*, the breadth can be replaced by the more general notion of *deviation* (see [243]).

For every $t \in [a, b]$, such that $\mathrm{B}(t, b) \geq \epsilon$, let

$$q_\epsilon(t) = \inf\{s > t \text{ s.t. } \mathrm{B}(t, s) \geq \epsilon\}.$$

The arc of extremities $\gamma(t)$, $\gamma(q_\epsilon(t))$ have a breadth equal to ϵ.

One can construct a sequence $t_0 = a$, t_1, ..., $t_{N(\epsilon)}$ such that $t_{i+1} = q_\epsilon(t_i)$. The sequence stops at a value $t_{N(\epsilon)}$ such that $\mathrm{B}(t_{N(\epsilon)}, b) < \epsilon$. The polygonal curve C_ϵ of vertices $x_0 = \gamma(a)$, x_1, ..., $x_{N(\epsilon)}$, where $x_i = \gamma(t_i)$, is an approximation to Γ. It is easy to verify that $\text{dist}(\Gamma, C_\epsilon) \leq \epsilon$. Let

$$\mathcal{U}_\epsilon = \{u \text{ s.t. } u \text{ is maximal, and breadth}(u) = \epsilon\}.$$

The corresponding length is denoted

$$\mathcal{L}(\mathcal{U}_\epsilon) = L_\epsilon(\Gamma; \text{breadth}).$$

Then Proposition 1 implies that

$$L_\epsilon(\Gamma;\text{breadth}) \simeq \sum_{i=1}^{N} \mathrm{S}(t_i, t_{i+1})\,. \tag{14.18}$$

For evaluating $L_\epsilon(\Gamma;\text{breadth})$, one can now apply Proposition 14.10, with local velocities as follows:

$$\begin{aligned} v(t,\mathcal{U}_\epsilon) &= \inf\{\frac{\mathrm{S}(s,q_\epsilon(s))}{q_\epsilon(s)-s} \text{ for all } s \text{ such that } s \le t \le q_\epsilon(s)\ \} \\ V(t,\mathcal{U}_\epsilon) &= \sup\{\frac{\mathrm{S}(s,q_\epsilon(s))}{q_\epsilon(s)-s} \text{ for all } s \text{ such that } s \le t \le q_\epsilon(s)\ \}\,. \end{aligned}$$

When the breadth is *uniform* over Γ (see Definition 14.3), Equation 14.15 gives

$$L_\epsilon(\Gamma;\text{breadth}) \simeq \frac{1}{\tau(\epsilon)} \int_a^{b-\tau(\epsilon)} \mathrm{S}(t, t+\tau(\epsilon))\,dt\,. \tag{14.19}$$

Remark 4 The method of constant breadth have been used with success for the simplified design of curves ([194, 193]). One can replace a complicated curve Γ (a geographical coastline for instance) by a polygonal curve using arcs of a given breadth. This process is fast in regular (almost rectifiable) parts of Γ, and slow in the irregular parts. The precision depends on the initial value given to the breadth. The new curve is simpler, but it presents all the essential irregularity features as the old one. In particular, the fractal dimension is preserved within suitable scales. The compass method does not have this advantage.

14.9 Comparison of Lengths

Theorem 3 *Let $\Gamma = \gamma([a,b])$ be a parametrized curve. Then*

$$L_\epsilon(\Gamma;\text{Minkowski}) \preceq L_\epsilon(\Gamma;\text{breadth}) \preceq L_\epsilon(\Gamma;\text{diameter})\,. \tag{14.20}$$

The first inequality is obtained by constructing a polygonal curve C_ϵ, so that $\text{dist}(\Gamma, C_\epsilon) \le \epsilon$, and the segments of C_ϵ are chords of arcs of breadth ϵ: Then $L_\epsilon(\Gamma;\text{breadth}) \simeq L(C_\epsilon)$. Equation 14.3 gives $L_\epsilon(\Gamma;\text{Minkowski}) \le L(C_\epsilon)$.

For the second inequality, use the fact that $p_\epsilon(t) \preceq q_\epsilon(t)$ for all t.

Remark 5 But these three functions are not equivalent, as it is shown in the examples of Section 14.11.

Let us recall a definition used in [243]:

Definition 14.4 *The curve Γ is* expansive *if there exists a constant $c > 0$, and for all ϵ, smaller than the breadth of Γ, a covering $\Gamma = \cup_k \Gamma_k^\epsilon$ by arcs, such that*

- *For all k, $\epsilon/c \le \text{breadth}(\Gamma_k^\epsilon) \le \epsilon$;*

- $\sum_k \mathcal{A}(\mathcal{K}(\Gamma_k^\epsilon)) \simeq \mathcal{A}(\cup_k \mathcal{K}(\Gamma_k^\epsilon))$.

In other words, Γ is expansive if the local convex hulls do not have too much in common: For example, if they have disjoint interiors, or if the index of the covering $\{\mathcal{K}(\Gamma_k^\epsilon))\}$ is bounded. If it is expansive, Γ does not come back too often on itself – a behaviour which explains the word "expansive".

Now we can give a much simpler characterization of expansivity:

Theorem 4 *The curve* Γ *is* expansive *if*

$$L_\epsilon(\Gamma; \text{Minkowski}) \simeq L_\epsilon(\Gamma; \text{breadth})\,. \tag{14.21}$$

In the family of the simple, expansive curves, are: The curves of finite length, the self-similar, or self-affine, curves, all graphs of continuous functions [243].

14.10 Fractal Dimension of Curves

Equation 14.6 gives the relation between $L_\epsilon(\Gamma; \text{Minkowski})$ and the fractal dimension $\Delta(\Gamma)$. We can deduce from 14.20 that

$$\begin{aligned}\Delta(\Gamma) &= \limsup_{\epsilon\to 0}\Big(1 + \frac{\log L_\epsilon(\Gamma;\text{Minkowski})}{|\log\epsilon|}\Big)\\ &\le \limsup_{\epsilon\to 0}\Big(1 + \frac{\log L_\epsilon(\Gamma;\text{breadth})}{|\log\epsilon|}\Big) \qquad (14.22)\\ &\le \limsup_{\epsilon\to 0}\Big(1 + \frac{\log L_\epsilon(\Gamma;\text{diameter})}{|\log\epsilon|}\Big)\,. \qquad (14.23)\end{aligned}$$

These inequalities may be strict. If Γ is expansive, the relation 14.22 is an equality. A special mention must be made about the curves such that the breadth is *uniform* (Definition 14.3 in Section 14.6) which is a property of the parametrization rather than Γ. From Equation 14.19, one deduces the following:

Proposition 6 *If* Γ *is an expansive curve, and if the breadth is uniform over* Γ*, then*

$$\Delta(\Gamma) = \limsup_{\epsilon\to 0}\Big(1 + \frac{\log\tau(\epsilon)}{\log\epsilon} - \frac{\log\int_a^{b-\tau(\epsilon)} \mathrm{S}(t, t+\tau(\epsilon))\,dt}{\log\epsilon}\Big)\,. \tag{14.24}$$

Using the function g such that $g(\tau(\epsilon)) = \epsilon$, Equation 14.24 may be written as

$$\Delta(\Gamma) = \limsup_{\tau\to 0}\Big(1 + \frac{\log\tau}{\log g(\tau)} - \frac{\log\int_a^{b-\tau} \mathrm{S}(t, t+\tau)\,dt}{\log g(\tau)}\Big)\,. \tag{14.25}$$

A short form of 14.25 is obtained when

$$\int_a^{b-\tau} \mathrm{S}(t, t+\tau)\,dt \simeq \tau^\alpha$$

and

$$g(\tau) \simeq \tau^\beta\,.$$

Then, 14.25 becomes

$$\Delta = \frac{\beta + 1 - \alpha}{\beta}. \tag{14.26}$$

The numbers α and β must fulfill the conditions

$$0 < \alpha \leq 1\,,\, \alpha \leq \beta\,,\, \alpha + \beta \geq 1\,.$$

Figure 14.6 shows the corresponding (α, β, Δ) surface in $\Re^3$. Now let us study some particular cases.

14.11 A Few Typical Curves

14.11.1 Rectifiable curves

Rectifiable curves are expansive, with $\alpha = 1$, which implies $\Delta = 1$, independantly of the value of β. A two-times continuously differentiable curve has uniform breadth, with $\beta = 2$. It would be interesting to construct a curve of finite length, uniform breadth, for any value of $\beta \leq 1$. All finite-length curves verify

$$L_\epsilon(\Gamma; \text{Minkowski}) \simeq L_\epsilon(\Gamma; \text{breadth}) \simeq L_\epsilon(\Gamma; \text{diameter}) \simeq 1\,.$$

14.11.2 Spirals

The *locally rectifiable* curves can be covered with arcs of finite length, but they may have infinite length altogether. A good example is the spiral defined in polar coordinates by the following parametrization:

$$\left\{ \begin{array}{l} \rho(t) = t^\delta \\ \theta(t) = \frac{2\pi}{t} \end{array} \right. \quad 0 < t \leq 1$$

where $0 < \delta \leq 1$. The breadth is not uniform over Γ, so that Equation 14.26 cannot be used. A direct calculation of the Minkowski sausage area gives

$$\Delta(\Gamma) = \frac{2}{\delta + 1}$$

(see [243]). By looking at Figure 14.3 one can easily realize that the length $L_\epsilon(\Gamma; \text{Minkowski})$ may be much smaller than $L_\epsilon(\Gamma; \text{breadth})$, obtained by following the curve in the sense of the parametrization. Indeed, such a curve is a good example of a *non-expansive* curve. One obtains

$$L_\epsilon(\Gamma; \text{Minkowski}) \simeq \epsilon^{\frac{\delta - 1}{\delta + 1}}\,,$$

whilst $L_\epsilon(\Gamma; \text{breadth})$ is approximately the length of Γ for $t_\epsilon \leq t \leq 1$, where $\rho(t_\epsilon) = \epsilon$. One gets

$$L_\epsilon(\Gamma; \text{breadth}) \simeq \epsilon^{1 - \frac{1}{\delta}}\,.$$

In this case, $L_\epsilon(\Gamma; \text{breadth})$ is equivalent to $L_\epsilon(\Gamma; \text{diameter})$.

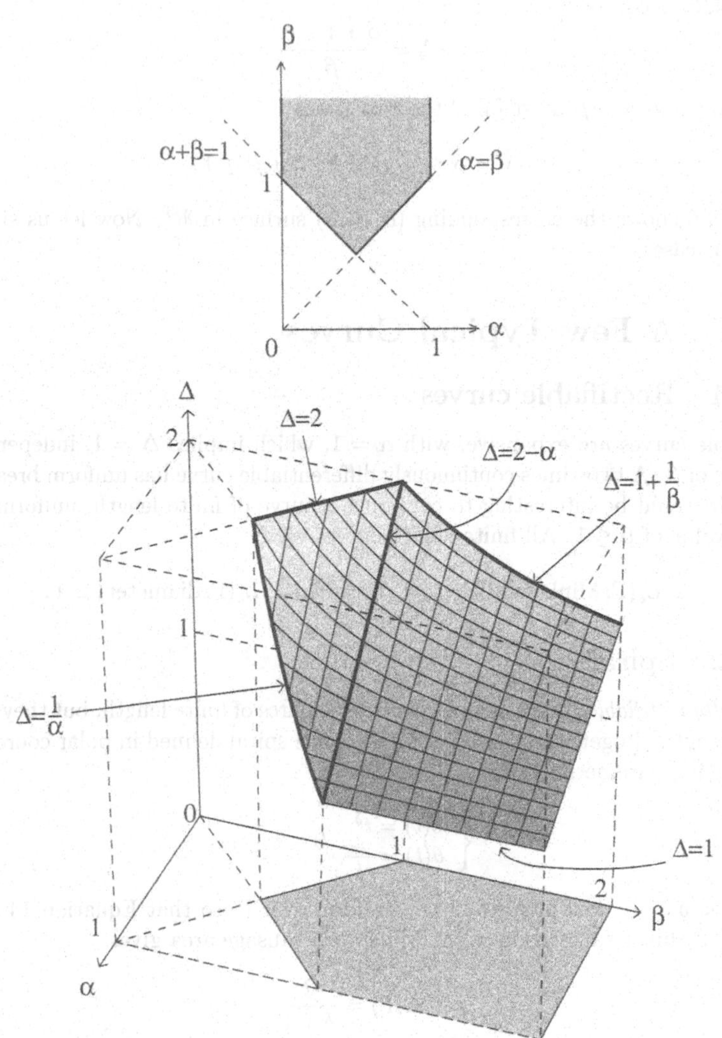

Figure 14.6: The admissible values of (α, β, Δ) form a part of an hyperboloid.

14.11.3 Self-similarity

Self-similar fractal curves are attractors of an iterated system of similitudes. They verify the equality

$$\Gamma = \cup_{i=1}^{N} F_i(\Gamma)\,,$$

where the transformations F_i can be written $F_i = \rho_i\, M_i\, x + b_i$, $0 \leq \rho_i < 1$, $1 < \sum_i \rho_i$, M_i an orthogonal linear function, b_i a translation vector. It is well known that, if Γ satisfies an *open set condition*, then the equation

$$\sum_{i=1}^{N} \rho_i^{\Delta} = 1$$

has a unique solution which is equal to the fractal dimension. The weights ρ_i^{Δ} induce a measure on Γ, which itself determines the parametrization in a canonical way. Following this parametrization, one can verify that the breadth is uniform over Γ. Actually Γ is expansive, and every subarc of such a curve has a breadth equivalent to its diameter. Therefore $\alpha = \beta < 1$, and

$$\Delta = \frac{1}{\alpha}\,.$$

This gives

$$\Delta = \lim_{\tau \to 0} \frac{\log \tau}{\log \int_a^{b-\tau} \mathrm{S}(t, t+\tau)\, dt}\,,$$

a very accurate method to evaluate the fractal dimension of a self-similar curve. For such curves,

$$L_\epsilon(\Gamma; \text{Minkowski}) \simeq L_\epsilon(\Gamma; \text{breadth}) \simeq L_\epsilon(\Gamma; \text{diameter})\,.$$

14.11.4 Self-affine graphs

Let $z(t)$ be a continuous function whose graph Γ is the attractor of an iterated system of contractions. This curve verifies the equality

$$\Gamma = \cup_{i=1}^{N} F_i(\Gamma)\,.$$

We suppose moreover that the transformations F_i can be written as $F_i = M_i\, x + b_i$, where M_i is a diagonal linear function, with spectral radius less than 1. Such a curve is represented in Figure 14.7. Like any graph of a continuous function, it is expansive. With the natural parametrization $\gamma(t) = (t, z(t))$, the breadth is uniform over Γ, and $\beta = 1$. Equation 14.26 yields

$$\Delta = 2 - \alpha\,. \tag{14.27}$$

Let us denote by

$$\mathrm{osc}_\tau(t) = \sup\{z(t') - z(t'') \text{ where } t \leq t' \leq t + \tau,\, t \leq t'' \leq t + \tau\,\} \tag{14.28}$$

the τ-oscillation of z at point t. For such a function,

$$\mathrm{S}(t, t+\tau) \simeq \max\{\mathrm{osc}_\tau(t), \tau\}\,.$$

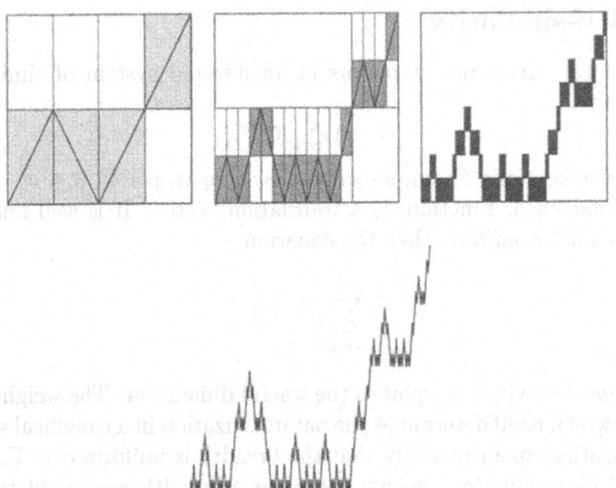

Figure 14.7: This function has Hölder exponant $\alpha = 1/2$. Since $\beta = 1$, the fractal dimension is 3/2.

Using 14.19, with $\tau(\epsilon) \simeq \epsilon$,

$$\alpha = \lim_{\tau \to 0} \frac{\log \int_a^{b-\tau} \mathrm{osc}_\tau(t)\, dt}{\log \tau}.$$

One may interpret α as the *local Hölder exponant* of this function.

14.11.5 Graphs of continuous functions

A nowhere differentiable function need not be self-affine (e.g. the Weierstrass function), but it still verifies many properties of self-affine graphs. In particular, if

$$\alpha = \liminf_{\tau \to 0} \frac{\log \int_a^{b-\tau} \mathrm{osc}_\tau(t)\, dt}{\log \tau},$$

then Equation 14.27 is valid. This result provides an excellent method for the evaluation of the graph dimension, for any non-constant, continuous function (see [244, 66, 243]).

14.11.6 Other curve with uniform breadth

Self-similar curves are such that $\alpha = \beta$, self-affine curves described in 14.11.4 are such that $\beta = 1$. It is interesting to exhibit some other example of curve, with $\alpha < \beta < 1$. Such a curve cannot be constructed as the attractor of an iterated system of affine functions. Our example is geometrically described in Figure 14.8. It depends upon a

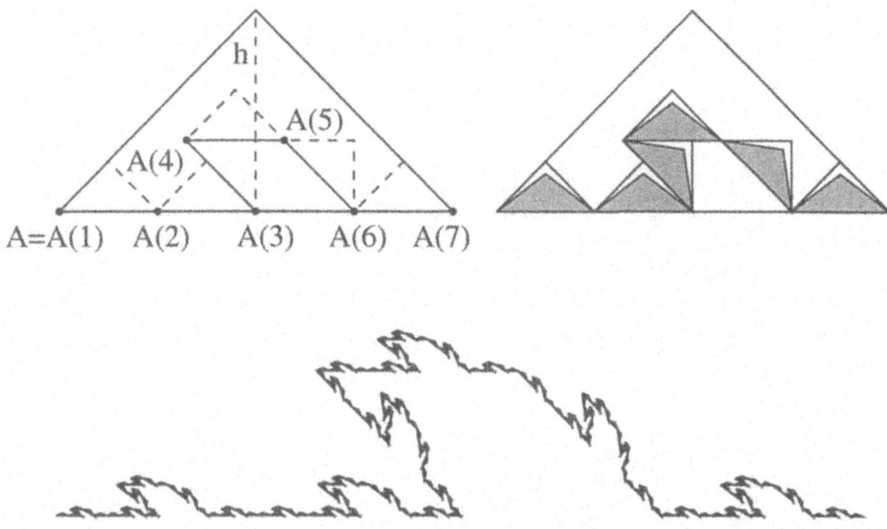

Figure 14.8: We start from an isocele triangle with base 1, height h, and an acute angle $< \pi/3$. We draw a polygonal curve C consisting of 6 segments of length 1/4 with vertices $A(i)$, $i = 1$, ..., $A(7)$. Let b be a parameter, $0 < b < 1/4$. The first operation consists of replacing D with 6 isocele triangles of basis $A(i)A(i+1)$, height hb. Each of these triangles is then replaced, at a second stage, by 2 smaller triangles, whose base length is 4^{-2} and height is hb^2. And so forth. The n-th stage is made with 6^n triangles whose base length is 4^{-n} and height is hb^n. The limit curve Γ is simple and expansive. When $b \to 1/4$, Γ tend to a self-similar curve whose generator is the curve C. For $b < 1/4$, Γ is not even self-affine.

parameter b, $0 < b < 1/4$. The breadth is uniform over Γ, and simple calculations show that

$$\Delta(\Gamma) = 1 + \frac{\log 3/2}{|\log b|}, \alpha = \frac{\log 4}{\log 6}, \beta = \frac{|\log b|}{\log 6}.$$

Equation 14.26 is satisfied. The computation of the ϵ-lengths gives

$$L_\epsilon(\Gamma; \text{Minkowski}) \simeq L_\epsilon(\Gamma; \text{breadth}) \simeq \epsilon^{\frac{\log 3/2}{\log b}},$$

but

$$L_\epsilon(\Gamma; \text{diameter}) \simeq \epsilon^{\frac{\log 3/2}{\log 1/4}}.$$

For $b < 1/4$, $L_\epsilon(\Gamma; \text{breadth})$ and $L_\epsilon(\Gamma; \text{diameter})$ are not equivalent.

Chapter 15

A Local Multiscale Characterization of Edges Applying the Wavelet Transform

Carl J.G. Evertsz, Kathrin Berkner, Wilhelm Berghorn

Abstract: A multiscale detection and characterization of edges in images is presented. The method is based on ideas involving scale-space filtering in the field of human vision, and on more recent developments in the theory of wavelets, singular functions and the analysis of fractal time series. Edges and their properties on all scales of observation are looked for, at the same time avoiding a proliferation of redundant information and computation. This is achieved by only looking at modulus maxima lines in the wavelet transform. In the end three descriptive parameters are determined for each pixel in the image from which a modulus maxima line starts. These three edge-parameters make it possible to select edges according to the largest distance at which they are still visible, their strength, and their sharpness. In generic cases, this makes it possible to, e.g., separate edges due to noise from real ones. The numerical implementation and the applications to test-images and medical-images are discussed. The latter involve the detection of boundaries of bone- and liver-tumors in X-ray images and CT-scans.

15.1 Introduction

In image and time-series analysis there is often a need for scale dependent analysis and description. For the characterization of texture the small scales are typically more important, while large objects are more easily segmented and analyzed by considering the larger scales. The importance of multiscale analysis of images, in particular their edges, has long been recognized in the field of human vision (see e.g. [181]). In the last decade, the multiscale approach to images got new impetus from both the introduction of fractal geometry [173] and the concept of wavelets [50]. The discovery of the fractal geometry of nature showed that many naturally occurring geometries, and therefore also their images are self-similar to some degree. The wavelet transform decomposes a signal or image in functions which are translations and dilations of one and the same localized mother function called a wavelet.

In recent years some important connections have been made between wavelets, fractals, and image analysis. This paper discusses an edge detection algorithm that is based on both scaling ideas from fractal geometry and the theory of one-dimensional wavelets. It goes back to a paper by Witkin [257] which addressed the problem of how events found in a smoothed version of a signal, can be located in the original signal. The solution proposed there was to track the extrema in the convolution of the signal with a Gaussian smoothing kernel, as a function of decreasing kernel width [257, 12, 260]. One of the main aims there was to reduce the proliferation of redundant data inherent in scale-space considerations, by means of a "scale-space filtering."

In this paper, we combine Witkin's tracking of extrema, with recent developments using modulus-maxima lines [131, 129, 127, 128, 166, 191] of wavelet transforms to estimate the local properties of singularities in functions. Such singularities give rise to a scaling behavior of the wavelet transform of the signal as a function of scale. The corresponding scaling exponent, a Hölder exponent, is a single number that provides a concise characterization of a singularity, and has been widely applied in the characterization of fractal signals and measures [9, 191, 105, 71]. Using the derivatives of the Gauss function as analyzing wavelets, one is able to find, and to follow edges across scales. Furthermore, the theory of modulus-maxima lines allows one to characterize the scale, strength, and through the local Hölder exponent also the scaling behavior of the edge-singularities. Applications to medical images and a grey-scale digitized picture of Lena are discussed. Also the effects of noise are discussed.

15.2 Gaussian Kernel Smoothing

The convolution of a signal $f(t)$ with a Gaussian function of width a can be interpreted as coarse graining at scale a. It acts like a low bandpass filter in the frequency domain. Therefore, the result of this transform is a new signal where all details on time- or spatial-scales smaller than a are removed. Denoting the Gaussian kernel $\theta(x) = \frac{1}{\sqrt{2\pi}} e^{-\frac{1}{2}x^2}$ and the convolution in the point b at scale a by

$$T_\theta[f](b,a) = \frac{1}{a}\int_{-\infty}^{\infty} \theta(\frac{x-b}{a}) f(x)dx, \tag{15.1}$$

we see that for fixed a, $T_\theta[f]$ is a smoothed version of the original signal, with a determining the degree of smoothing.

If the one-dimensional signal $f(t)$ is a cut through the intensity field of a grey-scales image, then it is well-known (e.g. [166]) that edges are located at, or near those points in the signal where the intensity field changes most rapidly. If the signal is continuously differentiable, then these points of rapid variation are those in which the modulus of the first derivative has a local maximum. Such a point is called a modulus maximum point. The second derivative is always zero in such points. Therefore, the modulus maxima points of the first derivative are always *zero-crossing points* of the second derivative. However, as is illustrated in the next figure, zero-crossing points are not necessarily modulus maxima points.

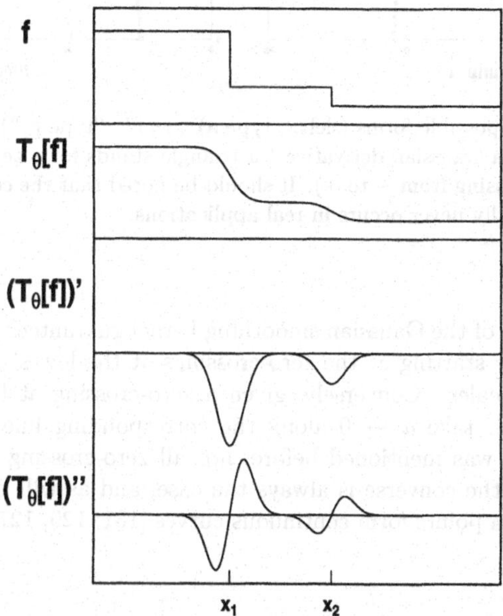

Figure 15.1: Transforms $T_\theta[f]$, $(T_\theta[f])'$ and $(T_\theta[f])''$ at a given scale a of a signal f with two sharp variation points.

Modulus maxima points and the zeros of the second derivative can be determined in all smoothed versions $T_\theta[f](.,a)$ of the signal. A zero-crossing point at scale a is then a point b where $\frac{\partial^2 T_\theta[f](b,a)}{\partial b^2} = 0$. Figure 15.1 shows the smoothed version of a signal with two step edges, together with its first and second derivative. The two outer zero-crossings of the second derivative at x_1 and x_2, are modulus maxima points, while the center one is not. Only the modulus maxima points relate to edges in the "image."

For the Gaussian smoothing kernel, it has been shown by Witkin and Baboud et al. and Yuille and Poggio [257, 12, 260] that in regular points zero-crossing points form continuous curves as a function of the smoothing scale a. The possible shapes of these curves are shown in Figure 15.2.

It should be noted that the configuration shown in the right plot, practically never occurs in real applications. However, they do occur in functions with special symmetry properties, like the devil-staircase function associated with the cantor-set [191]. This

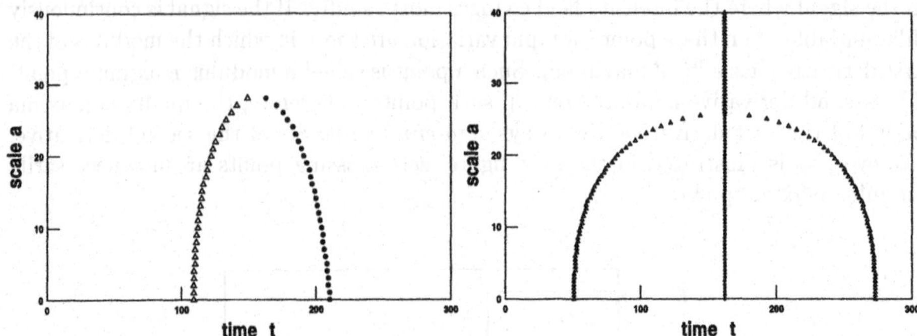

Figure 15.2: The only possible forms (left: "type a)", right: "type b)") of zero-crossing lines of a convolution with a Gaussian derivative (a triangle stands for a zero-crossing from + to −, a dot for a zero-crossing from − to +). It should be noted that the configuration shown in the right plot, practically never occurs in real applications.

very special property of the Gaussian smoothing kernel guarantees that one can find all zero-crossing lines by starting at the zero-crossings at the lowest scale, and following them up to higher scales. Conversely, given a zero-crossing at large scale a, one is able to zoom in, i.e., take $a \to 0$ along the corresponding line, and find its more precise location. As was mentioned before, not all zero-crossing points are modulus maxima points, but the converse is always the case, and as will be discussed shortly, also modulus maxima points form continuous curves [131, 129, 127, 128, 166, 191] as a function of scale a.

15.3 Edge Detection and Filtering with Wavelets

15.3.1 The Wavelet Transform

Is is not difficult to show that the partial derivatives of order n of the smoothed signal $T_\theta[f]$ can also be obtained by smoothing f with the nth derivative of the Gaussian. More precisely:

$$\frac{\partial^n T_\theta[f]}{\partial b^n}(b,a) = (-\frac{1}{a})^n T_{\theta^{(n)}}[f](b,a)$$

In practice one therefore uses the right-hand side to numerically estimate the partial derivatives of smoothed signals. How this provides the link with the wavelet transform, and ultimately to the estimation of local regularity properties, is discussed next. We first review some basic properties of wavelets. A wavelet is a localized function of zero integral, out of which a whole family of functions can be obtained through translations and dilations, which under certain general conditions [50, 53, 48], form an orthonormal bases in $L^2(\mathbb{R})$. More precisely,

Definition: $\psi \in L^2(\mathbb{R})$ is called a *wavelet*, if its Fourier-transform $\hat{\psi}$ satisfies:

$$\int_0^{\infty} \frac{|\hat{\psi}(\omega)|^2}{|\omega|} d\omega = \int_{-\infty}^{0} \frac{|\hat{\psi}(\omega)|^2}{|\omega|} d\omega = c_\psi < \infty$$

The convolution

$$W_\psi[f](b,a) = \frac{1}{\sqrt{C_\psi |a|}} \int_{-\infty}^{\infty} \psi(\frac{x-b}{a}) f(x) dx, \tag{15.2}$$

is called the *wavelet transform* of f. The parameter a controls the dilation of the analyzing wavelet ψ; its inverse being a measure of frequency. The parameter b controls the location of the dilated wavelet. One can show (see next lemma) that the derivatives

$$\theta^{(n)} = \frac{\partial^n \theta(x)}{\partial x^n} \tag{15.3}$$

of the Gaussian are wavelets for $n > 0$. Up to a normalization factor, the second derivative of the Gaussian is the well-known Mexican hat wavelet.

Up to a factor $\sqrt{C_\psi |a|}$, the wavelet transform, Equation 15.2, is the same as the convolution in Equation 15.1 with θ replaced by ψ. Therefore, the zero-crossings discussed before, are zero-crossings in the wavelet transform of the signal. For more details about wavelet transforms we suggest the reader to consult, e.g., [50, 53, 48].

15.3.2 Hölder Exponents of Singularities and Edges

The main reason for discussing this connection with wavelets in the present context, is that we would like to use some recent results due to Holschneider & Tchamitchian [114, 113], Jaffard [131, 129, 127, 128] and Mallat & Hwang [166], which link the scaling behavior of the wavelet transform, especially along modulus maxima lines ([166]), with Hölder exponents of singularities in signals. We first discuss the concept of regularity of functions that will be used for the characterization of edges.

Definition: Let $n > 0$ an integer, $\alpha \in \mathbb{R}_{>0}$, $n \leq \alpha \leq n+1$. A function $f : \mathbb{R} \to \mathbb{R}$ is called *Lipschitz α in x_0*, if there are constants $A \in \mathbb{R}$ and $\delta \in \mathbb{R}_{>0}$ and a polynomial $P_n(x)$ of order n , that for all $h \in \mathbb{R}$ with $|h| < \delta$ holds:

$$| f(x_0 + h) - P_n(h) | \leq A|h|^\alpha \tag{15.4}$$

$\alpha^*(x_0) = \sup\{\alpha |\ f(x) \text{is Lipschitz } \alpha \text{ in } x_0\}$ is called *Hölder exponent of f in x_0*.

A point where f has a Hölder exponent $\alpha < \infty$ is a singularity of f. Points where f is discontinuous are related to Hölder exponent 0. For example the image-row in figure 15.1 with two "edges" can be interpreted as a function which has two singularities with the same Hölder exponent $\alpha = 0$. The results due to Holschneider & Tchamitchian [114, 113], Jaffard [131, 129, 127, 128] and Mallat & Hwang [166] to be discussed now, provide a method to determine such Hölder exponents using wavelets, and can be used to make numerical estimates. It should be noted that in the following applications of the wavelet transform for the computation of Hölder exponents, it is more convenient [9, 13, 191] to use the alternative renormalization, $|a|^{-1}$, in Equation 15.1, instead of $|a|^{-\frac{1}{2}}$ in the wavelet transform, Equation 15.2. Therefore, throughout the rest of the paper, we use the convolution $T_\psi[f]$, in Equation 15.1 for the wavelet transform.

First we state some useful definitions, and a lemma ([161]).

Definition: Let $n \in \mathbb{N}$. A wavelet ψ *has n vanishing moments (is of order n,)* if for all integers $k < n$

$$\int_{-\infty}^{\infty} \psi(x)x^k dx = 0 \quad \text{and}$$

$$\int_{-\infty}^{\infty} \psi(x)x^n dx \neq 0.$$

Lemma: Let ϕ a n-times differentiable function and $\phi, \phi^{(n)} \in L^2(\mathbb{R}), \phi^{(n)} \neq 0$. Then it follows that $\psi = \phi^{(k)}$ is a wavelet.

Note that applying this lemma to the Gaussian function θ yields that all its derivatives are wavelets. It is easy to prove that these derivatives of the Gaussian function are wavelets with a number of vanishing moments equal to the order of the derivative.

We can now state the main results from [166] using the scaling properties of the wavelet transform of functions with local Hölder exponents from [113, 114, 131, 129], which are relevant for this paper. Let ψ be a wavelet with n vanishing moments and compact support, which is the nth derivative of a n-times continuously differentiable smoothing function.

1. If all points of an interval $(b-\epsilon, b+\epsilon)$ $(\epsilon > 0)$ are not origins of modulus-maxima lines of the transform $T_\psi[f]$, then f is Lipschitz n in all points of $(b-\epsilon, b+\epsilon)$.
2. If f has singularities with Hölder exponents $\alpha < n$, then these points are starting points of modulus-maxima lines of $T_\psi[f]$.
3. Estimation of local Hölder exponents via modulus maxima lines in non-oscillating isolated singularities:
   Let $\Gamma(a)$ be the parameterization of a modulus-maxima line as a function of scale a. If f has Hölder exponent $\alpha < n$ in $x_0 = \Gamma(0)$ and there exists a constant C and a scale a_0, such that for all $a < a_0$, $|\Gamma(a) - x_0| < Ca$, then it follows:
   f is Lipschitz α in x_0, if and only if there exists a constant B, such that for all $a < a_0$
   $$|T_\psi[f](\Gamma(a), a)| \leq Ba^\alpha.$$
   If f has non-isolated singularities, it is necessary to make some extra assumptions (see [166]) on the wavelet in order to get the Lipschitz exponents from the decay of

the wavelet transform on the modulus-maxima lines. For example, problems arise when on tries to measure the Lipschitz-exponent e.g. in an oscillating singularity. However, this type of singularities almost never occurs in images.

Result (2) shows that each singularity yields modulus-maxima lines, as long as the analyzing wavelet has enough vanishing moments. Therefore singularities and edges can be located by finding the modulus-maxima lines. Furthermore, result (3) shows that the Hölder exponent α of a singularity can be estimated from the scaling behavior of the wavelet transform on the modulus-maxima line as a function of $a \to 0$.

15.3.3 Edge-Parameters

We now discuss the significance of three parameters related to modulus maxima lines, that can be used to describe and to selectively segment edges in images. For convenience we denote the parameterization of a modulus maxima line as a function of scale, by $\Gamma(a)$. The *edge-parameters* are

- *the scale of the edge*
- *the strength of the edge*
- *the Hölder exponent associated with the regularity of the edge*

When the smoothing parameter a increases, the image becomes increasingly blurred, and edges of small spatial extend tend to become dominated by those of larger extend. As a consequence, at large enough scales, the contribution of small edges diminishes to the extent, that they stop producing modulus-maxima lines. Therefore, a measure for the distance from which an edge in an image is still distinguishable, is provided by the magnitude of the scale at which the modulus-maxima line stops existing.

The size of the wavelet transform, $|T_\psi[f](\Gamma(a), a)|$, at the smallest numerically available scale $a = 1$ of a modulus maxima line $\Gamma(a)$, is a measure for the variation in the intensity field. For example, in case the wavelet ψ is the derivative of the Gaussian, this value is the local derivative of the signal, and therefore characterizes the *strength* of the edge.

The Hölder exponent at the pixel located at position $\Gamma(1)$, describes the sort of singularity, if any, associated with the intensity field in its vicinity. Small Hölder exponents are related to sharp edges and large Hölder exponents to smooth edges. It is important to remark that one can only detect singularities with Hölder exponents smaller than the number of vanishing moments of the analyzing wavelet (see point 3 in previous section). A simple analytical and numerical example illustrating the use of these edge-parameters is discussed next.

15.4 Numerical Implementation

15.4.1 Finding Modulus-Maxima Lines

Our applications involved digitized images $I(x, y)$ of size $n_x \times n_y$ pixels, with $n_b < 12$ bits quantization of the intensity $I(x, y)$. In order to get 1-dimensional signals, we scanned the images in the four different ways shown in Figure 15.3. These are a *horizontal scan*, a *vertical scan*, and two *diagonal scans*. Each of these scans, yields a 1-dimensional signal of size $n_x n_y$.

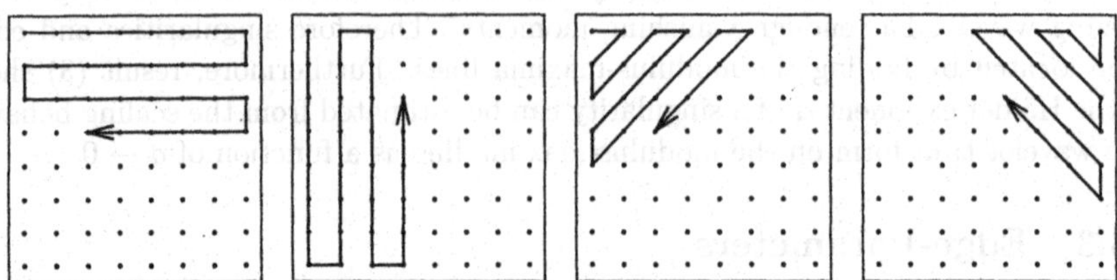

Figure 15.3: The four serpentine scans used in the edge detection. From left to right these are the horizontal, the vertical, and two diagonal scans.

It is essential to scan the image in these various ways, because, e.g. horizontal edges are missed in the horizontal scan. Also, experience showed that the horizontal and vertical scan are not sufficient, and that, in addition diagonal scans are needed. In each of these 1-d signals, modulus-maxima points are located at the smallest scale, and a special procedure tracks each one of them to the largest scale at which it still exists. This has the advantage that the time consuming wavelet transform only needs to be computed in the direct vicinity of the modulus maxima lines, instead of for all values of a and b. In this way the amount of computations grows as $Na_{\max}$, where N is the length of the signal and a_{max} is the maximum scale a.

The reason for transforming 2-dimensional images into 1-dimensional signals in this paper, is that it is much easier to find and parameterize the modulus maxima lines. In the 2-dimensional case, complications arise due to the fact that the modulus maxima points can form surfaces. However, as is discussed in, e.g., [166] and [167], it is possible to apply the two-dimensional wavelet transform to edge detection, denoising, and enhancement of images.

Figure 15.4 shows the result of the tracking of a modulus maxima line for a simple signal with isolated singularities with various Hölder exponents, $\alpha(x_1) = 0.7$, $\alpha(x_2) = \alpha(x_3) = 0.3$, $\alpha(x_4) = 1.5$, $\alpha(x_5) = 0$. The upper-left plot shows the modulus maxima lines tracked, using the wavelet $\theta^{(1)}$, and the upper-right plot those using $\theta^{(2)}$. According to the result (1) in the previous section, the absence of modulus maxima lines in between the singularities, confirms that there the signal is at least Lipschitz 1 in the case $\theta^{(1)}$, and at least Lipschitz 2 in the case $\theta^{(2)}$.

The number of modulus maxima lines emanating from singularities grows linearly with the number of vanishing moment in the wavelet [166]. This proliferation of lines is not desirable, because it increases the number of lines to be tracked. Also, the errors in the numerical (discrete) approximation of the wavelets, increases at smaller scales because of the increasing number (n) of maxima in the wavelet. Nevertheless, depending on the application, one may need higher orders. In the example at hand, the singularity in x_4 is missed by the wavelet of order $n = 1$, because the Hölder exponent is larger than 1 there. Increasing the order to 2 in the upper-right figure

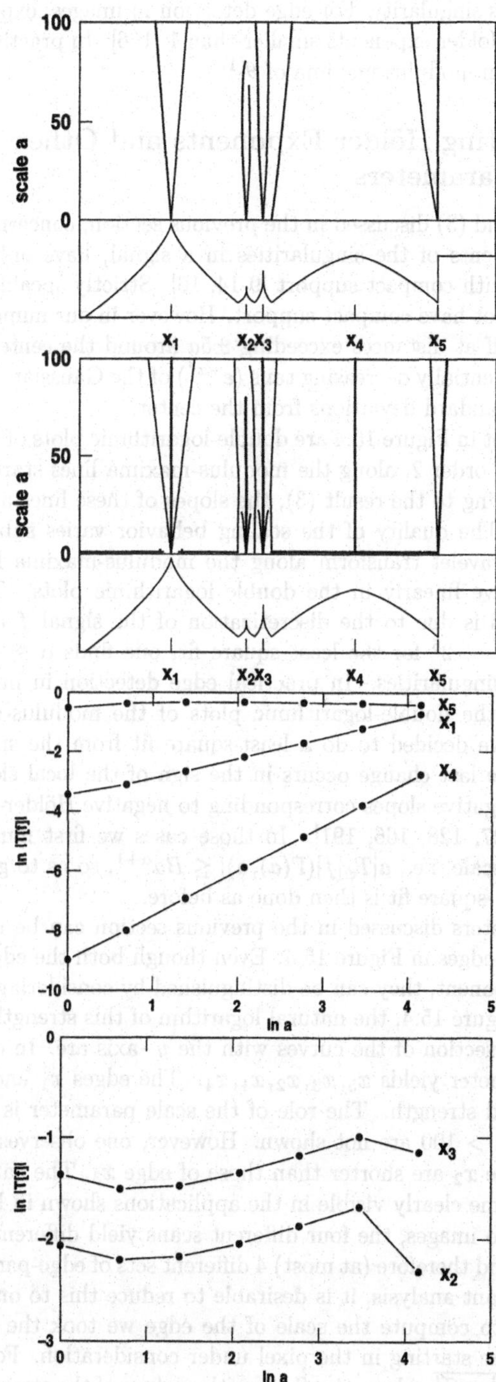

Figure 15.4: Top: Modulus maxima lines in isolated singularities with analyzing wavelets $\theta^{(1)}$ (left) and $\theta^{(2)}$ (right). Bottom: log-log plot of modulus maxima lines of $T_{\theta^{(2)}}[f]$ starting in the singularities.

suffices to detect this singularity. For edge detection in images, experience shows that it suffices to detect Hölder exponents smaller than 1 [166]. In practice we therefore use the wavelet-transform modulus maxima of $\theta^{(1)}$.

15.4.2 Estimating Hölder Exponents and Other Edge-Parameters

The results (1),(2) and (3) discussed in the previous section, concerning the estimation of the Hölder exponents of the singularities in a signal, have only been proved for analyzing wavelets with compact support [9-14, 16]. Strictly speaking, the derivatives of the Gaussian do not have compact support. However in our numerical applications, the support is cut off at distances exceeding $\pm 5a$ around the center b of the wavelet. Because of the exponentially decreasing tails (e^{-x^2}) of the Gaussian, there is essentially no mass outside 5 standard deviations from the center.

The lower-left plot in Figure 15.4 are double-logarithmic plots of the modulus of the wavelet transform of order 2, along the modulus-maxima lines starting at the various singularities. According to the result (3), the slopes of these lines are estimates of the Hölder exponents. The quality of the scaling behavior varies substantially between singularities. The wavelet transform along the modulus-maxima lines related to x_2 and x_3 do not behave linearly in the double logarithmic plots. The upward swing at the smaller scales is due to the discretization of the signal f around x_2 and x_3. Using the scales $2^2 - 2^5$ for the least square fit, one finds $\alpha \approx 0.3$ for the Hölder exponents of these singularities. In practical edge detection in images, one can not visually inspect all the double-logarithmic plots of the modulus-maxima lines. As an approximation, we decided to do a least-square fit from the smallest scale up to that scale, where the last change occurs in the sign of the local slope in the double-logarithmic plot. Negative slopes corresponding to negative Hölder-exponents ($\alpha < 0$) are also possible [127, 128, 166, 191]. In those cases we first multiply the wavelet-transform with the scale, i.e., $a|T_\psi[f](\Gamma(a), a)| \leq Ba^{\alpha+1}$, so as to get a positive slope $\alpha + 1 > 0$. The least-square fit is then done as before.

The edge-parameters discussed in the previous section can be used to distinguish between the various edges in Figure 15.4. Even though both the edges x_2 and x_3 have the same Hölder exponent, they can be distinguished by considering their strength. In the lower plots of Figure 15.4, the natural logarithm of this strength $|T_\psi[f](x_i, a = 1)|$ is given by the intersection of the curves with the $y-$axis are. In order of decreasing strength, this parameter yields x_5, x_3, x_2, x_1, x_4. The edges x_1 and x_5 differ both in Hölder exponent and strength. The role of the scale parameter is lost in these plots because the scales $a > 100$ are not shown. However, one observes that the modulus maxima lines of edge x_2 are shorter than those of edge x_3. The variation in the scale parameter will become clearly visible in the applications shown in Figure 15.6.

In applications to images, the four different scans yield different modulus maxima lines in each pixel, and therefore (at most) 4 different sets of edge-parameter values. For a rotationally invariant analysis, it is desirable to reduce this to one set of parameter values. Therefore, to compute the scale of the edge we took the size of the longest modulus-maxima line starting in the pixel under consideration. For the strength, we took $\sqrt{S_1^2 + S_2^2 + S_3^2 + S_4^2}$, where the S_i's are the values of the strengths along the four different directions of scanning through the pixel. To estimate the Hölder exponent we took the smallest value, because it is associated with the scanning path through the pixel along which the variation is sharpest. Then, by thresholding each of the parameters one can filter the image in multiple ways.

15.4.3 Noise

The three edge-parameters can also be used, to filter out noise in signals. In a signal with additive Gaussian white noise, typically the noise produces shorter modulus-maxima lines with relatively lower strength than those of modulus-maxima lines of events related to the original signal. In addition, the points related to the Gaussian white noise tend to have lower Hölder exponents. These characteristics of Gaussian white noise in terms of the edge-parameters can be used to separate the information coming from the underlying signal from the noise, by appropriately thresholding the three parameters.

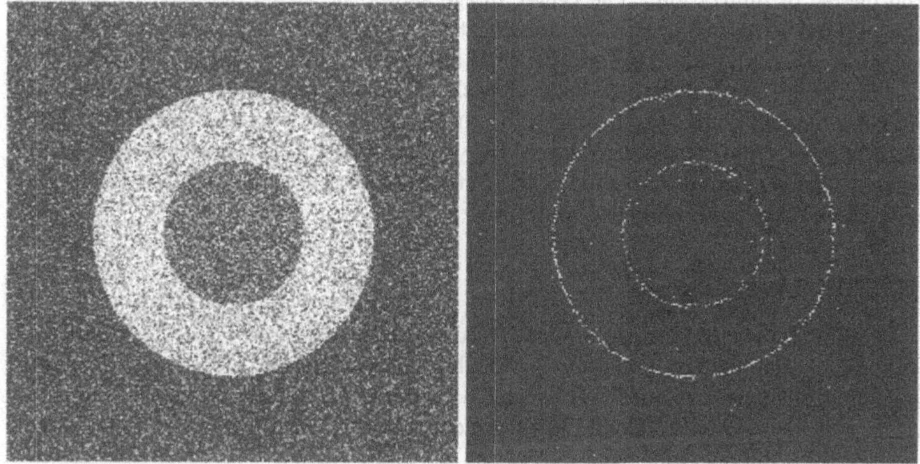

Figure 15.5: *left)* Test image for studying effects of noise. The inner disc has grey-scale value $I_i = 64$, the outer one $I_o = 192$, and the background $I_b = 0$. The diameter of the inner disc is 80, and that of the outer is 160 pixels. *right)* Edges found with appropriate thresholding of the edge parameters.

Figure 15.5 left, shows the grey-scale image of two discs, on to which has been added Gaussian white noise of standard deviation $\sigma = 96$. The inner disc has grey-scale value $I_i = 64$, the outer one $I_o = 192$, and the background $I_b = 0$. The diameter of the inner disc is 80, and that of the outer is 160 pixels. For the inner edge one has a signal to noise ratio $(I_o - I_i)/\sigma = 1.33$, and for the outer one $(I_o - I_b)/\sigma = 2$. The right plot in the same figure, shows the output of the filter after thresholding the edge parameters. All pixels are colored black, except those with strength bigger than 78, modulus-maxima line length bigger than 23, and Hölder exponent bigger than -0.8; these are colored white. With this parameter-setting we can detect points which are only related to the boundaries of the discs and not to the noise.

This test also shows that the detection of edges in this case is independent of the direction of the edges in the images, and that the four scanning directions, one horizontal, one vertical and two diagonal, are sufficient.

15.5 Applications

In this section we will discuss the application of the edge-parameters in the segmentation of edges of bone tumors, liver tumors and of edges in the picture of Lena.

15.5.1 Bone Tumors

The uppermost image in Figure 15.6 shows the X-ray image of a human bone with a tumor. The task is to find the boundary of the tumor so that the physician can assess the degree of infiltration. The center image shows the signal obtained by scanning the original X-ray along the line crossing the tumor shown in the upper plot. This signal is superimposed on the graph of the modulus-maxima lines. The boundaries of the tumor are marked with x_l and x_r, both of which are starting points of modulus maxima lines. Trying a segmentation by thresholding the length of modulus-maxima lines keeping only those with sizes > 15, yields the result shown in the bottom left image. The segmentation shown in the bottom right figure was obtained by keeping only those pixels with strengths above 3.2.

15.5.2 Liver Tumor

Figure 15.7 shows the original slice of a computed tomography scan (CT-scan) of the abdominal region of the human body. The big organ to the left is the liver. The white regions in the liver are blood vessels which are enhanced because of the application of a contrast medium in the blood. The large dark region in the liver is a section through a large liver tumor. The right plot shows all pixels having a modulus maxima line starting in them.

The left and right plots in Figure 15.8 show the result for different settings of the edge-parameters. In the left plot only the strength and the scale have been thresholded. The right plot shows how many irrelevant boundaries can eliminated by not taking into account the very singular boundaries demarcating e.g. the spine. This is realized by an additional thresholding of the Hölder exponent, in this case only keeping those with values larger than 0.

15.5.3 Lena

The 512x512 grey-scale image of Lena is shown in the left part of Figure 15.9. The right plot shows the significance of the length-parameter. All pixels with modulus-maxima line lengths larger than 30 are shown. These correspond to edges that are still visible

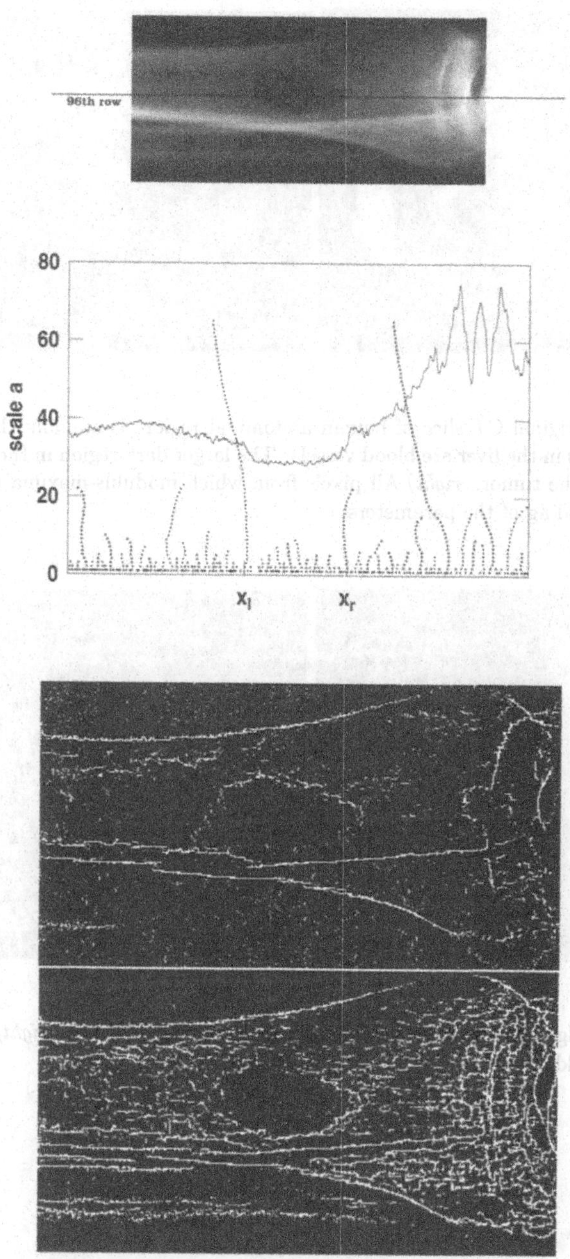

Figure 15.6: *top)* X-ray of a human bone with tumor. *middle)* Modulus maxima lines of $T_{\theta'}[f]$ from the signal obtained by scanning along of the row market in the top figure. *bottom left)* Pixels (white) with modulus maxima lines with length greater than 15 *bottom right)* Modulus maxima with strengths larger than 3.2.

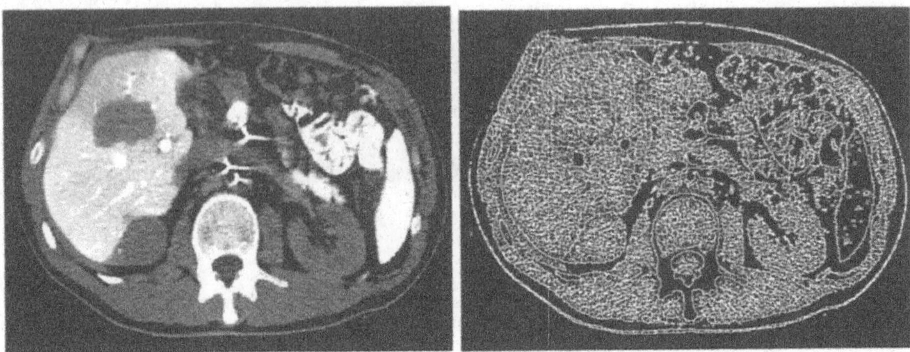

Figure 15.7: *left)* Original CT slice of human abdominal region, containing the liver on the left. The white spots in the liver are blood vessels. The larger dark region in the central upper part of the liver is the tumor. *right)* All pixels from which modulus-maxima lines emanate. That is: No thresholding of the parameters.

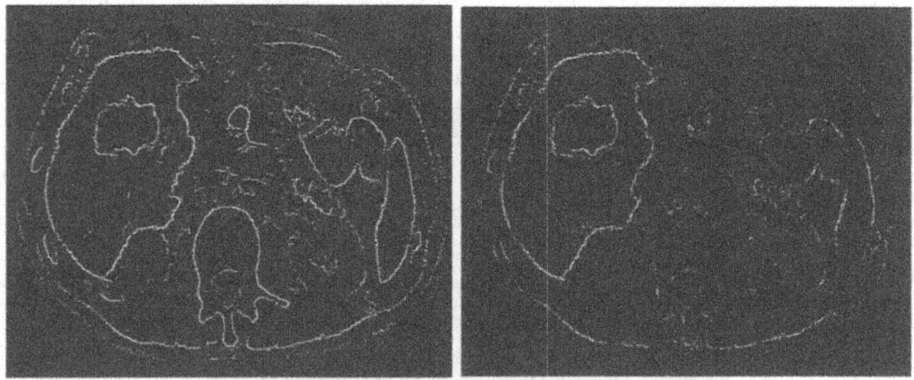

Figure 15.8: *left)* Edge-parameter setting: Length > 30, strength > 12.8. *right)* Length > 30, strength > 12.8, Hölder > 0.

Figure 15.9: *left)* Original Lena (512 × 512). *right)* Only those pixels with a modulus-maxima line length > 30 are shown in white. The rest is made black.

after extreme blurring. The effects of a reduction of this parameter is shown in the left plot in Figure 15.11.

Figure 15.10 shows the significance of the other two parameters, the strength and the Hölder exponent. In the left plot all pixels with a strength larger than 5 are shown. These correspond to edges which have a higher contrast as most of the noise-pixels, which therefore are removed. The right plot in Figure 15.10 shows all pixels with Hölder exponent smaller or equal to 0. These represent the very sharp edges, and since all lengths and strengths are allowed, one finds that the noise becomes very dominant. The left plot in Figure 15.11 shows an optimum setting of the edge parameters obtained by interactive tuning of the parameters. A segmentation of the feathers is obtained by keeping pixels with shorter modulus-maxima line lengths < 14, and larger strengths > 22. The result is shown in the right part of Figure 15.11.

15.5.4 Conclusions and Summary

The edge detection and segmentation method discussed in this paper combines information obtained from a multiscale analysis. Three tuning parameters involving the scale, the strength and the Hölder exponent can be associated to each pixel from which a modulus maxima line emanates. The number of edge parameters could be further expanded or reduced, depending on the application at hand. Data redundancy is avoided by only considering the modulus maxima lines in the wavelet transform. In addition, the amount of computations is reduced by tracking the modulus maxima lines from fine to coarse scales. Typically the amount of computations grows as the size of the signal times the maximum scale of interest. For a 512x512 image up to scale 100, precomputing and storing the three edge parameter values of the image allows real-time

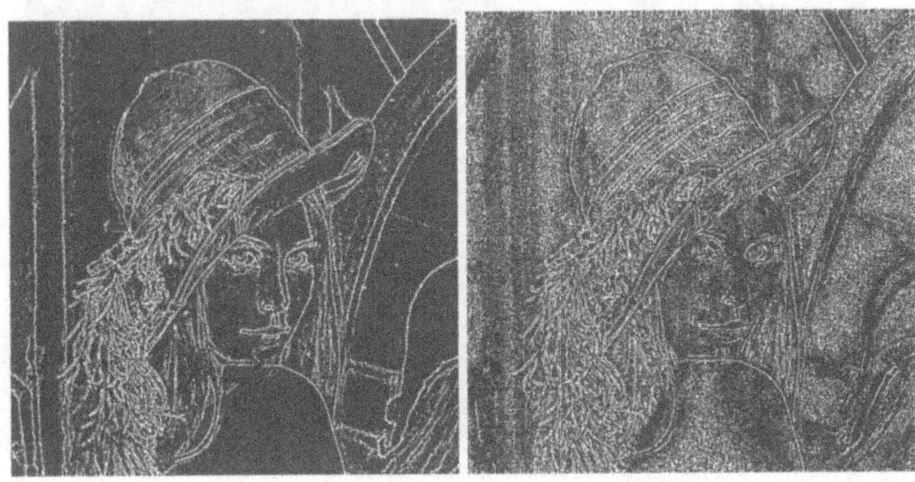

Figure 15.10: *left)* Strength > 5. *right)* Hölder < 0.

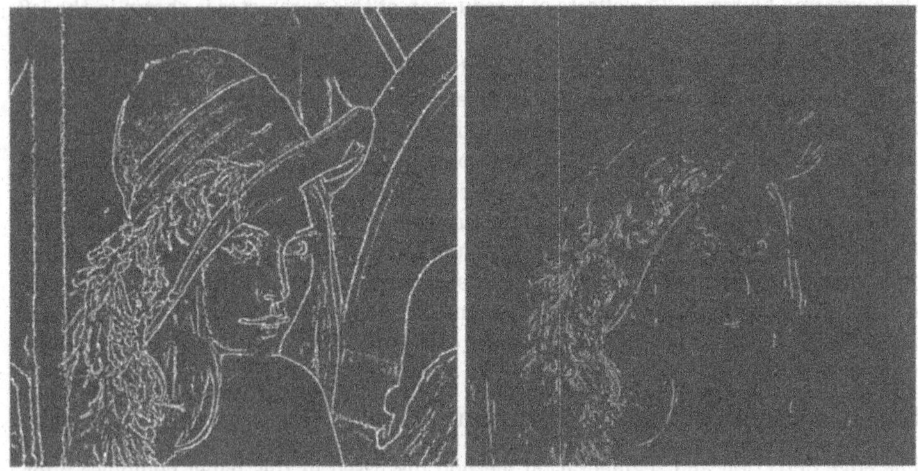

Figure 15.11: *left)* A subjectively optimal setting of the edge parameters: strength > 7, modulus maxima line length > 2. *right)* Obtaining the feathers with modulus-maxima line length < 14, strength > 22.

manipulation of sliders, which facilitates the interactive study of the effects of various parameter settings. Once the right range of setting has been found for a particular type of image, one can optimize the algorithm for that specific task.

Acknowledgment

We would like to thank Thomas Netsch for various discussions and computer support. Also, we would very much like to thank Heinz-Otto Peitgen for discussions and his support of this research. We thank K.J. Klose for the liver CT and J. Freyschmidt for bone X-ray.

Chapter 16

Local Connected Fractal Dimension Analysis of Early Chinese Landscape Paintings and X-Ray Mammograms

Richard F. Voss

Abstract: Local and global applications of multifractals to the analysis of digitized image intensities $I(x, y)$ is discussed. The magnitude of the local slope $|\nabla I(x, y)|$ is shown to be a more useful measure than $I(x, y)$. A global fractal dimension D may be estimated from the spectral density $S(\vec{k})$, the angle-averaged pair correlation $C(r)$, and mass-radius $M(R)$. For image enhancement, the concept of *local fractal dimension* can be used to construct a color-coded dimensional image. Applications to the classification of early Chinese landscape paintings and x-ray mammograms, however, suggest that the local *connected* fractal dimension provides the best agreement with the human eye for highlighting and discriminating between images.

16.1 Fractals: From Self-Similarity to Self-Affinity and Multifractals

The term *fractal* was invented and developed by IBM mathematician Benoit Mandelbrot [171] in 1975 to characterize a previously disjoint and little known collection of mathematical techniques and models applicable to irregular shapes and time series [171, 172, 173]. Since that time, fractal geometry has become a scientific discipline in its own right and has achieved tremendous impact on other fields of science, technology, and art.

The initial exposition of fractals [171, 172] focussed on the concept of *self-similarity* or *dilation symmetry* in which a small part of a shape mimicked the characteristics of the whole. When a shape can be described as consisting of N sub-shapes, each scaled down by a factor $r < 1$ from the original, it can be characterized by a *fractal dimension* D where

$$D = \frac{\log N}{\log 1/r}. \tag{16.1}$$

Although extremely useful, the concept of fractal dimension is strictly speaking only applicable to *sets*, collections of points or regions that may be specified according to some membership rule. In other words, use of a fractal dimension assumes a *black and white* or *binary* view of the world. A specific point is either a member of the *set* of interest, or it isn't.

Most views of the world (and most scientific problems) are not so clear-cut, and require continuous *shades of gray* in their description. Fractal geometry has spawned two strategies for dealing with this problem. *Self-affinity* is used to describe *functions* of time or space [179, 174, 252, 253]. Models for self-affine functions $X(t)$, such as fractional Brownian motion (fBm) [179], typically involve the *addition* of perturbations $f(t/\tau)$ each having a typical size τ,

$$X(t) = \sum_{\tau} a(\tau) f(t/\tau). \tag{16.2}$$

Self-similar shapes are constructed from sub-shapes scaled equally in all directions. With self-affinity the sub-shapes may be scaled differently in different directions. For a Brownian motion or random walk, the displacement ΔX scales with time change Δt as $\Delta X \propto \Delta t^{1/2}$. When the time scale is changed by 4, the distance scale is changed by 2. Consequently, the original concept of *fractal dimension*, which was based on self-similarity, must be used with great care for self-affine functions [174, 252, 253]. Different procedures for *estimating* D that give the same answer for self-similar shapes will, in general, give different answers for self-affine shapes. A single procedure, moreover, may give different D for different spatial scales.

Multifractals, on the other hand, are used to describe mathematical *measures* [170, 175, 178]. Models for multifractals, such as the *Besicovitch measures*, typically involve the *multiplication* of perturbations $f(\tau)$ each having a typical size τ,

$$X(t) = \prod_{\tau} a(\tau) f(t/\tau). \tag{16.3}$$

As can be seen from Eqs. (16.2) and (16.3) there can be a great deal of similarity between *self-affine* and *multifractal* descriptions of a problem. A *measure* may be converted to an arbitrary number of *sets*. For example, one set includes all points for which $X > X_0$. Since each X_0 produces a different set, with a possibly different D, multifractals are typically characterized by their *distributions* or *histogram* of D values.

16.2 Measures and Mass Dimension

The fractal analysis techniques described here will first be illustrated in 1-D then extended to the case of 2-D images. The basic problem is the fractal description of a *measure* or local *mass density* $m(x)$.

As shown in Fig. 16.1 even a discrete distribution of point masses becomes approximately continuous under coarse-grained averaging. Most experimental data, and digitized images in particular, represent a coarse-grained average of some underlying point process (such as the actual locations of photons reaching a CCD or film).

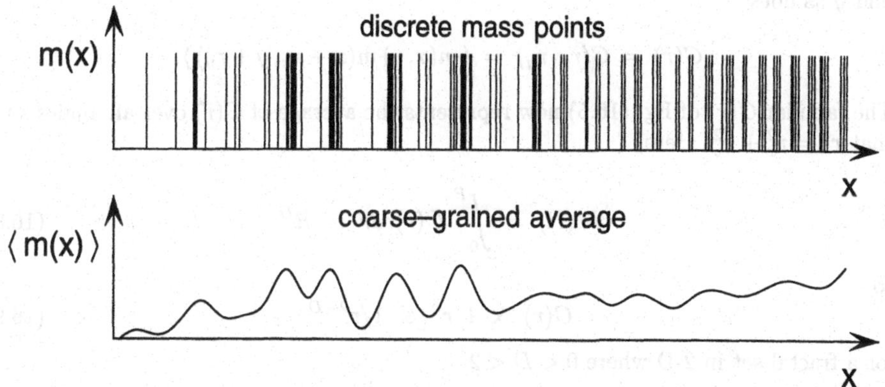

Figure 16.1: A sample discrete point mass distribution and its coarse-grained average.

There are a number of standard techniques for characterizing random functions $m(x)$ and their correlations [215]. The *autocorrelation* or *pair-correlation* function $C(r)$ is a quantitative measure of how the fluctuations in a $m(x)$ are related at x and $x + r$:

$$C(r) = \langle m(x)\, m(x+r)\rangle$$

where the brackets $\langle \ldots \rangle$ denote sample or ensemble averages. $C(r)$ is that probability of finding another mass point a distance r from an existing mass. Similar information is provide by the *mass-radius* relation. $M(R)$ is the conditional total average "mass" within a distance r of an existing point,

$$M(R) = \int_0^R C(r)dr \propto R^D \tag{16.4}$$

for a *set* of mass points characterized by a fractal dimension $0 < D < 1$. Thus,

$$C(r) \propto 1/r^{\eta} \propto 1/r^{1-D} \tag{16.5}$$

for a fractal set in 1-D.

An alternate characterization is provided by the Fourier coefficients $m(k) \propto \int e^{-2\pi i k x} dx$ and the spectral density $S(k) \propto |m(k)|^2$ in frequency or k-space. $S(k)$ and $C(r)$ are not independent. In most cases they are related by the Wiener-Khintchine relations [215]:

$$S(k) \propto \int C(r) \cos 2\pi k r\, dr \tag{16.6}$$

and

$$C(r) \propto \int S(k) \cos 2\pi k r\, dk. \tag{16.7}$$

Fast Fourier Transform (FFT) algorithms allow efficient direct computation of $S(k)$ from sample sequences and the estimation of $C(r)$ and D from $S(k)$ via Eqs. (16.5) and (16.7).

For the 2-D case of image analysis, the *measure* $m(x, y)$ becomes a function of x and y as does

$$C(\vec{r}) = C(r_x, r_y) = \langle m(x,y)\, m(x + r_x, y + r_y) \rangle.$$

The familiar $C(r)$ of Eg. (16.5) now represents the average of $C(\vec{r})$ over all angles such that $r^2 = r_x^2 + r_y^2$. Here,

$$M(R) = \int_0^R C(r) r dr \propto R^D \tag{16.8}$$

so

$$C(r) \propto 1/r^{\eta} \propto 1/r^{2-D} \tag{16.9}$$

for a fractal set in 2-D where $0 < D < 2$.

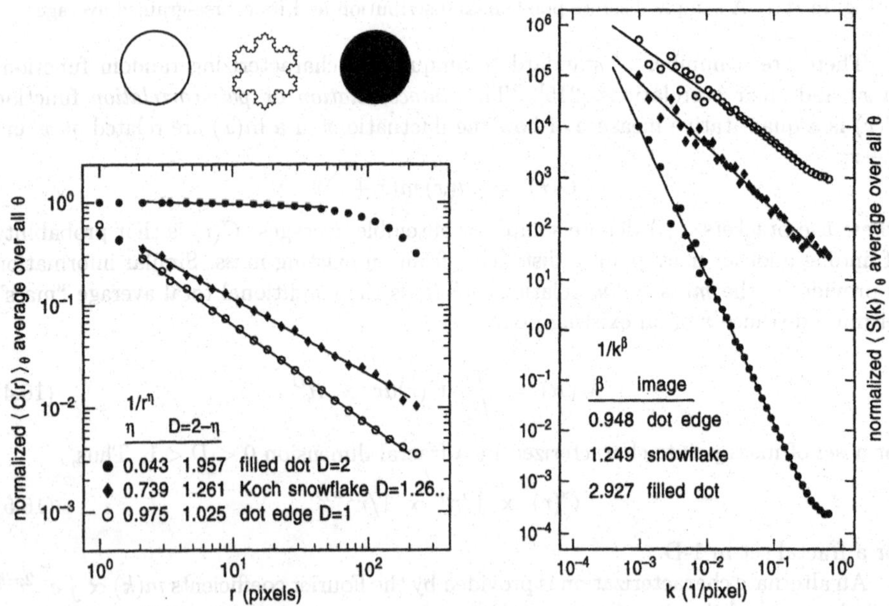

Figure 16.2: $C(r)$ image analysis of test shapes with known fractal dimension D. $S(\vec{k})$ was estimated for each 512 by 512 binary image via 2-D FFT. The 2-D cosine FFT of $S(\vec{k})$ gives $C(\vec{r})$ and averages over all angles produce $C(r)$. Least squares power-law fits to $C(r) \propto 1/r^{\eta}$ and $S(k) \propto 1/k^{\beta}$ are shown as solid lines.

As with the 1-D case above, 2-D FFT algorithms allow efficient computation of the 2-D $S(\vec{k})$ from $m(x,y)$, $C(\vec{r})$ via the 2-D equivalent of Eg. (16.7), and the estimation of D from $C(r)$ or $M(R)$ by averaging over angles. This procedure is illustrated in Fig. 16.2 which shows the results of 2-D FFT estimation of $C(r)$ for some simple test shapes on a 512 by 512 image. The angle average should be the last step in the procedure. Angle averaging the 2-D $S(\vec{k})$ to produce the 1-D $S(k)$ and using Eq. (16.7) for $C(r)$ may introduce errors.

16.3 Local Image Gradient and the Statistics of Natural Images

As shown in Fig. 16.2 this 2-D *global* characterization by a single fractal dimension D does work well for test images of well-defined fractal *sets*. There are, however, numerous problems in trying to apply this procedure directly to digitized images. What *measure* $m(x,y)$ should be measured? If $I(x,y)$ represents the coarse-grained image intensity (photon flux striking the sensor at x,y), any non-zero background, sensitivity changes across the image, or sharp edges can seriously corrupt $S(\vec{k})$ and $C(\vec{r})$.

Although Field [80] describes typical "natural scenes" as having $S(k) \propto 1/k^2$, his choice of test scenes avoided the common *light - dark edge* such as the land - sky boundary of many natural scenes. As shown in Fig. 16.3, natural scenes that include such a prominent boundary (the Norwegian coast scenes) have $S(k)$ that is indistinguishable from an simple test edge (sharp transition from white to black) with $S(k) \propto 1/k^3$. Only the gumtree image in Fig. 16.3, which was cropped to eliminate large-scale boundaries, has $S(k) \propto 1/k^2$ as reported by Field [80].

The major difficulties with $I(x,y)$ as the image measure $m(x,y)$ for analysis can be eliminated by using the *local image gradient* $|\nabla I(x,y)|$. $|\nabla I(x,y)|$ emphasizes the *edges* in an image[1]. Figure 16.4, which shows $|\nabla I(x,y)|$ spectral analysis for the same images as Fig. 16.3, demonstrates that the effects of large-scale gradients, uniform background, and prominent edges are all greatly reduced. All of the natural scenes now show $S_{|\nabla|}(k) \propto 1/k^{\beta}$ with $\beta \approx 1.5 - 1.6$.

Indeed, perceptual experiments on random fractal images [219] show that human estimates of fractal characteristics are based primarily on edge characteristics. Solid-filled and outline shapes have the same effect.

16.4 Local Fractal Mass Dimension

The above procedures provide a *global* characterization of an image by a single fractal dimension D. In many cases, particularly when the object(s) of interest represent only a subset of the original image, this is inappropriate. A more accurate and useful data reduction follows the multifractal lead and studies the distribution of *local* fractal dimensions.

[1] $|\nabla I(x,y)|$may be estimated by a local fit of $I(x,y)$in the neighborhood of x,y to the form $a\Delta x + b\Delta y + I(x,y)$. This gives $|\nabla I(x,y)| = (a^2+b^2)^{1/2}$.

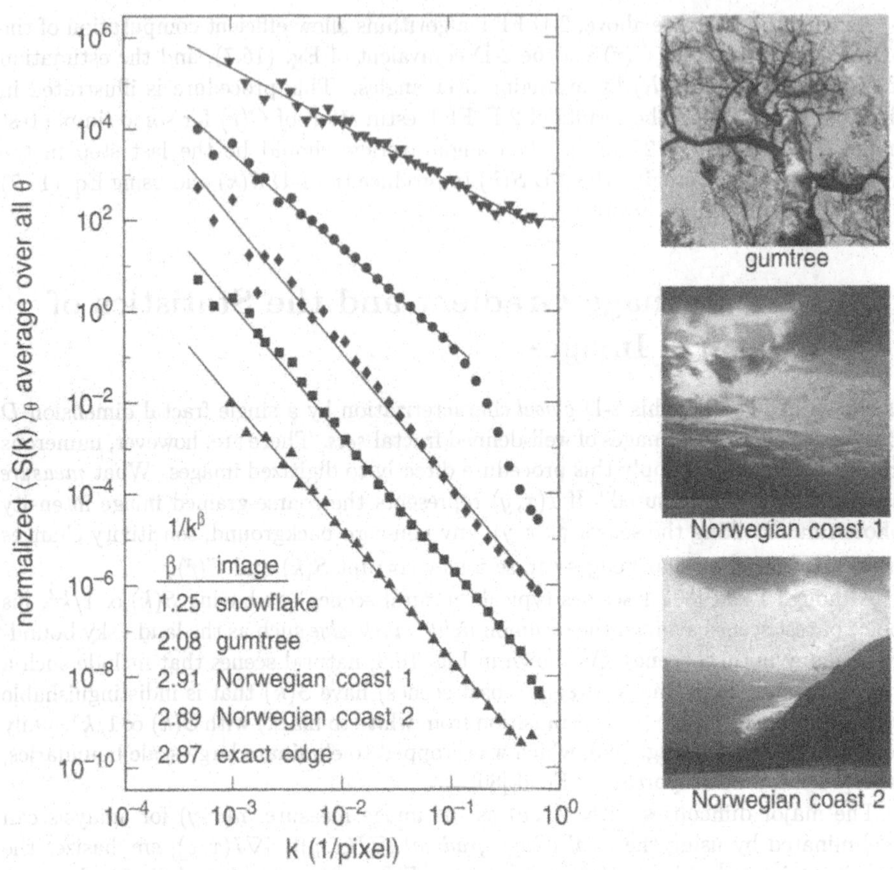

Figure 16.3: $S(k)$ for $I(x,y)$ image analysis of test shapes and natural scenes that contain a prominent edge such as land - sky boundary. In this case the natural scenes are indistinguishable from the $S(k) \propto 1/k^3$ of a test edge. $1/k^\beta$ power-law fits to $S(k)$ are shown as solid lines.

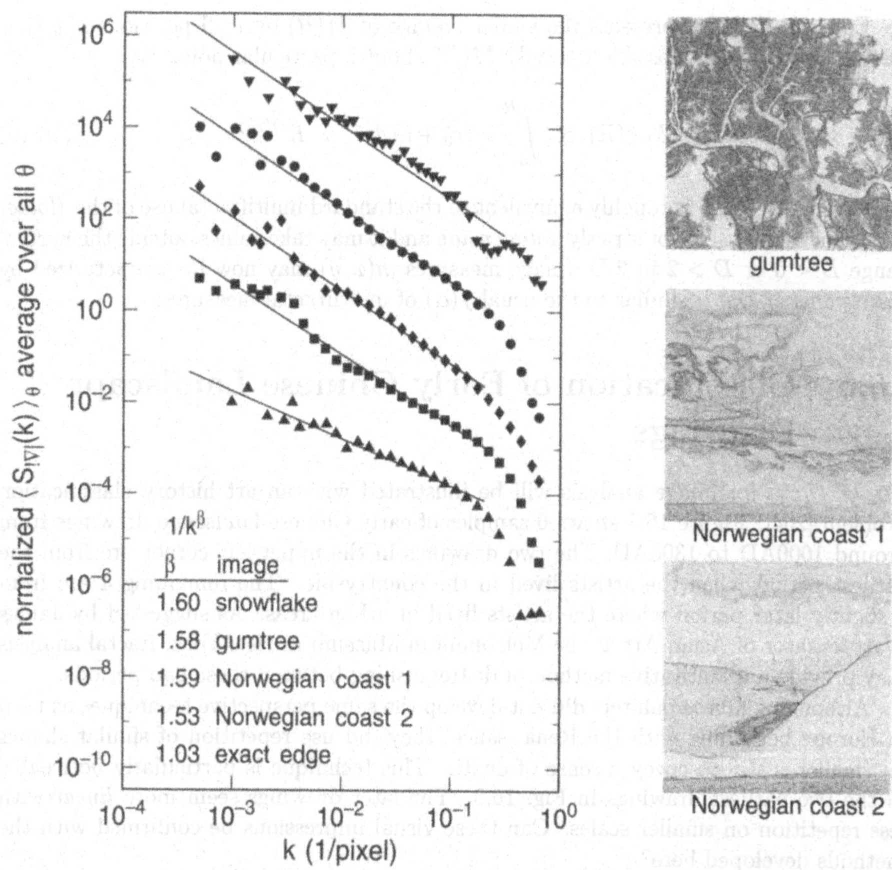

Figure 16.4: $S_{|\nabla|}(k)$ for $|\nabla I(x,y)|$ analysis of the same images as Fig. 16.3. The use of $|\nabla I(x,y)|$ rather than $I(x,y)$ allows even natural scenes with prominent edges to be distinguished from the $S_{|\nabla|}(k) \propto 1/k$ of a test edge. $1/k^\beta$ power-law fits to $S_{|\nabla|}(k)$ are shown as solid lines.

Equation (16.8) represents the global average of $M(R)$ over all positions x, y that have mass. It is also possible to study $M(R)$ about a particular point $\vec{r_0}$

$$M_{\vec{r_0}}(R) \propto \int_0^R m(\vec{r_0} + \vec{r})\, d^2\vec{r} \propto R^{D(\vec{r_0})} \tag{16.10}$$

This $D(\vec{r_0})$ or D_{local} is roughly equivalent to the standard multifractal use of the *Hölder* exponent α. D_{local} is not strictly a *dimension* and it may take values outside the normal range $D < 0$ or $D > 2$ in 2-D. Image measures $m(x, y)$ may now be characterized by histograms of D_{local} similar to the usual $f(\alpha)$ of multifractal measures.

16.5 Classification of Early Chinese Landscape Drawings

Use of D_{local} for image analysis will be illustrated with an art history classification problem [254]. Figure 16.5 shows 6 samples of early Chinese landscape drawings from around 1000AD to 1300AD. The two drawings in the upper-left corner are from the earliest period when the artists lived in the countryside. The remaining 4 are from a slightly later period where the artists lived in urban areas. As suggested by James Watt, curator of Asian Art at the Metropolitan Museum in New York, fractal analysis may provide a quantitative method of distinguishing between these two periods.

Although Chinese painters did not develop the same perspective techniques as used in Europe beginning with the Renaissance, they did use repetition of similar shapes on smaller scales to covey a sense of depth. This technique is particularly noticeable in the two earliest drawings in Fig. 16.5. The later drawings seem more *linear* with less repetition on smaller scales. Can these visual impressions be confirmed with the methods developed here?

Figure 16.6 which shows the local image slope $|\nabla I(x, y)|$, provides the starting measure $m(x, y)$. Figure 16.7 shows the corresponding D_{local} image where each pixel is shaded according to the least squares estimate of $M_{local}(R) \propto R^{D_{local}}$. Such a transform yields few differences between the images. This is confirmed with the D_{local} histogram in Fig. 16.10(a). Each image has $\langle D_{local} \rangle$ in the range 1.6 - 1.7.

16.6 Local Connected Fractal Dimension

To improve the classification it was necessary to take a lesson from percolation theory [251, 234]. Here the important fractal shapes require a consideration of local *connectivity*. As illustrated in Fig. 16.8, connectivity highlights important fractal shapes. The left image shows randomly occupied (white) sites with probability $p = 0.48$. At larger scales, for example when viewed from a distance, the image becomes a uniform gray with $D_{local} = 2$. The right image, however, shows the corresponding largest connected clusters which are known to be fractal with $D \approx 1.89$ [251, 234]. This concept may be applied to image analysis by including in $M(R)$ only those mass points that are *connected to* $\vec{r_0}$. In practice this involves selecting some minimum threshold m_0.

Figure 16.5: Sample digitized images $I(x, y)$ of 6 early Chinese landscape drawings. The two drawings in the upper-left corner are from the earliest period when the artists lived in the countryside. The remaining 4 are from a slightly later period where the artists lived in urban areas.

Figure 16.6: Local slope $|\nabla I(x,y)|$ for the Chinese landscape drawings of Fig. 16.5. This $|\nabla I(x,y)|$ is used as the fundamental image measure $m(x,y)$ for later analysis.

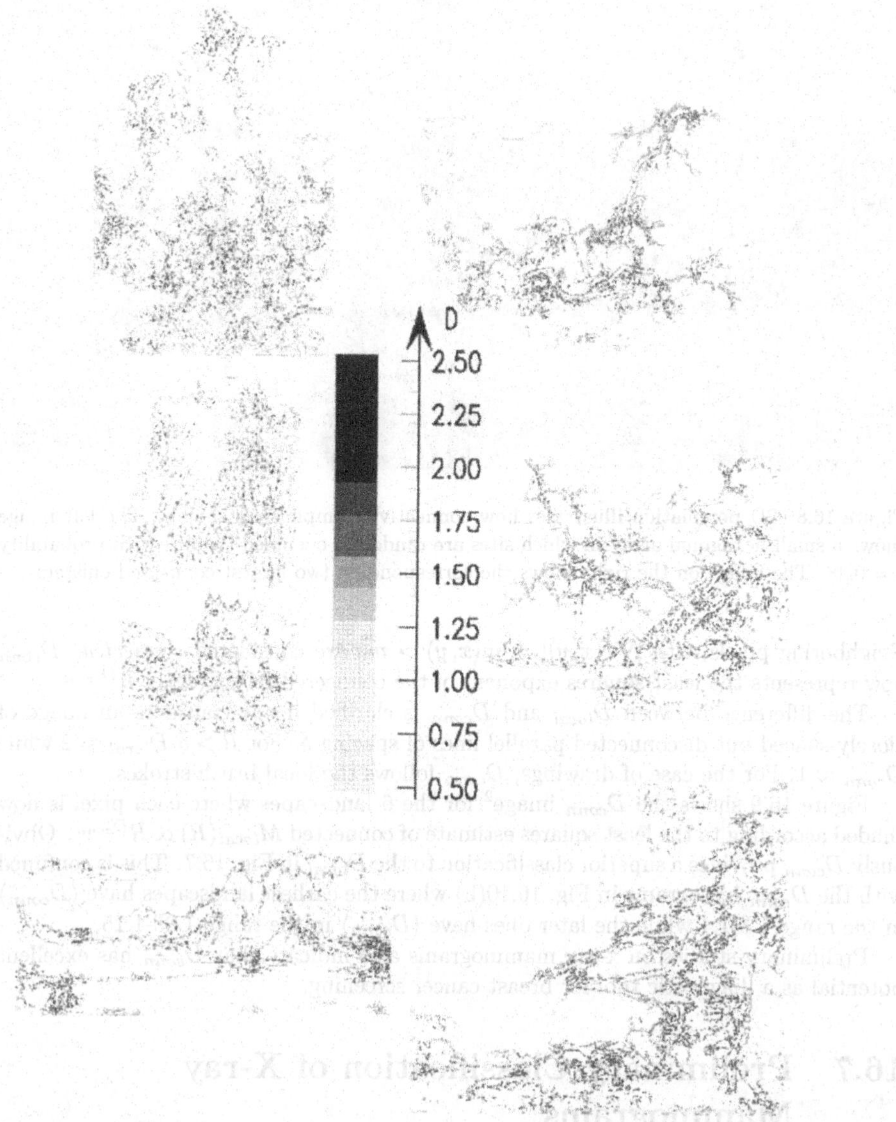

Figure 16.7: D_{local} transformed images for the Chinese landscape drawings of Fig. 16.5. Different shades represent different D_{local} from local fits of $M_{local}(R)$ to R^D including all $|\nabla I(x,y)|$ within $R < 32 pixels$.

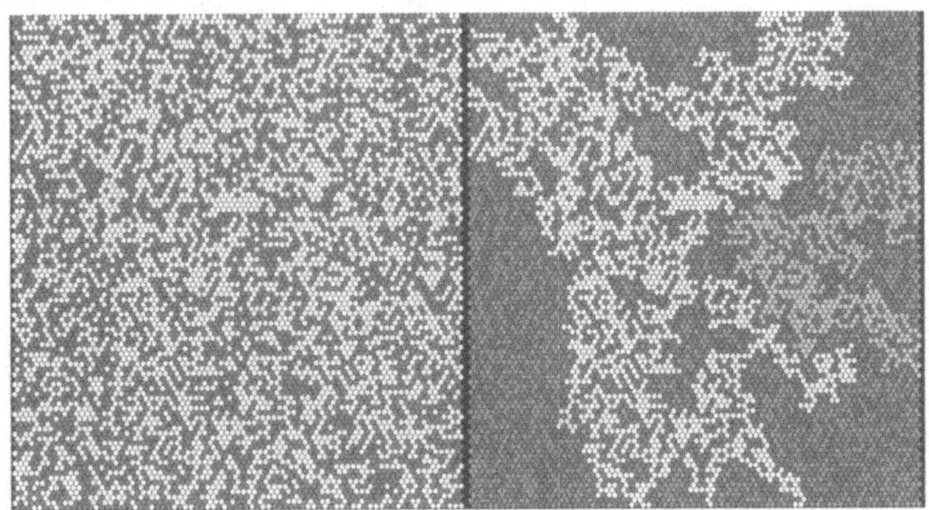

Figure 16.8: 2D percolation illustrates how connectivity emphasizes shapes. The left image shows a small hexagonal array in which sites are randomly occupied (white) with probability $p = 0.48$. The image on the right shows the corresponding two largest connected clusters.

Neighboring pixels with $|\nabla I(x,y)| = m(x,y) > m_0$ are considered *connected*. D_{conn} now represents the least squares exponent of the connected $M_{conn}(R) \propto R^{D_{conn}}$.

The difference between D_{local} and D_{conn} is clarified if one considers an image of closely spaced but disconnected parallel lines of spacing δ. For $R > \delta$ $D_{local} \approx 2$ while $D_{conn} \approx 1$. For the case of drawings, D_{conn} follows the local brush strokes.

Figure 16.9 shows the D_{conn} image[2] for the 6 landscapes where each pixel is now shaded according to the least squares estimate of connected $M_{local}(R) \propto R^{D_{conn}}$. Obviously D_{conn} provides a superior classification to the D_{local} in Fig. 16.7. This is confimed with the D_{conn} histograms in Fig. 16.10(b) where the earliest landscapes have $\langle D_{conn} \rangle$ in the range 1.3-1.4 while the later ones have $\langle D_{conn} \rangle$ in the range 1.05-1.15.

Preliminary studies on x-ray mammograms also indicate that D_{conn} has excellent potential as a diagnostic tool for breast cancer screening.

16.7 Preliminary Classification of X-ray Mammograms

Approximately 12 x-ray mammograms were obtained from Prof. Barry Thornton (Mathematics Dept., University of Technology, Sydney, Australia) and Dr. Cherrell Hirst (Breast Clinic, Wesley Hospital, Brisbane, Australia) that contained readily distinguishable malignant and benign calcifications.

[2]The choice of threshold m_0 may be based on the image statistics for a particular class. Given the quantized $|\nabla I(x,y)|$histogram H_m for the images in Fig. 16.6, m_0 was defined as the weighted average, $m_0 = \sum mH_m \,/\, \sum H_m$.

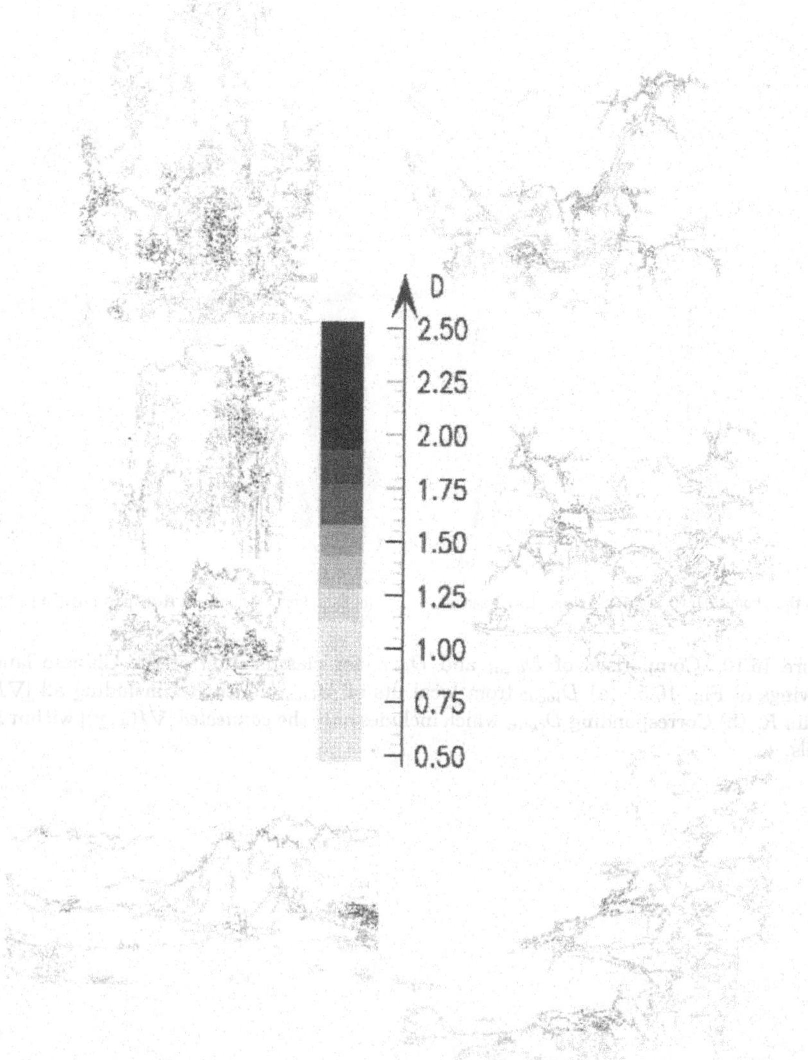

Figure 16.9: D_{conn} shaded images for the Chinese landscape drawings of Fig. 16.5. Different shades represent different D_{conn}. $M_{local}(R)$ includes only the *connected* $|\nabla I(x,y)|$ within R.

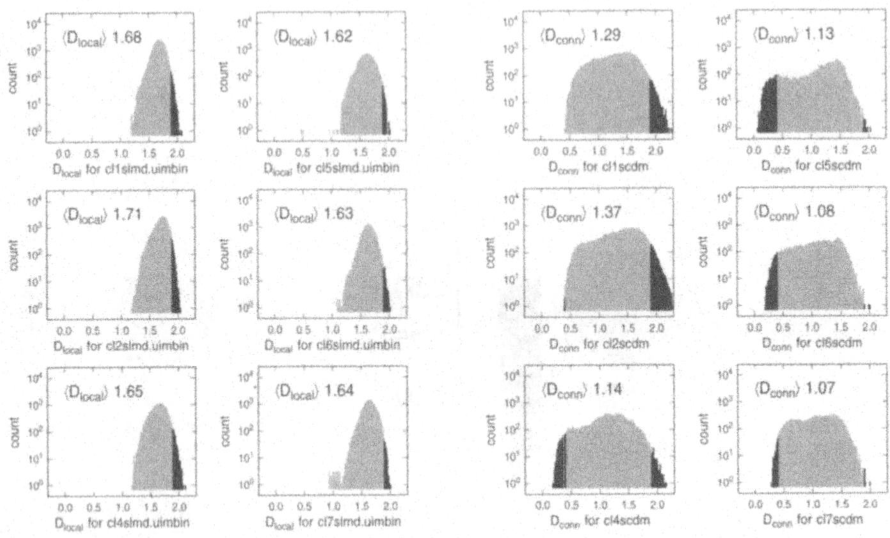

(a) D_{local} from M(R) fits with R<32 pixels D>1.9 darkest

(b) D_{conn} from M(R) fits with R<32 pixels D<0.4 D>1.9 darkest

Figure 16.10: Comparison of D_{local} and D_{conn} for classification of the Chinese landscape drawings of Fig. 16.5. (a) D_{local} from local fits of $M_{local}(R)$ to R^D including all $|\nabla I(x,y)|$ within R. (b) Corresponding D_{conn} which includes only the *connected* $|\nabla I(x,y)|$ within $R < 32$ pixels.

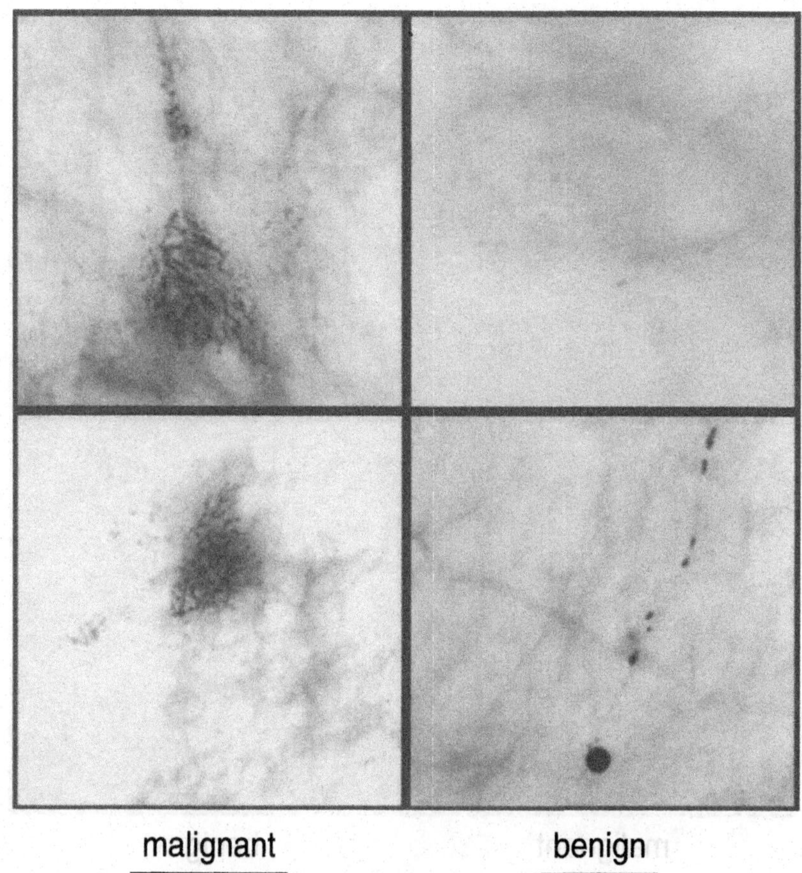

Figure 16.11: Sample digitized x-ray mammograms.

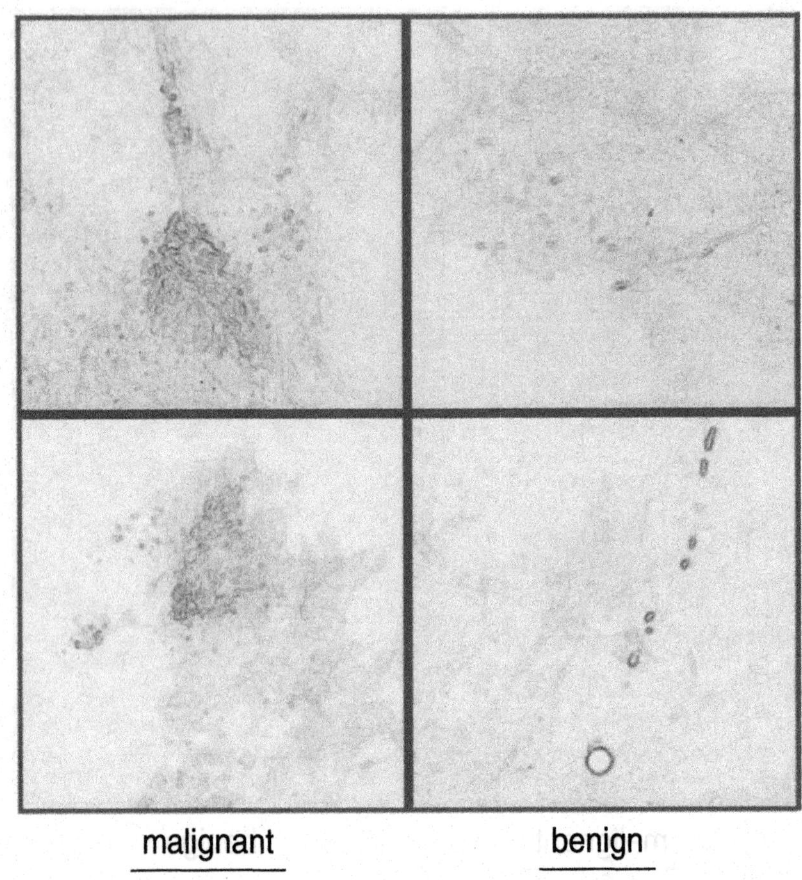

Figure 16.12: Local slope $|\nabla I(x, y)|$ for edge enhancement of the x-ray mammograms from Fig. 16.11.

Sample of easily diagnosed x-ray mammograms and their edge enhanced versions are shown in Fig. 16.11 and Fig. 16.12. All samples were digitized to 8 bits (0-255) at a resolution of 512 by 512 from 35mm slides of the original x-rays. The two malignant samples on the left are easily distinguished from the two benign samples on the right. Nonetheless, traditional methods of local fractal analysis in which D_{local} includes all clusters (not shown here) gives little consistent distinction.

Fig. 16.13 shows the corresponding histograms of the local connected fractal dimensions D_{conn} for the samples. The distinction is now obvious. One quantitative measure of the analysis is the average $\langle D_{conn}(x,y)\rangle$ shown as $\langle D\rangle$ above each of the histograms. $\langle D\rangle$ is higher for the malignant samples. The benign sample in the upper right is most similar to the malignant samples and its $\langle D\rangle$ is correspondingly closest.

Fig. 16.14 shows corresponding D_{conn} of the enhanced images. Each pixel is shaded according to its local $D_{conn}(x,y)$ and the malignant regions are clearly emphasized.

16.8 Discussion

The human perceptual system has evolved over millions of years in a natural fractal environment. Only recently, by evolutionary time scales have we found ourselves in a primarily Euclidean environment of straight lines and few spatial scales. Our visual system shows a particular response to random fractals [219]. Nevertheless, it has proven extremely difficult to quantitatively mimic with fractal analysis the classification and feature detection that the eye performs with ease.

As demonstrated here with early Chinese landscape drawings and x-ray mammograms, significant image processing is required to reliably distinguish between the two classes of images. Care must be taken in the decision of which *measure* to measure. For many natural scenes the image *edges* $|\nabla I(x,y)|$ are far less prone to artifacts than the original intensities $I(x,y)$.

For *local* enhancement and feature detection, the concept of *local fractal dimension* D_{local} may be defined analogous to the multifractal $f(\alpha)$. Even with $|\nabla I(x,y)|$, however, D_{local} detects few of the differences perceived by eye.

The critical step here is the use of local *connected* fractal dimension D_{conn} analogous to percolation theory. D_{conn} only considers the local connected portions of the measure. Histograms of D_{local} are the best fractal tool for automatically classifying the early Chinese landscapes. These preliminary results suggest D_{conn} may also be of significant practical use in the quantification, automatic screening, and enhancement of medical images.

Acknowledgments

The author wishes to thank Dr. James Watt, curator of Asian Art at the Metropolitan Museum, New York for suggesting the problem of fractal analysis of early Chinese landscape drawings and for providing image samples; and Prof. Barry Thornton, at the University of Technology, Sydney for suggesting the use of fractals in mammogram analysis and Dr. Cherrell Hirst for providing the x-ray samples.

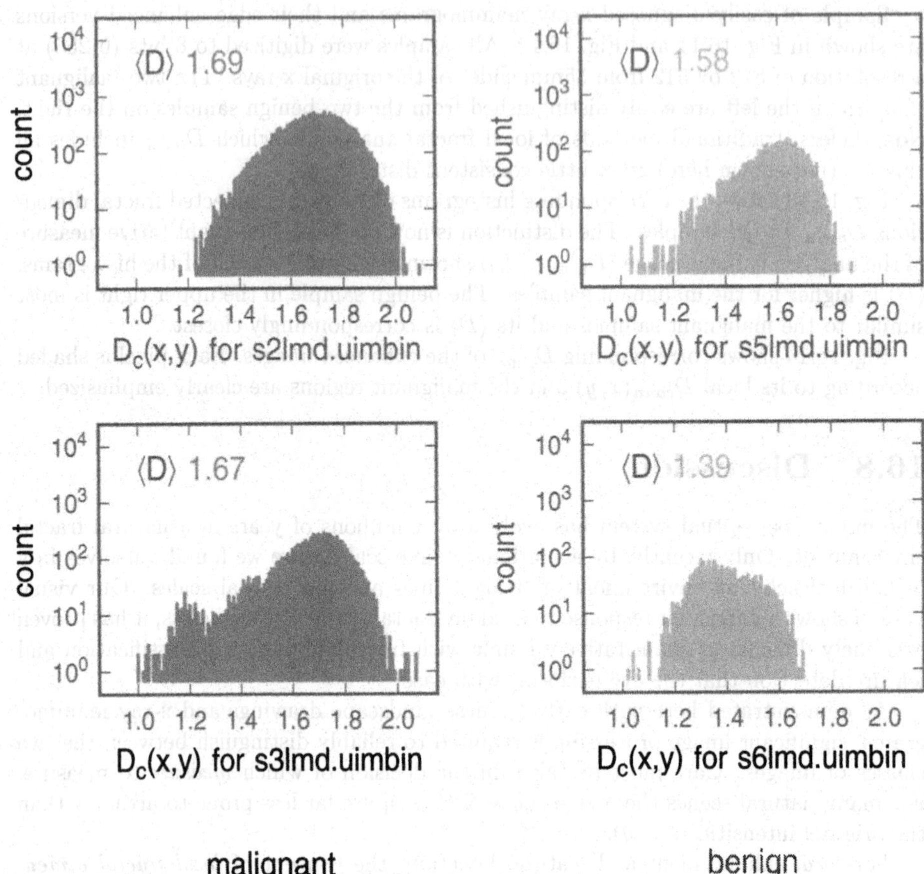

Figure 16.13: Histograms of D_{conn} from the x-ray mammograms. D_{conn} from local fits of $M_{local}(R)$ to R^D including only the *connected* $|\nabla I(x,y)|$ within $R < 32$ pixels.

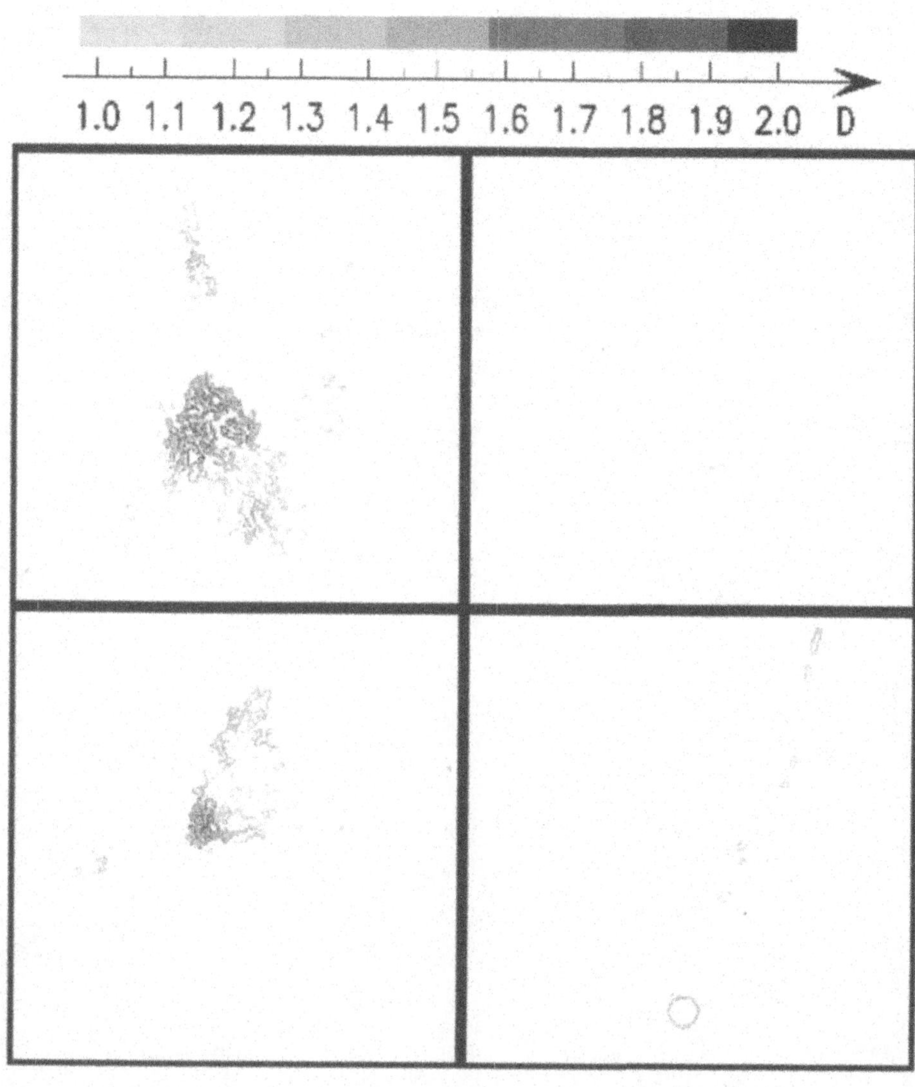

Figure 16.14: Transformed images based on D_{conn} of the x-ray mammograms. D_{conn} from local fits of $M_{local}(R)$ to R^D including only the *connected* $|\nabla I(x,y)|$ within $R < 32$ pixels.

Chapter 17

Introduction to the Multifractal Analysis of Images

Jacques Lévy-Véhel

Abstract: After a brief review of some classical approaches in image segmentation, the basics of multifractal theory and its application to image analysis are presented. Practical methods for multifractal spectrum estimation are discussed and some experimental results are given.

17.1 Introduction

Image Analysis deals with teaching a computer how to "see". It has applications in many fields, including robotics, medical imaging (e.g. automatic detection of diseases from images), satellite imaging (e.g. monitoring vegetation evolution), and a lot more. In most cases, the aim is either to let the computer process data which are too numerous to be analyzed by human beings, or to be able to detect features in images that are not easily found with the eye.

A first step towards these goals is to detect and localize objects and features in images. A considerable body of research has been devoted to this problem in the past twenty years. It has raised difficult problems, some of which have lead to elegant mathematical developments, even though several issues are still unsolved.

In this paper, we describe the multifractal approach to image analysis. The essential difference between this approach and "classical" ones lies in the way they handle irregularities. In order to clearly illustrate this and other important differences, we briefly recall in section 2 some of these classical methods. Section 3 presents the principles of the multifractal theory, whose application to image analysis is detailed in section 4. Practical issues and experiments are described at the end of the paper.

17.2 A Quick Review on Some "Classical" Methods

17.2.1 Canny's Edge Detector

Canny's detector is based on a gradient approach [41]. The image is modeled by a function I from $\mathbb{R}^2$ to $\mathbb{R}^+$. Edges are assumed to correspond to sharp variations of the grey levels, i.e. of I. These sharp variations may be detected by looking at the maxima of the gradient of I. Two problems arise:

- Because the data are discrete, it is not straightforward to compute a gradient.
- In many cases, noise is present in the image, which induces spurious edges.

The basic idea is to filter the image in order to get rid of these problems. More precisely, a linear filter f is defined and edges are detected as follows:

- Filter the image I with f: $I * f$.
- Compute the gradient of the filtered image: $(I * f)' = I * f'$.
- Compute the norm of the gradient $|(I * f)'|$.
- Select the points where the norm of the gradient is maximum in the gradient's direction.

The problem of edge detection is thus reduced to the one of finding a linear filter f. f should satisfy the following conditions:

Detection: the edge detector should react to edges.

Localization: edges should be precisely localized.

Unicity of response: an edge should trigger the detector only once.

Different types of edges may be considered:

- Step edge or 0th order discontinuity (figure 17.1).
- Roof edge or 1st order discontinuity (figure 17.2) .

Other types of edges include corners, lines, ... Some assumptions are then made about the nature of the noise in the image (additive/multiplicative, Gaussian/Rayleigh/K-distributed, ...), and, for some simple cases, it is possible to derive an explicit optimal filter f w.r.t. the three criteria above. For instance, for an ideal step edge with additive Gaussian noise, the optimal filter is:

$$\begin{aligned} f(x) &= a_1 e^{\alpha x} \sin wx + a_2 e^{\alpha x} \cos wx \\ &\quad + a_3 e^{-\alpha x} \sin wx + a_4 e^{-\alpha x} \cos wx \qquad \text{(Canny)} \end{aligned}$$

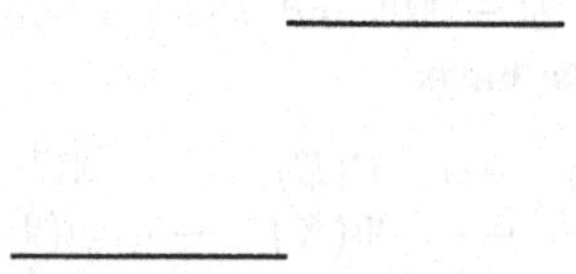

Figure 17.1: Model of a step edge.

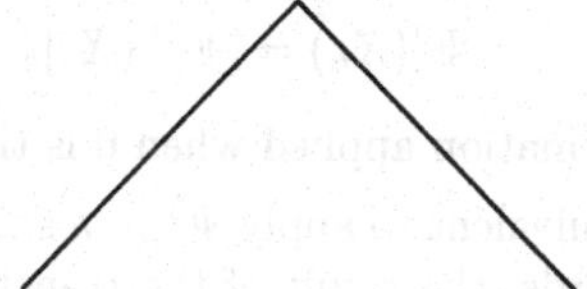

Figure 17.2: Model of a roof edge.

Usual filters in this approach are:

$$
\begin{aligned}
f(x) &= ce^{-\alpha|x|} && \text{(Shen–Castan)} \\
f(x) &= (1 + c|x|)e^{-\alpha|x|} && \text{(Deriche)}
\end{aligned}
$$

where α is a parameter that tunes the smoothing of the signal. It allows to adjust the trade-off between localization and robustness.

17.2.2 Mathematical Morphology

Mathematical Morphology was first introduced by G. Matheron in the 1960's and then developed by a number of authors ([231], [228])

The basic idea underlying Mathematical Morphology is that the notion of a geometrical structure or texture is not purely objective. It does not lie in the phenomenon itself, nor in the observer, but somewhere between the two. Mathematical Morphology quantifies this intuition by introducing the concept of structuring elements. The structuring elements are chosen by the analyst, and they interact with the studied object to modify its shape and transform it into a simpler one. The result thus reflects information both on the structuring elements and on the object. An important issue is then the choice of the structural elements. This choice is based on several principles which can be loosely stated as:

- "We only see what we want to look at."
- "To perceive an image, is to transform it."

More precisely, given a set X, a morphological operation on X is by definition the composition of a transformation Ψ of X into a new set $\Psi(X)$ followed by a measure

μ of $\Psi(X)$. Thus, if we note $E = \mathbb{R}^n$ and $\mathcal{P}(E)$ a family of subsets of E satisfying certain technical conditions, we have:

$$\begin{array}{ccccc} \mathcal{P}(E) & \longrightarrow & \mathcal{P}(E) & \longrightarrow & \mathbb{R}^+ \\ X & \longmapsto & \Psi(X) & \longrightarrow & \mu(\Psi(X)) \end{array}$$

A morphological operation must satisfy four constraints:

1. **Compatibility under translation:**

   Let 0 be an "origin" in E and $X_h = \{x + h, x \in X\}$. Then we must have:

   $$\Psi^0(X_h) = [\Psi^{-h}(X)]_h$$

   where Ψ^0 is the transformation applied when 0 is the origin.

   This means that it is equivalent to apply Ψ^0 to a shifted version of X or to apply Ψ^{-h} to X. In other words, the result of the transformation should not depend on the origin.

2. **Compatibility under change of scale:**

   Let $X_\lambda = \{\lambda x, x \in X\}$. Then, we must have

   $$\Psi_\lambda(X) = \lambda\Psi\left(X_{\frac{1}{\lambda}}\right)$$

   where Ψ_λ is the transformation applied when the coordinate axes are scaled by λ. In other words, the transformation must be independent of an arbitrary magnification/reduction of the object.

3. **Local knowledge:**

   $\forall Z'$ bounded, $\exists Z$ bounded such that:

   $$(\Psi(X \cap Z)) \cap Z' = \Psi(X) \cap Z'$$

   In other words, for any bounded set Z' in which we want to know $\Psi(X)$, we can find a bounded set Z in which the knowledge of X is sufficient to locally perform the transformation. This requirement stems from the fact that we generally do not see the whole object X but only a part of it, $X \cap Z$.

4. **Semi Continuity:**

   This principle insures that, if we have a decreasing sequence of closed sets $(X_n)_{n\geq 1}$ converging to $X \neq \emptyset$, then $(\Psi(X_n))_{n\geq 1}$ should converge to $\Psi(X)$, as soon as Ψ is increasing (i.e. $A \subset B \Longrightarrow \Psi(A) \subset \Psi(B)$).

   To be more precise, we have to define a fundamental tool of Mathematical Morphology, the "Hit or Miss" topology. Let $\mathcal{F}$ be the space of all closed subsets of $\mathbb{R}^n$. Given two finite sequences $(G_1, \ldots, G_m)$ of open sets and $(K_1, \ldots, K_p)$ of compact sets of $\mathbb{R}^n$, the class of all the closed sets that hit every G_i and miss every K_j defines an open neighborhood in $\mathcal{F}$. The induced topology is called the Hit or Miss topology (see figure 17.3).

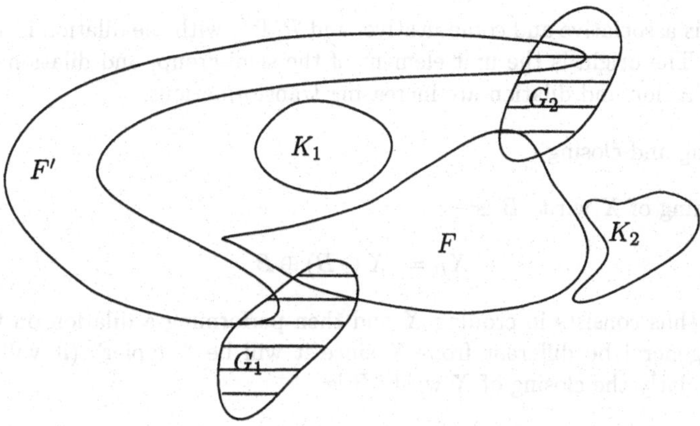

Figure 17.3: F and F' belongs to the same neighborhood defined by K_1, K_2 and G_1, G_2.

The following result is fundamental.

Theorem: *The space $\mathcal{F}$ equipped with the Hit or Miss topology is compact and has a countable base.*

The fourth principle now reads: Ψ satisfies the fourth principle iff it is semi continuous w.r.t. the Hit or Miss topology.

This last principle implies that morphological operations are ideally performed in the continuous space, and not in the discrete space of the pixels.

Basic morphological transformations include:

- Erosion

  Let B be a structuring elements. The eroded set of X is defined as:

$$Y = \{x/B_x \subset X\} = \cap_{y \in B_0} X_{-y}$$

  Let:

$$\check{B} = \cup_{y \in B}\{-y\} \text{ and } X \ominus B = \cap_{b \in B} X_b$$

  Then: $Y = X \ominus \check{B}$ ($X \ominus B$ is the classical Minkowski substraction).

  Thus Y is the set of points where we can center the structuring element with the constraint that it is included in X.

- Dilation

  It is the dual operation of erosion and is defined through Minkowski addition:

$$X \oplus \check{B} = \{x/B_x \cap X \neq \emptyset\} = \cup_{y \in \check{B}} X_y$$

  The dilated set is thus the set of points where we can center B with the constraint that it intersects (or hits) X.

Dilation is associative and commutative, and $\mathcal{P}(\mathbb{R}^2)$ with the dilation is an Abelian semi-group. The origin is the unit element of the semi-group, and dilation distribute the union. Erosion and dilation are increasing transformations.

- Opening and closing

The opening of X w.r.t. B is:

$$X_B = (X \ominus \check{B}) \oplus B$$

An opening thus consists in eroding X and then performing a dilation on the result. X_B will in general be different from X since it will be "simpler" (it will have less details). Similarly the closing of X w.r.t. B is:

$$X^B = (X \oplus \check{B}) \ominus B$$

i.e. a dilation followed by an erosion.

Intuitively, an opening smoothes the contours and suppresses small "islands". Conversely, a closing blocks up small "lakes".

The main properties of opening and closing are:

- $(X^c)_B = (X^B)^c$ and $(X_B)^c = (X_c)^B$
  (the opening of the complement of X is the complement of the closing of X, and conversely: opening and closing are dual w.r.t. the complementation.)
- The morphological opening transform satisfies the three properties defining an opening in algebra. It is:
  - anti-extensive: $X_B \subset X$,
  - increasing: $X \subset X' \Longrightarrow X_B \subset X'_B$,
  - idempotent: $(X_B)_B = X_B$.
- The closing transform verifies:

$$X^B \supset X$$

$$X \subset X' \Longrightarrow X^B \subset X'^B$$

$$(X^B)^B = X^B$$

To conclude, let us say a few words about the choice of the structuring element. Different structuring elements lead to different information on the studied set: this is illustrated on figure 17.4, where B has given some information relative to the size of X, as B' has given some information about the spatial distribution of X.

As noted above, it thus makes no sense to speak of the characteristic features of X, because we "only see what we want to look at": the information we get depends on the structuring element used. Some basic rules to choose the structuring elements are:

- simplicity: B should have in any case a much simpler geometric structure than X,

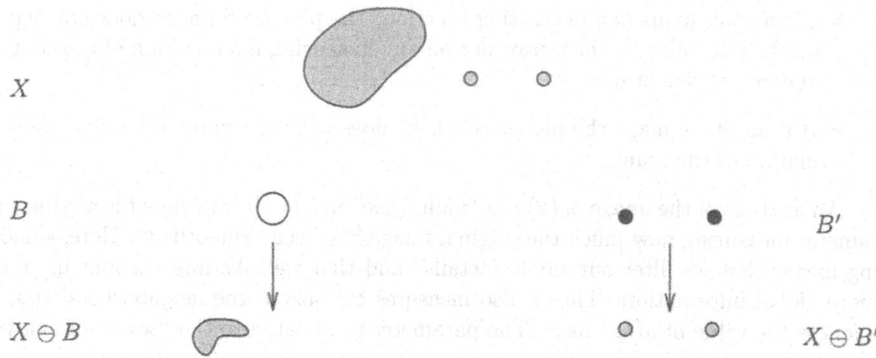

Figure 17.4: Erosion with two different structuring elements.

- B should be bounded,
- B should have "extreme" properties: if an isotropic filtering is needed, B should consist in a union of circular annuli with same center. If we want to perform an anisotropic filtering, B should be a union of aligned segments rather than elongated ellipses.
- Convex structures should be used to investigate size distributions, and clusters of isolated points should be used to measure set covariances.

17.2.3 Image Multiscale Analysis

This approach formalizes some basic assumptions that are usually made in image analysis and shows that, under these assumptions, the analysis must obey a simple equation. The merits of this approach are:

- to clearly state which assumptions are made,
- to discuss the validity of these assumptions and their relative importance,
- to derive rigorously the consequences of these assumptions and to obtain the unexpected result that in general only one equation is compatible with them.

This approach has been extensively studied in recent years. Useful references are [141], [6], [209]. Here, an image is a "brightness" function $u_0(x)$, for x in a subset of $\mathbb{R}^2$. u_0 is assumed to be continuous and to verify :

$$\exists N, C / \forall x, ||(1+|x|)^{-N} u_0(x)|| \leq C.$$

The set of all such functions will be denoted by $\mathcal{F}$. u_0 has no "absolute" meaning, but rather is an element of an equivalence class. More precisely, an image is an equivalence class of functions $g(u_0(Ax))$ where:

- g is a continuous non decreasing function: the perceived image does not depend on the sensibility of the sensor nor on any increasing modification of the contrast and/or the brightness.

- A is an affine map: the perceived image does not depend on any affine transformation of the plane.

An analysis of the image $u_0(x)$ is a "multiscale" image $u(t, x)$ where t is a "time" parameter measuring how much the original image has been "smoothed". Here, smoothing means that we filter out small "details" and that we take into account more and more global information. Thus t also measures the size of the neighborhood that influences the value of $u(t, .)$ at x. The parameter t is related to the "scale" of analysis.

The series of operation leading from $u_0(x) = u(0, x)$ to $u(\infty, x)$ is sometimes referred to as the Visual Pyramid. The basic principles that a visual pyramid must obey are the following ones:

- Causality:

  This principle states that what happens at scale t cannot influence what happens at scale $t' < t$. Furthermore, the output at scale t can be computed from the output at scale $t - h, h > 0$.

  Let:

$$T_t \quad : \quad \begin{array}{ccc} \mathcal{F} & \rightarrow & \mathcal{F} \\ u_0 & \longmapsto & u(t, .) \end{array}$$

  Then

$$\exists\ T_{t+h,t} \quad : \quad \begin{array}{ccc} \mathcal{F} & \longrightarrow & \mathcal{F} \\ u(t, .) & \longmapsto & u(t+h, .) \end{array}$$

  such that:

$$T_{t+h} = T_{t+h,t} T_t, \qquad \text{with} \qquad T_0 = Id$$

- Local Comparison Principle

  This principle states that $T_{t+h,t}$ acts "locally", i.e. that $T_{t+h,t}u(x)$ should depend upon the values of u only in a neighborhood of x. Let u and v be two images in $\mathcal{F}$. If for some $\varepsilon > 0$, $u(y) > v(y)$ for $y \in B(x, \varepsilon), y \neq x$, then for all h small enough,

$$(T_{t+h,t}u)(x) \geq (T_{t+h,t}v)(x)$$

  This implies in particular that no new boundaries are created when the scale increases.

- Regularity:

  This principle states that a smooth image should evolve in a smooth way: let u be a quadratic form of $\mathbb{R}^2$. Then there exists a function $F(u, x, t)$ continuous w.r.t. u such that:

$$\frac{(T_{t+h,t}u - u)(x)}{h} \longrightarrow F(u, x, t) \text{ when } h \longrightarrow 0$$

The last principles deal with the fact that an image is really an equivalence class. The operators should thus commute with some perturbations of the data:

- Morphological Invariance:

  Let g be a continuous non decreasing function.

  Then:

  $$gT_{t+h,t} = T_{t+h,t}g$$

- Euclidean Invariance:

  Let A be an isometry. Then:

  $$AT_{t+h,t} = T_{t+h,t}A$$

- Affine Invariance:

  Here we only impose a weak commutation property, because A may reduce or enlarge the image. For instance, if A is a zoom by a factor $\lambda > 0$: $Au(x) = u(\lambda x)$

  The affine invariance principle is stated as follows: there exists a C^1 function $t'(t, A) \geq 0$ such that:

  $$AT_{t'(t,A),t'(s,A)} = T_{t,s}A$$

  Moreover, $\frac{\partial t'}{\partial \lambda}(t, \lambda Id)$ is positive for $t > 0$.

  This means that the result of the multiscale analysis T_t is independent of the size and the position in space of the analyzed features.

Although the most general invariance should be projective, the affine approximation is valid if the studied objects are sufficiently small or sufficiently far away from the sensor. The condition on $\frac{\partial t'}{\partial \lambda}$ means that the analysis scale increases with the size of the picture. It may be shown that, if $t \longrightarrow T_t$ is a one-to-one family of operators satisfying the affine invariance principle and the causality principle, then:

- $t'(t, B)$ only depends on t and $|detB|$
- t' is increasing w.r.t. t
- there exists an increasing differentiable function

  $$\sigma : [0, \infty] \longmapsto [0, +\infty]$$

  such that:

  $$t'(t, B) = \sigma^{-1}(\sigma(t)|detB|^{1/2})$$

σ acts as a rescaling function: if $S_t = T_{\sigma^{-1}(t)}$ then $t'(t, B) = t|detB|^{1/2}$ for the rescaled analysis S_t.

The following results provide a complete solution to the problem of image analysis under the principles and assumptions presented above:

- If a multiscale analysis is causal, local, Euclidean invariant and linear, then $u(t, x) = (T_t u)(x)$ obeys the heat equation:

$$\frac{\partial u}{\partial t} = \Delta u$$

- If a multiscale analysis is causal, local and regular, then $u(t, x)$ is a viscosity solution of:

$$\frac{\partial u}{\partial t} = F(D^2u, Du, u, x, t)$$

  Where Du is the gradient of u, F is defined in the regularity principle and is non decreasing w.r.t. to D^2u. If u_0 is a bounded uniformly continuous image, this equation has a unique viscosity solution.

- If a multiscale image analysis is causal, local, regular, morphological and Euclidean invariant, then:

$$\frac{\partial u}{\partial t} = |Du| F\left(div\left(\frac{Du}{|Du|}\right), t\right)$$

  where F is non decreasing w.r.t. its first argument. The quantity $div\left(\frac{Du}{|Du|}\right)(x)$ can be interpreted as the curvature of the level line of the image $u(t, x)$ passing by x. In the particular case where if $F = +1$ is constant, we get:

$$\frac{\partial u}{\partial t} = |Du|$$

  This corresponds to a morphological dilation with a ball $B(x, t)$ as a structuring element. Likewise, if $F = -1$, we obtain the morphological erosion.

The "most invariant" model is the following.

Theorem: *There is a single causal, local, regular morphological and affine invariant multiscale analysis, whose equation is:*

$$\frac{\partial u}{\partial t} = |Du| \left(t.div\left(\frac{Du}{|Du|}\right)\right)^{1/3}$$

17.3 The Multifractal Approach

17.3.1 Overview

In this approach, the image is not modeled by a function but by a measure μ. This modeling allows to emphasize the fundamental role played by resolution. In this framework, the basic assumptions are:

- The relevant information for the analysis can be extracted from the Hölder regularity of μ.
- Three levels contribute to the perception of the image:
  - the pointwise Hölder regularity of μ at each point,
  - the variation of the Hölder regularity of μ in local neighborhoods.
  - the global distribution of the regularity in the whole scene.
- The analysis should be translation and scale invariant.

Compared to the approaches described above, the multifractal one has the following specific features:

- As in mathematical morphology (MM) and image multiscale analysis (IMA), translation and scale invariance principles are fulfilled.
- There is no "local comparison principle" or "local knowledge principle". On the contrary, information about whole parts of the image is considered essential to analyze each point (see section 17.4.1).
- The most important difference between classical and multifractal methods lies in the way they deal with regularity. While the former aim at obtaining smoother versions of the image (possibly at different scales) in order to get rid of irregularities, the latter tries to extract information directly from the singularities. Edges, for instance, are not considered as points where large variations of the signal still exist after smoothing, but as points whose regularity is different from the "background's" regularity in the raw data. Such an approach makes sense for "complex" images, in which the relevant structures are irregular in nature. Indeed, an implicit assumption of MM and IMA is that the useful information lies at boundaries between originally smooth regions, so that it is natural to filter the image. But there are cases (e.g. in medical imaging, satellite or radar imaging) where the meaningful features are essentially singular.
- As in MM, and contrarily to IMA, the multifractal approach does not assume that there is a universal scheme for image analysis. Rather, depending on what one is looking for, different measures may be used.
- Both MM and IMA consider the relative values of the grey levels as the basic information ("morphological invariance"). Here the Hölder regularity is considered instead. This again is justified in situations where the important information lies

in the singularity structure of the image. As an example, figure 17.5 displays the lumping of two 2D Weierstrass functions with different exponents. On this image, the use of classical methods with standard parameters would result in the detection of several edges. The multifractal approach, on the other hand, will detect only one vertical line in the middle. This case of a fractal texture is somewhat extreme. Experiments on real images are shown at the end of the paper.

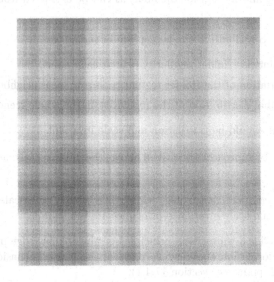

Figure 17.5: Lumping of two 2D Weierstrass functions.

The natural theoretical framework that allows to perform an analysis based on the above ideas is an extension of multifractal analysis that deals with multifractal correlations and sequences of Choquet capacities. A complete presentation of these extensions goes beyond the scope of this paper, and only the basics principles are presented below.

Before concluding this paragraph, let us note that the multifractal approach does not assume that the image is a "fractal" object or a "multifractal" measure, whatever meaning is given to these words. Instead, a multifractal analysis of the image is performed, without making any assumptions on its structure or regularity.

17.3.2 Choquet Capacities

Let E be a set, $\mathcal{P}(E)$ the power set of E. A paving on E is a set $\mathcal{E}$ of subsets of E containing the empty set and stable under finite union and finite intersection. The pair $(E, \mathcal{E})$ is called a paved space. A Choquet $\mathcal{E}$-capacity c is a function:

$$\mathcal{P}(E) \to \overline{\mathbb{R}}$$

such that:

- c is non decreasing,
- if (A_n) is an increasing sequence of subsets of E, then:

$$c(\cup_n A_n) = \sup_n c(A_n),$$

- if (A_n) is a decreasing sequence of elements of $\mathcal{E}$, then:

$$c(\cap_n A_n) = \inf_n c(A_n)$$

In image analysis, it is useful to consider sequences of Choquet capacities. Let us give a few examples. Let $E := [0,1)$, $\mathcal{P} := ((I_k^n)_{0\le k<\nu_n})_{n\ge 1}$ a sequence of partitions of $[0,1)$, $x_k^n := \inf(I_k^n)$ and for all n, $\zeta_n : \mathcal{P}(E) \longrightarrow [0,1]$. The following set functions defined on $\mathcal{P}(E)$ are Choquet capacities:

$$c_n(A) \;:=\; \max\{\zeta_n(I_k^n); x_k^n \in A\}$$

$$\forall p \ge 1 \;\; c_n^{(p)}(A) \;:=\; \left(\sum_{x_k^n \in A} \zeta_n(I_k^n)^p \right)^{1/p}$$

$$c_n(A) \;:=\; \max_t \;\# \; \{k \,/\, \zeta_n(I_k^n) = t \,; x_k^n \in A\}$$

Remark : if $A \subseteq I_k^n$, then

$$c_n(A) = 0 \quad \text{or} \quad c_n(A) = c_n(I_k^n) = \zeta_n(I_k^n)$$

Such capacities are called myopic capacities.

17.3.3 Multifractal Analysis of Sequences of Choquet Capacities

Multifractal analysis has recently drawn much attention as a tool for studying the structure of singular measures, both in theory and in applications ([110, 177, 96]). Efforts have mainly been focused on special cases such as self-similar and self-affine measures, both in the deterministic [216, 207, 38] and in the random case [176, 138, 112, 76, 8, 206]. Multifractal analysis has been extended to deal with Choquet capacities in [155]. Studies have also been conducted to extend the analysis to point functions [132, 130].

In the multifractal scheme, the pointwise structure of a singular measure is analyzed through the so-called "multifractal spectrum", which gives either geometrical or probabilistic information about the distribution of points having the same singularity. Several definitions of a multifractal spectrum exist. Some are related to measure theory (Hausdorff and packing measures), other to large deviation theory (Rényi exponents and Legendre spectrum). The so-called "multifractal formalism" assesses that, in some situations, all the spectra coincide. The motivation for studying the multifractal formalism stems from the fact that both Hausdorff and packing dimensions are in general difficult to compute. The other definitions are more suited to applications, as they are easier to evaluate. The multifractal formalism is known to hold for the class of multiplicative measures, but it fails in general.

Let $c := (c_n)$ be a sequence of Choquet capacities defined on $[0,1)$, with values in $[0,1]$. Let μ be a non-atomic probability measure, called the "reference measure". Let $\mathcal{P} := ((I_j^n)_{0\leq j<\nu_n})_{n\geq 1}$ be a sequence of partition of $[0,1)$ such that:

(C1) $\lim_{n\to\infty} \max_{0\leq j<\nu_n} |I_j^n| = 0$,

(C2) $\forall n, k,\ \ I_k^n$ is an interval, semi-open to the right.

Define:

$$\begin{aligned}
\mathcal{H}^s_{\mu,\delta}(E) &:= \inf\{\sum_{i=1}^{+\infty} \mu(E_i)^s \ / E \subset \bigcup_i E_i,\ \ \mu(E_i) \leq \delta, E_i \in \mathcal{P}\ \ \forall i\} \\
\mathcal{H}^s_{\mu}(E) &:= \lim_{\delta\to 0} \mathcal{H}^s_{\mu,\delta}(E) \\
\dim_\mu(E) &:= \inf\{s \ / \ \mathcal{H}^s_\mu(E) = 0\} = \sup\{s \ / \ \mathcal{H}^s_\mu(E) = +\infty\}
\end{aligned}$$

Under additional conditions on $\mathcal{P}$, $\dim_\mu(E)$ coincides with the Hausdorff dimension of E w.r.t. to μ.

Let $I^n(x)$ denote the interval I_j^n which contains x. The Hölder coarse grain exponent is defined as:

$$\alpha_n(x) := \frac{\log c_n(I^n(x))}{\log \mu(I^n(x))}$$

and the Hölder pointwise exponent as:

$$\alpha(x) = \lim_{n\to\infty} \alpha_n(x)$$

when the limit exists. Let

$$E_\alpha := \{x \in [0,1) \ / \ \alpha(x) = \alpha\}$$

The Hausdorff singularity spectrum of c is defined as:

$$f_h(\alpha) := \dim_\mu E_\alpha$$

Let $n \in \mathbf{N}$ and $\varepsilon > 0$. Note:

$$K_\varepsilon^n(\alpha) := \left\{ k \in \{0,\ldots,\nu_n - 1\} \ / \ \alpha_k^n := \frac{\log c_n(I_k^n)}{\log \mu(I_k^n)} \in (\alpha-\varepsilon, \alpha+\varepsilon) \right\}$$

and

$$N_\varepsilon^n(\alpha) := \# K_\varepsilon^n(\alpha)$$

Let, for $\beta > 0$,

$$\begin{aligned} S_\varepsilon^n(\alpha,\beta) &:= \sum_{k \in K_\varepsilon^n(\alpha)} \mu(I_k^n)^\beta \\ S_\varepsilon(\alpha,\beta) &:= \limsup_{n\to+\infty} S_\varepsilon^n(\alpha,\beta) \end{aligned}$$

There exists a real $f_g^\varepsilon(\alpha)$ such that:

$$\begin{aligned} \beta < f_g^\varepsilon(\alpha) &\implies S_\varepsilon(\alpha,\beta) = +\infty \\ \beta > f_g^\varepsilon(\alpha) &\implies S_\varepsilon(\alpha,\beta) = 0 \end{aligned}$$

The large deviation spectrum is:

$$f_g(\alpha) := \lim_{\varepsilon\to 0} f_g^\varepsilon(\alpha)$$

When all intervals have the same size and μ is the Lebesgue measure $\mathcal{L}$,

$$f_g(\alpha) = \lim_{\varepsilon\to 0} \limsup_{n\to+\infty} \frac{\log N_\varepsilon^n(\alpha)}{\log \nu_n}$$

Finally, let:

$$X_n(x,y) := {\sum_{j<\nu_n}}' c_n(I_j^n)^x \mu(I_j^n)^{-y}$$

and

$$X(x,y) := \limsup_{n\to\infty} n^{-1} \log X_n(x,y)$$

where $\sum'$ means that the summation runs through the indices j such that $c_n(I_j^n)\mu(I_j^n) \neq 0$. X is convex function, non decreasing in y and non increasing in x. Let: $\Omega := \{(x,y)/X(x,y) < 0\}$. There exists a concave function τ such that:

$$\overset{\circ}{\Omega} = \{(x,y) \in \mathbb{R}^2 \ / \ y < \tau(x-0)\}$$

The Legendre multifractal spectrum is defined as τ^*, the Legendre transform of τ:

$$f_l(\alpha) := \tau^*(\alpha) := \inf_q [q\alpha - \tau(q)]$$

When all intervals have the same size and $\mu = \mathcal{L}$,

$$\tau(q) = \limsup_{n\to+\infty} \frac{\log {\sum_k}' \mu(I_k^n)^q}{\log \nu_n^{-1}}$$

As mentioned above, a central concern of the multifractal theory is to relate f_l, f_g and f_h. This is the so called multifractal formalism. It has important applications in our case of image analysis:

- f_g measures, loosely speaking, the probability of finding a given value of α_n at resolution n. It gives a probabilistic description of the distribution of the singularities.
- f_h measures the Hausdorff dimension of the set of points having a given α. It gives a geometrical description of the distribution of the singularities.

However, both f_g and f_h are difficult to compute on real discrete data. On the other hand, f_l is much easier to compute, because it involves only the evaluation of average quantities. As a counterpart, f_l generally contains less information, because it is always a concave function. The following results hold.

Theorem: *Under conditions (C1) and (C2),*

$$f_h \leq f_g \leq f_l$$

Theorem: *Under conditions (C1) and (C2),*

$$f_l = f_g^{**}$$

17.3.4 An Example: the Binomial Measure

Define a sequence of measures μ_n on $[0,1)$ as follows: let $m_0 \in (0,1), m_1 = 1 - m_0$. For $n \in \mathbb{N}^*$ and for $k \in \{0, \ldots 2^{n-1}\}$ let:

$$\mu_n[k\ 2^{-n}, (k+1)2^{-n}) = m_0^{n\phi_0^n} m_1^{n\phi_1^n}$$

where ϕ_0^n is the proportion of 0's in the base-2 expansion of $x \in [k\ 2^{-n}, (k+1)2^{-n})$ up to rank n, and ϕ_1^n the proportion of 1's:

$$x = \sum_{i=1}^{\infty} x_i 2^{-i} \qquad x_i \in \{0,1\}$$

$$\phi_0^n(x) = \frac{\sum_{i=1}^{n}(1-x_i)}{n} \qquad \phi_1^n(x) = \frac{\sum_{i=1}^{n} x_i}{n}$$

The sequence $(\mu_n)_{n\geq 1}$ has a weak* limit μ, which is called a binomial measure. It is easy to see that, if $\phi_0(x) = \lim_{n\to\infty} \phi_0^n(x)$ exists, then $\alpha(x) = -\phi_0 \log_2 m_0 - \phi_1 \log_2 m_1$.

As for the multifractal spectra, it can be proved that:

$$\begin{cases} \alpha(\phi_0) = -\phi_0 \log_2 m_0 - \phi_1 \log_2 m_1 \\ f_h(\phi_0) = f_g(\phi_0) = f_l(\phi_0) = -\phi_0 \log \phi_0 - \phi_1 \log \phi_1 \end{cases}$$

and the Hausdorff dimension of the set of points "without" a ϕ_0 is 1. Figure 17.6 displays a 2D version of the above construction (a "quadrinomial measure"). Figures 17.5 and 17.6 illustrate a striking difference between classical and multifractal methods for image segmentation: as said above, the latter will detect only one edge in figure 17.5 but several ones in image 17.6. The former will in general detect several edges in both images. Edges detected by IMA and multifractal analysis on figure 17.6 do however coincide with an appropriate choice of the parameters.

A typical spectrum of a binomial measure is shown on figure 17.7. Figure 17.8 displays a 1D random version of a multinomial measure.

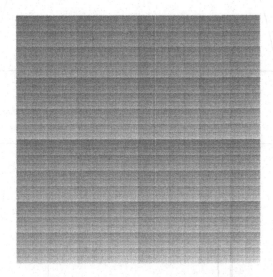

Figure 17.6: A 2D quadrinomial measure.

The above are simple cases where the multifractal formalism holds. Let us now briefly describe three examples where it fails.

1. Consider the following measure on $[0, 1)$:

$$\nu(A) = \frac{1}{2}(\nu_1(A) + \nu_2(A))$$

where:

- A is a Borel set in $[0, 1)$.
- ν_1 is an m_0-binomial measure on $\left[0, \frac{1}{2}\right)$.
- ν_2 is an m_0'-binomial measure on $\left[\frac{1}{2}, 1\right)$ and $m_0 \neq m_0'$.

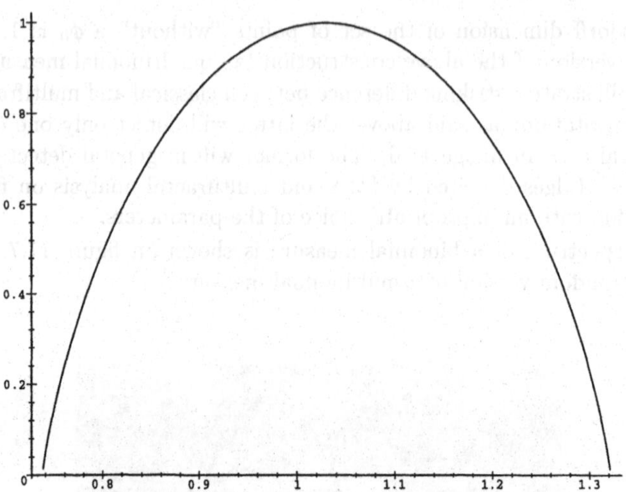

Figure 17.7: Spectrum of a binomial measure.

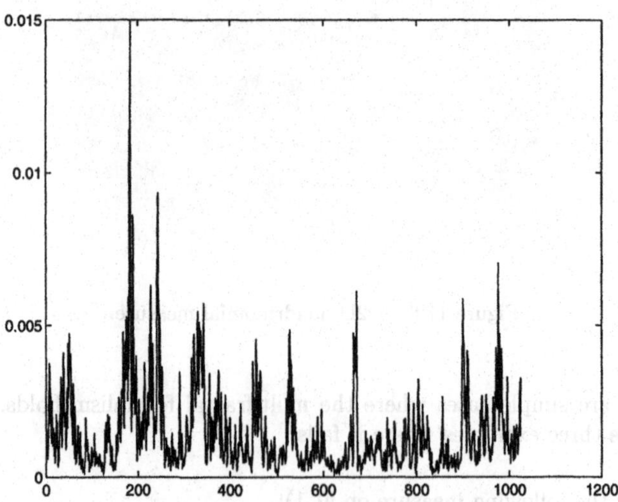

Figure 17.8: A random multinomial measure.

Then it is easy to show that (figure 17.9):

$$\begin{aligned} f_h^\nu = f_g^\nu &= \sup(f_h^{\nu_1}, f_h^{\nu_2}) \\ f_l^\nu &= (f_g^\nu)^{**} \end{aligned}$$

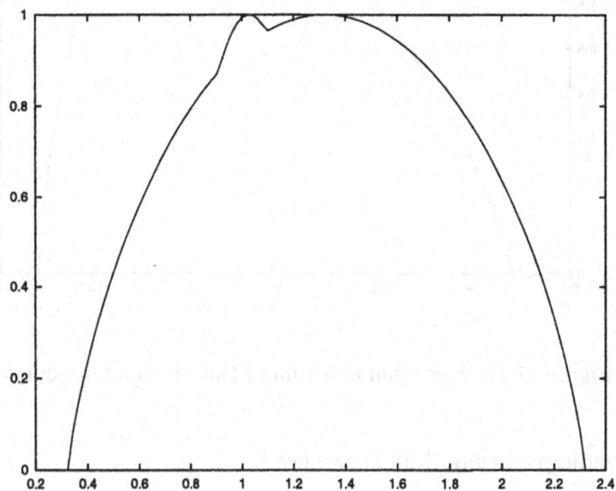

Figure 17.9: Spectrum of a lumping of two binomial measures.

2. Let now:

$$\nu(A) = \frac{1}{2}[\nu^1(A) + \nu^2(A)]$$

where:

- ν^1 is an m_0-binomial measure on $[0, 1)$,
- ν^2 is an m_0'-binomial measure on $[0, 1)$ and $m_0 \neq m_0'$.

Let $[\alpha^i_{min}, \alpha^i_{max}]$ be the support of $f_h^{\nu^i}$ (i.e. $f_h^{\nu^i} = 0$ outside $[\alpha^i_{min}, \alpha^i_{max}]$). Assume w.l.o.g. that $\alpha^1_{min} < \alpha^2_{min} < \alpha^2_{max} < \alpha^1_{max}$. Then (figure 17.10):

$$\begin{aligned} f_h^\nu = f_g^\nu &= f_h^{\nu^1} && on \quad [\alpha^1_{min}, \alpha^2_{min}] \\ f_h^\nu = f_g^\nu &= f_h^{\nu^2} && on \quad [\alpha^2_{min}, \alpha^2_{max}] \\ f_h^\nu = f_g^\nu &= -\infty && otherwise \\ f_l^\nu &= (f_g^\nu)^{**} \end{aligned}$$

3. The last example deals with a case where $f_h(\alpha) < f_g(\alpha) = f_l(\alpha)$ on an interval. Let

$$c_n(A) = \max\{2^{-n\, k\, 2^{-n}};\ k\, 2^{-n} \in A\}$$

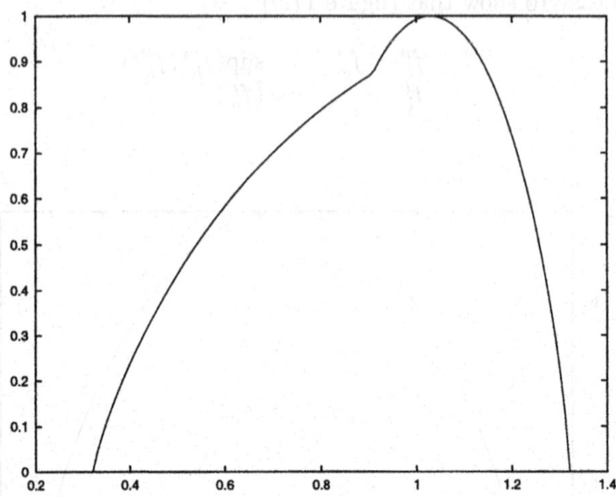

Figure 17.10: Spectrum of a sum of two binomial measures.

With a dyadic partition of $[0,1)$, we get:

$$f_h(\alpha) = \begin{cases} 0 & \text{if } \alpha \in [0,1] \\ -\infty & \text{otherwise} \end{cases}$$

and

$$f_g(\alpha) = f_l(\alpha) = \begin{cases} 1 & \text{if } \alpha \in [0,1] \\ -\infty & \text{otherwise} \end{cases}$$

Let us finally mention that, while f_l is always a concave function, it is possible to construct a sequence of capacities having a prescribed f_h spectrum under very weak assumptions on f_h. Figure 17.11 displays an example.

We end this section with a few words on the role of the reference measure. Let:

- ν_1 be a trinomial measure (0.1, 0.8, 0.1) on $\left[0, \frac{1}{3}\right]$
- ν_2 be a binomial measure $(p, 1-p)$ on a Cantor triadic set defined on $\left[\frac{2}{3}, 1\right]$
- $\nu = (\nu_1 + \nu_2)/2$

Let the reference measure, μ, equal $\mathcal{L}$ (the Lebesgue measure). Then it is possible to choose p such that:

$$f_h^\nu = \sup(f_h^{\nu_1}, f_h^{\nu_2}) = f_h^{\nu_1}$$

so that the singularities coming from ν_2 are "hidden" by those coming from ν_1. However, if we take $\mu = \nu_1$, the part of the spectrum f_h^ν due to ν_1 will reduce to the point (1,1) and the singularities coming from ν_2 will be "seen" on the spectrum.

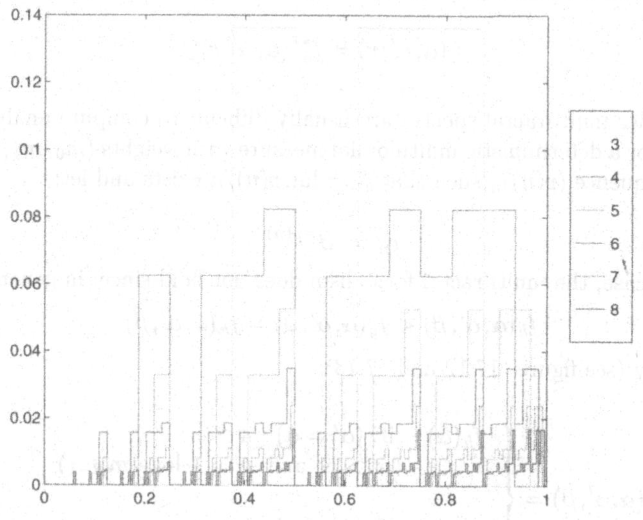

Figure 17.11: The first elements of a sequence of capacities whose spectrum is x^2 on an interval.

17.3.5 Multifractal Correlations

It is an obvious fact that the multifractal spectra do not uniquely characterize a measure. To obtain a finer analysis, one can compute "multifractal correlations". These correlations allow to describe the local interplay of singularities. This topic has been investigated for instance in [45, 78, 250].

Let:

$$\begin{aligned} E_{\alpha,\alpha',r} &= \{x \in [0;1[\ / \ \lim_{n\to+\infty} \alpha_n(x) = \alpha \ , \\ &\quad \limsup_{n\to\infty} \alpha_n(x+r_n) = \alpha'\} \end{aligned}$$

where $r := (r_n)_n$ is a sequence in $]0,1[$ such that $\lim_n r_n = 0$. Define:

$$f_h(\alpha,\alpha',r) := \dim_H E_{\alpha,\alpha',r}$$

The generalized definitions of f_g, τ and f_l, although conceptually simple, are a little involved and we don't write them here. The interested reader is referred to [250]. The following results hold:

Theorem: *Under conditions (C1) and (C2),*

$$f_h(\alpha,\alpha',r) \leq f_g(\alpha,\alpha',r) \leq f_l(\alpha,\alpha',r)$$

Theorem: *Under conditions (C1) and (C2),*

$$f_l(\alpha,\alpha',r) = f_g^{**}(\alpha,\alpha',r)$$

Higher order multifractal spectra are usually difficult to compute analytically. We give a result for a deterministic multinomial measure with weights $(m_0, m_1, \ldots, m_{B-1})$. Consider a sequence $(p(n))_n$ such that $\beta := \lim p(n)/n$ exists and let:

$$r_n := B^{-p(n)}$$

In such a case, the multifractal formalism does not hold since, in general,

$$f_h(\alpha,\alpha',\beta) < f_g(\alpha,\alpha',\beta) = f_h(\alpha,\alpha',\beta)$$

More precisely (see figures 17.12 and 17.13):

$$f_h(\alpha,\alpha',\beta) = \begin{cases} f_h(\alpha) & \text{if } (\alpha'=\alpha) \text{ or} \\ & (\alpha' > \alpha \text{ and } \alpha = -\log_B m_{B-1}) \\ & \\ -\infty & \text{if } (\alpha' < \alpha) \text{ or} \\ & (\alpha' > \alpha \text{ and } \alpha \neq -\log_B m_{B-1}) \end{cases}$$

$$\begin{aligned} \tau(q,q',r) &= -\lim_{n\to+\infty} \frac{1}{n} \log_B \sum_{k=0}^{B^n-1} \mu(I_k^n)^q \, \mu(I_k^n + B^{-p(n)})^{q'} \\ &= \tau(q+q') - \beta \min(0, L(q,q')) \end{aligned}$$

$$\begin{aligned} L(q,q') &:= \tau(q+q') + q\log_B m_{B-1} + q'\log_B m_0 \\ \lambda &:= \lambda(\alpha,\alpha') := \frac{\alpha'-\alpha}{\log_B \frac{m_{B-1}}{m_0}} \\ A &:= A(\alpha,\alpha') := \frac{\alpha + \lambda \log_B m_{B-1}}{1-\lambda} \end{aligned}$$

- If $m_0 \neq m_{B-1}$,

$$f_l(\alpha,\alpha',\beta) = \begin{cases} (1-\lambda) f_l(A) & \text{if } A \in]\alpha_{\min}, \alpha_{\max}[\\ & \text{and } \lambda \neq 1, \lambda \geq 0 \\ 0 & \text{if } \lambda = 1 \\ & \text{and } \alpha = -\log_B m_{B-1} \\ -\infty & \text{else} \end{cases}$$

- If $m_0 = m_{B-1}$,

$$f_l(\alpha,\alpha',\beta) = \begin{cases} f_l(\alpha) & \text{if } \alpha = \alpha' \\ -\infty & \text{if } \alpha \neq \alpha' \end{cases}$$

Finally:

$$f_g = f_l$$

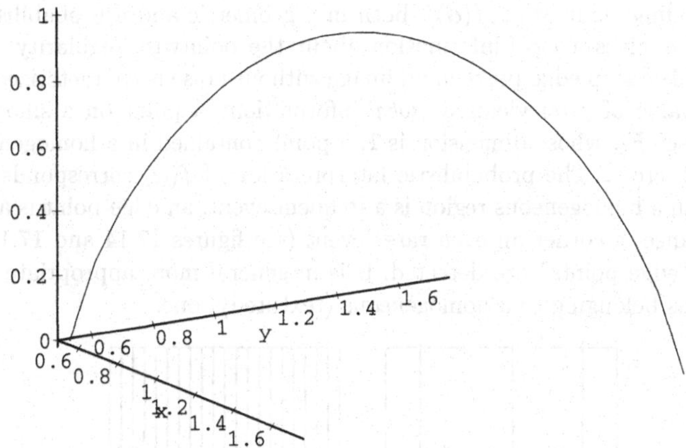

Figure 17.12: f_h second order spectrum of a multinomial measure.

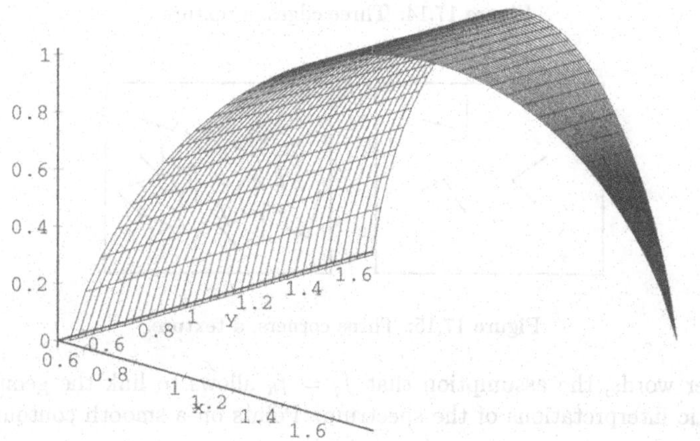

Figure 17.13: $f_l = f_g$ second order spectrum of a multinomial measure.

17.4 Application to Image Analysis

17.4.1 Introduction

Throughout the remaining of the paper, we make the following assumption:

$$f := f_h = f_g$$

The multifractal analysis of images consists in defining a sequence of capacities on the image, computing its multifractal spectrum, and classifying each point according to the corresponding value of $(\alpha, f(\alpha))$, both in a geometric and a probabilistic fashion. The value of α gives a *local* information about the pointwise regularity: for a fixed capacity, an ideal step edge point in an image without noise is characterized by a given value. The value of $f(\alpha)$ yields a *global* information: a point on a smooth contour belongs to a set E_α whose dimension is 1, a point contained in a homogeneous region has $f(\alpha) = 2$, etc ... The probabilistic interpretation of $f(\alpha)$ corresponds to the fact that a point in a homogeneous region is a frequent event, an edge-point is a rare event, and, for instance, a corner an even rarer event (see figures 17.14 and 17.15). Indeed, if too many "edge points" are detected, it is in general more appropriate to describe these points as belonging to a homogeneous (textured) zone.

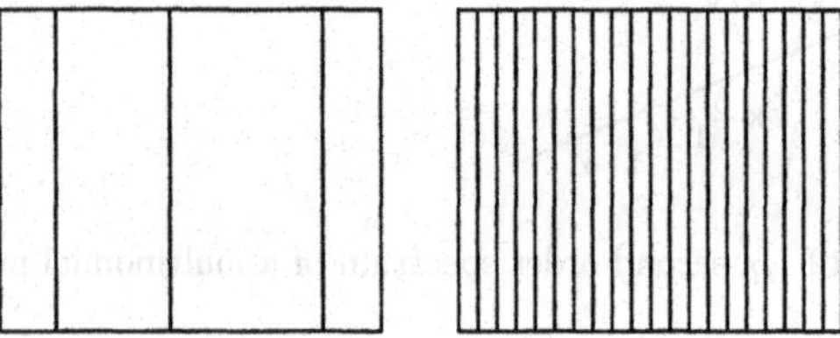

Figure 17.14: Three edges, a texture.

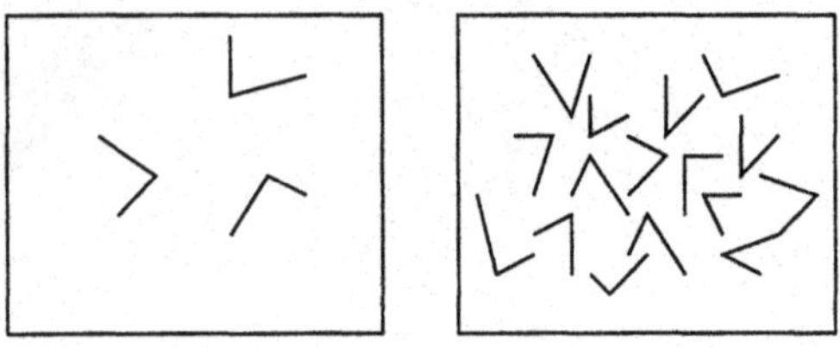

Figure 17.15: Three corners, a texture.

In other words, the assumption that $f_g = f_h$ allows to link the geometrical and probabilistic interpretations of the spectrum. Points on a smooth contour have an α such that:

- $f_h(\alpha) = 1$ because a smooth contour fills the space as a line.
- $f_g(\alpha) = 1$ because a smooth contour has a given probability to appear.

In fact, we may define the type of a point (i.e. edge, corner, smooth region ...) through its associated $f(\alpha)$ value:

Definition: *Let c be a sequence of capacities defined on the image. x is called a c-edge point if $f_h^c(\alpha(x)) = 1$. More generally, for $t \in [0, 2]$, x is called a point of type $c - t$ if $f_h^c(\alpha(x)) = t$. The following terminology is used:*

- *points of type $c - 2$ are called smooth points,*
- *points of type $c - 0$ are called rare points.*

A benefit of the multifractal approach is thus that it allows to define not only edge points, but a continuum of various types of points (see figure 17.27). Note that, in this setting, the type of a point has no absolute meaning. It depends on the sequence of capacities used to analyze the image.

Recall that in most "classical" approaches, an edge point is a point where some sort of "gradient" is maximal in its direction, or more generally, a point where the image exhibits a specific local behavior (even if the size of the neighborhood may evolve, as in IMA). In the multifractal approach, an edge point is defined through both a local ($\alpha(x)$) and a global property ($f(\alpha)$) of the image: one can not decide whether a point is on an edge only by looking at a neighborhood of this point (indeed, all edge points in an image are not characterized by the same value of α: the actual value depends on the contrast of the image and on the noise). Such a use of a global information is meaningful only if one assumes that the image is "homogeneous" in the following sense : the sequence c of capacities used to analyze the data is such that, for any subregions Ω_1 and Ω_2 in the image,

$$f_h^{\Omega_1}(\alpha) = f_h^{\Omega_2}(\alpha) \qquad \forall \alpha \in support(f_h^{\Omega_1}) \cap support(f_h^{\Omega_2})$$

where $f_h^{\Omega_i}$ is the spectrum of c computed on the restriction of the image to Ω_i. Note that we allow the possibility of finding a given value of α in only one of the two regions (i.e. if $f_h^{\Omega_1}(\alpha) = -\infty$, $f_h^{\Omega_2}(\alpha)$ may assume any value), which is essential for applications. This property can be verified by local computations of the spectrum. When it is not fulfilled, one needs to perform separate analysis for each homogeneous region of the image. Of course, from a practical point of view, this only makes sense if the whole image can be divided in a few sufficiently large homogeneous zones.

Another condition for the above definition to be relevant is that there exists a sequence of capacities that assigns different Hölder exponents to the set of points one wants to extract (the "foreground", for instance the edges) and to the "background". Remark that it is not necessary that all "foreground" points have the same exponent. Infinitely many different exponents may correspond to the same type t, as long as the corresponding subsets of points in the image all have the same dimension (see figures 17.16 and 17.17. The *sum* measure and *iso* capacity are defined below, and the procedure to extract the edges is explained in section 17.4.5).

As is probably clear to the reader by now, the core of the multifractal analysis of images lies in the choice of a relevant sequence of capacities for describing the scene. The following fact is easy to prove:

Proposition: *Let E_n be the set of edge points at time n derived from a multiscale analysis. Then there exist a sequence of capacities $c = (c_n)$ such that, for all n, $E_n = \{(x,y)/\alpha(x,y) = \alpha_0\}$.*

This result is however of little practical interest, and the problem of finding an optimal c in a general setting is still unsolved. Preliminary investigations are described in [43]. In practice, one uses myopic capacities because they allow to take into account the resolution in a simple way. The most basic functions are the following ones:

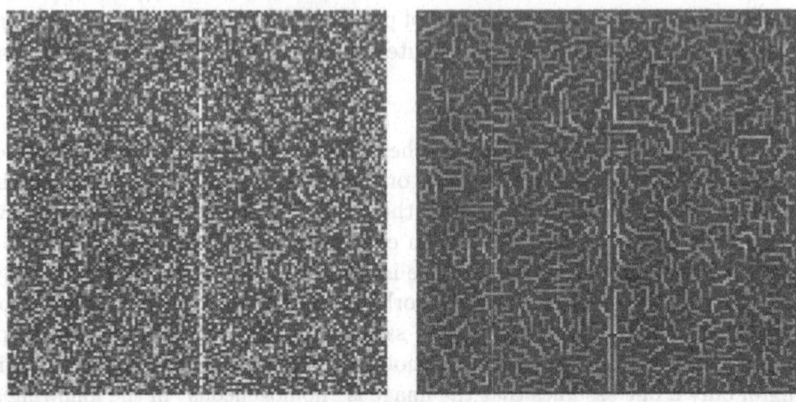

Figure 17.16: Original image (left) and edges obtained with Canny's detector.

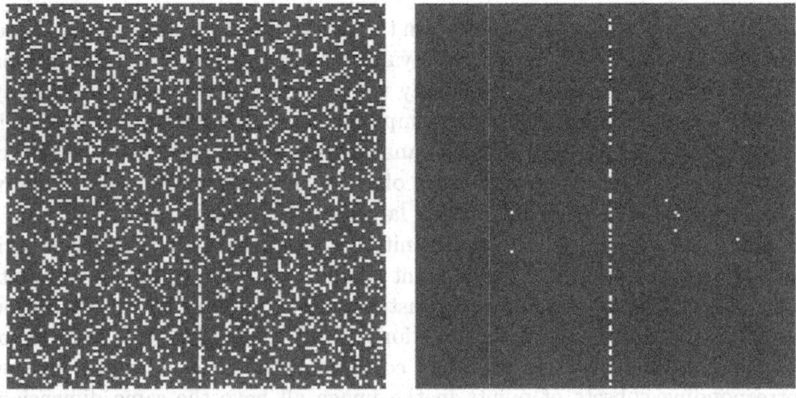

Figure 17.17: Edges obtained with the *sum* measure (left) and with the *iso* capacity(right).

Definitions: *Let the image be defined on* $[0,1] \times [0,1]$, $\mathcal{P} := ((I_k^n)_{0 \le k < \nu_n})_{1 \le n \le N}$ *be a sequence of partitions of* $[0,1] \times [0,1]$ *and* (x_k^n, y_k^n) *be any point in* I_k^n. *Each* I_k^n *is made of an integer number of pixels. Let* $L(I_k^n)$ *denote the sum of the grey levels in* I_k^n. *Let* (x, y) *denote a generic pixel in the image and* $L(x, y)$ *denote the grey level at* (x, y). *Let* $(p_n)_{1 \le n \le N}$ *be a fixed sequence of positive integers and* Ω *be a region in the image.*

sum **measure:**

$$c^s(\Omega) = \sum_{(x,y) \in \Omega} L(x, y)$$

max **capacities:**

$$c_n^m(\Omega) = \max_{I_k^{n+p_n} / (x_k^{n+p_n}, y_k^{n+p_n}) \in \Omega} L(I_k^{n+p_n})$$

$$c^M(\Omega) = \max_{(x,y) \in \Omega} L(x, y)$$

iso **capacities:**

$$c_n^i(\Omega) = \max_l \# \{k \,/\, L(I_k^{n+p_n}) = l, (x_k^{n+p_n}, y_k^{n+p_n}) \in \Omega\}$$

$$c^I(\Omega) = \max_l \# \{(x, y) \,/\, L(x, y) = l, (x, y) \in \Omega\}$$

Note that, since in practice resolution is finite, all sequences above are also finite (they are in fact finite versions of the myopic capacities defined in 17.3.2). It is easy to show that:

- $c^s(\Omega)$ depends on both the grey level values and their distribution in Ω,
- $c_n^m(\Omega)$ and $c^M(\Omega)$ only depend on the grey level values,
- $c_n^i(\Omega)$ and $c^I(\Omega)$ only depend on the grey level distribution.

Thus, (c_n^m, c^M) and (c_n^i, c^I) give in some loose sense "orthogonal" information about the image. Furthermore, it can be shown that they are more robust to noise than c^s.

17.4.2 Numerical Estimation of f_g

Before we apply the above ideas to specific problems in image analysis, let us say a few words about the numerical estimation of f_g (estimation of f_l is considered in [71] and estimation of f_h in [154]). The simplest method for estimating f_g involve the following steps ([71]):

1. for each n, compute all α_k^n;

2. compute $\alpha^n_{\min} = \min_k \alpha^n_k, \alpha^n_{\max} = \max_k \alpha^n_k$, and divide $[\alpha^n_{\min}, \alpha^n_{\max}]$ into N boxes
3. compute the number N^n_i of intervals I^n_k whose α^n_k falls in the i-th box, $i = 1, \ldots N$;
4. find $f_g(\alpha)$ by a linear regression on $(\log n, \log N^n_i)$.

Although, this "histogram method" may yield satisfactory results on strictly multiplicative cascades, it fails to estimate a good approximation in more complex situations, in particular when f_g is not a concave function. For instance, when dealing with a non stationary process, like the lumping of two binomial measures, or a compound process, like the sum of two binomial measures, the histogram method will tend to hide the non concavity of f_g (see figures 17.9 and 17.10). There are two types of problems:

- The choice of the number of "boxes" on the α axis: this choice is ad-hoc, although it influences a lot the result. In particular, it does not take into account the ε limit in the definition of f_g.
- The choice of the averaging procedure: when dealing with, say, a trinomial measure, averaging two adjacent boxes at each step produces strong oscillations with induce errors in the regression estimation step, and thus both in the exponent and dimension computations.

As said above, this method in fact introduces an implicit dependence between n and ε through step 2. Although it is desirable, in numerical methods, to pass from two limits on n and on ε to just one limit, it is clear that the dependency between n and ε should be carefully designed.

Indeed, more precise results may be obtained using a classical tool in density estimation: the kernel method, or more precisely the double kernel method ([63]). Kernel methods have been extensively studied and powerful theorems are known that assess the quality of the results. The difficulty here is that f_g is not the density corresponding to the α's, but rather a double logarithmic normalization of this density. Applying a kernel procedure to estimate f_g has the following advantages:

- The choice of the number of boxes is no longer ad-hoc but driven by the data. The dependence on ε is made explicit.
- No averaging needs to be done, since the spectrum is evaluated at only one resolution. Of course, in the general case, one needs to verify that the spectra estimated at each resolution are consistent. Such a verification should be performed carefully. Note that other authors have considered one resolution estimation methods ([101]).
- The use of smooth kernels allows to obtain more regular estimates.

For simplicity, we only consider the 1D case with $\mathcal{P}$, the sequence of partitions $\mathcal{P}_n$ of $[0, 1[$, being made of dyadic intervals, i.e.:

$$\mathcal{P}_n = \{I^n_k\}_{0 \le k < 2^n}, I^n_k = [k2^{-n}, (k+1)2^{-n}[$$

(recall section 17.3.3 for the notations.)

The starting point of the method is to write $N_\varepsilon^n(\alpha)$ as:

$$N_\varepsilon^n(\alpha) := \#\{\alpha_k^n / \alpha_k^n \in \mathcal{B}(\alpha,\varepsilon)\} = 2^{n+1}\varepsilon K_\varepsilon * \rho_n(\alpha)$$

where $\mathcal{B}(\alpha,\varepsilon)$ is the open ball centered at α of radius ε, ρ_n is the density of the $(\alpha_k^n)_k$, and K_ε is the rectangular kernel: $K(x) = 1$ for $x \in \left[-\frac{1}{2}, \frac{1}{2}\right]$, 0 outside, and $K_\varepsilon(x) = \frac{1}{\varepsilon}K\left(\frac{x}{\varepsilon}\right)$.

It is easily shown that K may be in fact chosen to be any compactly supported kernel. A classical problem in density estimation is the choice of an ε such that $K_\varepsilon * \rho_n$ is as close as possible to the limiting density ρ. In this setting, ε becomes a function of n. Adapting classical techniques allows to reduce the two limits in f_g to one limit, with an optimal choice of ε. However, one has to check under which conditions it is indeed possible to get rid of the limit on ε.

Let $f_g^\varepsilon(\alpha) = \overline{\lim}_{n\to\infty} \frac{\log N_\varepsilon^n(\alpha)}{n}$. The following results hold ([3]):

Theorem: *Let $\varepsilon_0 > 0$. The following statements are equivalent :*

- $f_g^\varepsilon(\alpha) = \sup_{x\in\mathcal{B}(\alpha,\varepsilon)} f_g(x) \qquad \forall\varepsilon \in (0,\varepsilon_0)$.
- $\varepsilon \mapsto f_g^\varepsilon(\alpha)$ *is left continuous for* $\varepsilon \in (0,\varepsilon_0)$.

Theorem: *Assume f_g is continuous on a neighborhood of α. Then, for sufficiently small ε,*

$$f_g^\varepsilon(\alpha) = \sup_{x\in\mathcal{B}(\alpha,\varepsilon)} f_g(x)$$

The following corollaries are then easy to prove:

Corollary: *Assume f_g is continuous on a neighborhood of α_0. Then $\alpha \mapsto f_g^\varepsilon(\alpha)$ is continuous at α_0.*

Corollary: *Assume f_g is continuous on $[\alpha_m, \alpha_M]$. Then $(f_g^\varepsilon)_\varepsilon$ converges uniformly to f_g when ε goes to 0^+.*

The two following results deal with the problem of passing from two limits to one limit in the computation of f_g.

Assume f_g is greater than or equal to 0 on $[a,b]$ and $-\infty$ outside. Then it is easy to see that $\overline{\bigcup_{n,k} \alpha_k^n} = [a,b]$. Let $(\beta_j^n)_j$ denote the ordered set of $(\alpha_k^n)_k$ at resolution n without multiplicity, i.e.

$$\beta_0^n = \min_k \alpha_k^n, \qquad \beta_1^n = \min_{k/\alpha_k^n \neq \beta_0^n} \alpha_k^n, \qquad \ldots.$$

With a slight abuse of notation, let $N_0^n(\alpha) := \#\{k \;/\; \alpha_k^n = \alpha\}$, and define $\widehat{\beta^n(\alpha)}$ as the smallest β_j^n that verifies:

$$|\beta_j^n - \alpha| < \varepsilon \qquad N_0^n(\beta_j^n) \geq N_o^n(\beta_k^n) \qquad \forall \beta_k^n \in \mathcal{B}(\alpha,\varepsilon)$$

Define:

$$\tilde{f}_g(\alpha) := \lim_{\varepsilon \to 0} \overline{\lim_{n \to \infty}} \frac{\log N_0^n(\widehat{\beta^n(\alpha)})}{n}$$

Finally, let $\varepsilon_\alpha(n) = |\alpha - \widehat{\beta^n(\alpha)}|$ and

$$\varepsilon_n := \sup_\alpha \varepsilon_\alpha(n) = 1/2 \max_j |\beta_{j+1}^n - \beta_j^n|$$

Define:

$$f_g^n(\alpha) := \frac{\log N_{\varepsilon_n}^n(\alpha)}{n}$$

The following results hold:

Proposition:

$$\forall \alpha, \quad \tilde{f}_g(\alpha) \leq \overline{\lim_{n \to \infty}} f_g^n(\alpha) \leq f_g(\alpha)$$

Proposition: *Assume that there exist positive reals a, b, ε_0 such that*

$$\forall \varepsilon < \varepsilon_0, \forall n, \forall \alpha, \; N_\varepsilon^n(\alpha) \leq b\, n^a \; N_0^n(\widehat{\beta^n(\alpha)})$$

Then

$$\forall \alpha, \quad \tilde{f}_g(\alpha) = f_g(\alpha)$$

Corollary: *Under the same hypothesis as the previous proposition,*

$$\forall \alpha, \quad \tilde{f}_g(\alpha) = \overline{\lim_{n \to \infty}} f_g^n(\alpha) = f_g(\alpha)$$

The last series of propositions are specific to the case of multinomial measures. Generalizing the above definition, let, for an arbitrary sequence $(\varepsilon_n)_n$ of positive reals:

$$f_g^{\varepsilon_n}(\alpha) := \frac{\log N_{\varepsilon_n}^n(\alpha)}{n}$$

Proposition: *Let μ be a binomial measure with weights (m_0, m_1), and $a = |\log_2 \frac{m_1}{m_0}|$. Let $(\varepsilon_n)_n$ be such that: $\varepsilon_n \geq \frac{a}{n^2}, \varepsilon_n \to 0$. Then, for all α, there exists a subsequence $s(n)$ (depending on α) such that*

$$\lim_n f_g^{\varepsilon_{s(n)}}(\alpha) = f(\alpha)$$

Moreover, if $\varepsilon_n \geq \frac{a}{2n}, \varepsilon_n \to 0$, then

$$\lim_n \sup_\alpha |f_g^{\varepsilon_n}(\alpha) - f(\alpha)| = 0$$

If we replace the rectangular kernel by a Gaussian kernel, we obtain

Proposition: *For a binomial measure, the optimal ε_n (in the sense of uniform convergence) for a Gaussian kernel is such that there exists $c > 0$ with $\lim_n(\varepsilon_n/c\sqrt{n}) = 1$.*

The double kernel method consists in choosing $(\varepsilon_n, \varepsilon'_n)$ such that the estimates $f_g^{\varepsilon_n}$ and $h_g^{\varepsilon'_n}$ with two different kernels are as close as possible. For instance, let $f_g^{\varepsilon_n}$ correspond to the rectangular kernel and $h_g^{\varepsilon'_n}$ correspond to the triangular kernel ($T(x) = 1 - x$ for $x \in [0,1], T(x) = 0$ for $x > 1, T(-x) = T(x)$). The following propositions are valid for finite sums or lumpings of binomial measures:

Proposition: *Let $(\varepsilon_n, \varepsilon'_n)$ be sequences that minimizes $\sup_\alpha(f_g^{\varepsilon_n}(\alpha) - h_g^{\varepsilon'_n}(\alpha))$. Then $f_g^{\varepsilon_n} \to f(\alpha)$ for all α.*

Proposition: *Let ε_n be such that there exists $c > 0$ with $\lim_n \frac{\varepsilon_n}{cn} = 1$. Then $\lim_n \sup_\alpha(f_g^{\varepsilon_n}(\alpha) - f(\alpha)) = 0$.*

It should be noted that the double kernel method does not in general give the optimal sequence ε_n. Some results using a modification of the above ideas in a $Max-$*plus* frame (see [4],[31]) are shown below. The theoretical f_g along with its kernel estimate for a lumping (resp. a sum) of two binomial measures is shown on figure 17.18 (resp. figure 17.19). A slightly more complex example is shown on figure 17.20, which is the estimated f_g spectrum of a trinomial measure $\mathcal{T}$ with overlapping: the first map sends $[0,1]$ to $[0,\frac{1}{2}]$ and has weight $m, 0 < m < \frac{1}{2}$, the second map sends $[0,1]$ to $[\frac{1}{4},\frac{3}{4}]$ and has weight $\frac{1}{2}$, and the third map sends $[0,1]$ to $[\frac{1}{2},1]$ and has weight $\frac{1}{2} - m$. The theoretical f_g spectrum is not known in this case. However, it is readily proven, using for instance Fourier transformation, that $\mathcal{T}$ is the convolution of the Lebesgue measure with a binomial measure with weights $(m, 1-m)$. In particular, all points have Hölder exponent greater than one, a fact which is reasonably well verified on figure 17.20. Further experiments may be found in [153].

17.4.3 Image Enhancement

The simplest application of multifractal theory in image analysis is *enhancement*. Consider for instance the *sum* measure. It is easily shown that points in smooth zones have $\alpha = 2$. Moreover, when one computes the spectrum of this measure on a "typical" image (for instance a non noisy optical image of an indoor scene), one finds that $f(\alpha = 2) = 2$: "most" points have exponent 2, because there exist large smooth regions in the signal. On the contrary, the spectrum computed on, say, a radar image has $f(\alpha_0) = 2$ for an α_0 strictly lower than 2 and $f(\alpha = 2) < 2$. A natural idea to "de-noise" the radar image is to transform it in such a way that its spectrum is shifted by an amount $\delta = 2 - \alpha_0$ along the α axis. The new image will then have a set of points in "smooth" zones whose dimension is 2. Other distortions of the spectrum may of course be considered. More on this topic can be found in [103]. We simply show here an example on a radar image (figure 17.21). Figure 17.22 displays a de-noising by a wavelet method (the wavelet coefficients soft thresholding of Donoho, [65]). Figure 17.23 shows the result of a spectrum shift. Note that, contrarily to the wavelet method,

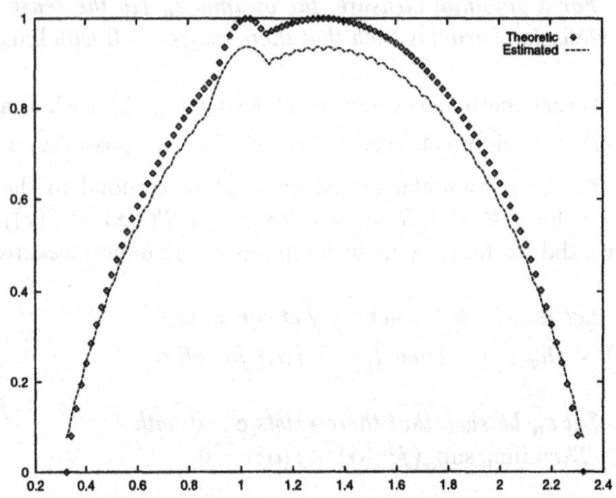

Figure 17.18: Theoretical and estimated f_g spectrum for the lumping of two binomial measures.

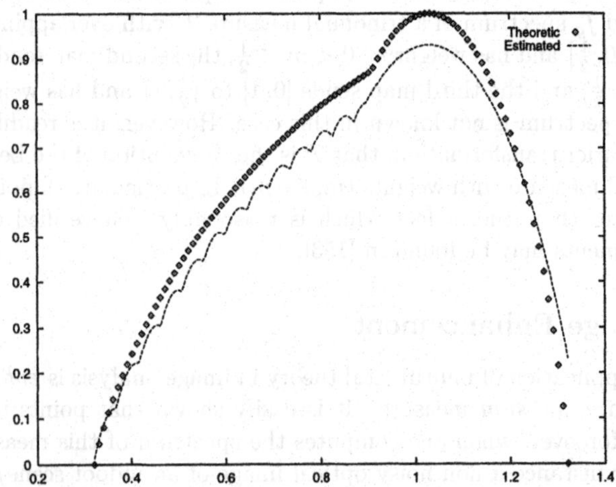

Figure 17.19: Theoretical and estimated f_g spectrum for the sum of two binomial measures.

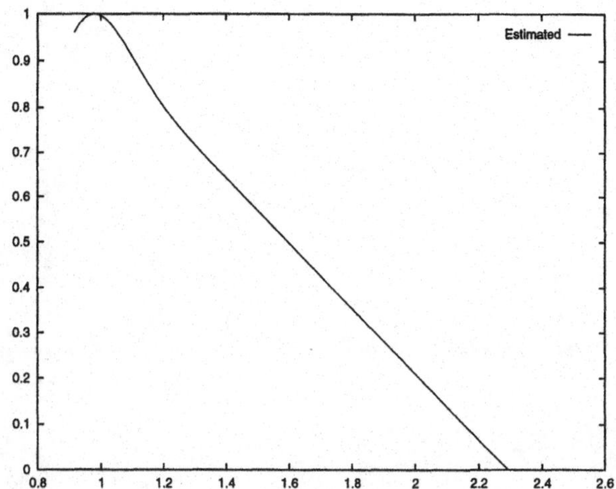

Figure 17.20: Estimated f_g spectrum for a trinomial measure with overlapping.

the multifractal one is not a filtering, since it is invertible. Figures 17.24, 17.25 and 17.26 display the result of Canny's edge detector as applied on the three images, the threshold on the norm of the gradient being set such that no spurious contours appear in the sea region.

17.4.4 Change Detection

An important application of image analysis is to provide means to monitor a scene over a period of time and to detect changes in the content of the scene. In photo-interpretation, change detection consists in finding significant differences – most of the time man-made changes in opposition to natural and/or seasonal changes – between a new image and site models derived from older images. In biomedical imagery, the aim is to control a disease evolution and its cure. In both contexts, most existing methods require a prior knowledge of the objects to be extracted in the new image.

The basic idea of change detection in the multifractal frame is to analyze the new image w.r.t. the older ones. More precisely, one defines a sequence of capacities on the new image, and compute its multifractal spectrum using the *sum* measure on some average of the old ones as the *reference* measure (μ with the notations of section 17.3.3). Intuitively, the spectrum will then highlight the importance of the changes in the new scene. More details on this method can be found in [42].

17.4.5 Edge Detection

The simplest procedure for extracting edges using multifractal analysis is as follows :

- Choose a sequence c of capacities.

Figure 17.21: A radar image, displaying sea in its middle part and land on the left and right sides (courtesy Alcatel Espace).

Figure 17.22: De-noising by soft thresholding of the wavelet coefficient.

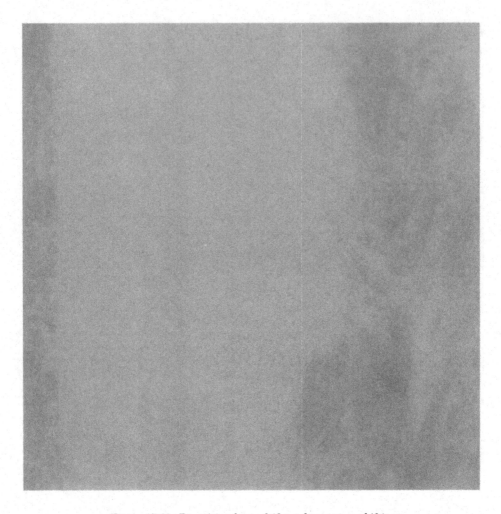

Figure 17.23: De-noising by multifractal spectrum shifting.

Figure 17.24: Edges obtained from image 17.21 with Canny's detector, thresholded such that no edges appear in the sea region.

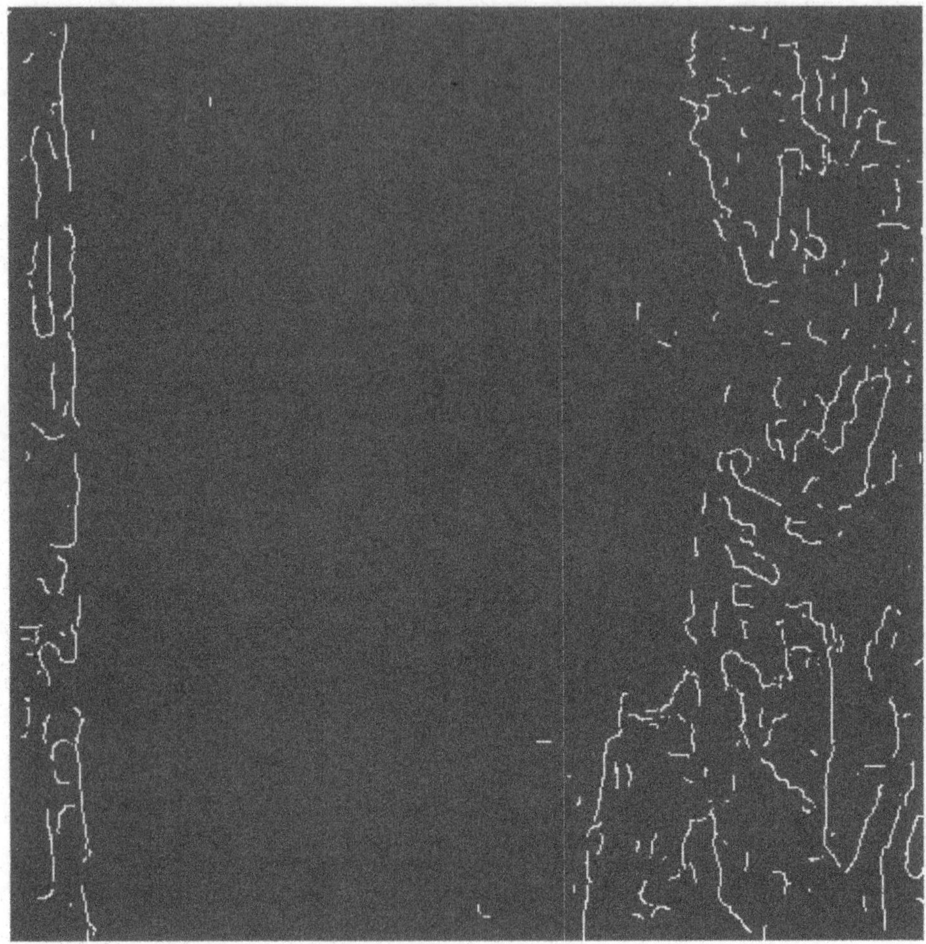

Figure 17.25: Edges obtained from image 17.22 with Canny's detector, thresholded such that no edges appear in the sea region.

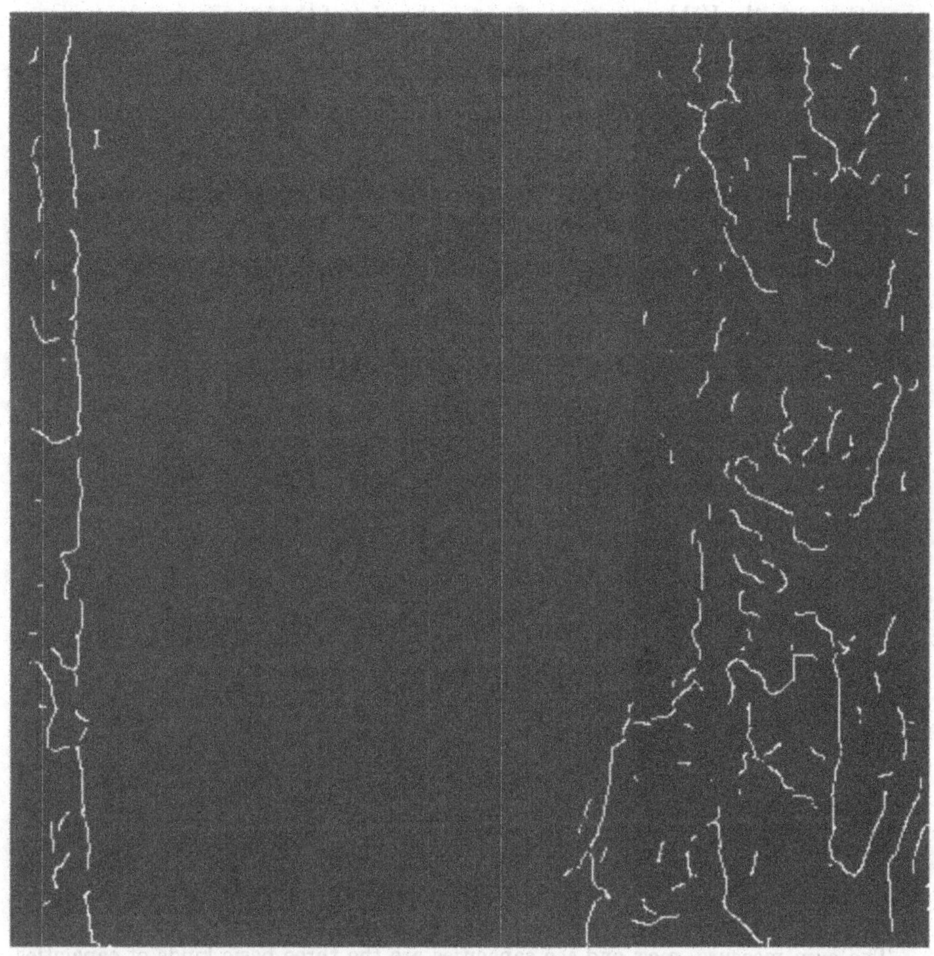

Figure 17.26: Edges obtained from image 17.23 with Canny's detector, thresholded such that no edges appear in the sea region.

- Compute the Hölder exponent of c at each point of the image.
- Compute the multifractal spectrum of c.
- Declare as "smooth" edge points those belonging to the set(s) E_α whose dimension is one.
- Declare as "irregular" edge points those belonging to the set(s) E_α whose dimension is between two fixed values, typically 1.1 and 1.5.

Results of segmentation using this approach are shown on figures 17.27, 17.28, 17.29 and 17.30.

Figure 17.27: Original image.

The *sum* measure, *max* and *iso* capacities are the three basic kinds of capacities used for analyzing images. There are cases where additional specific capacities have to be designed in order to get robust and precise results. This happens for instance in medical imaging. In certain situations, it is not enough to use one capacity (or one sequence of capacities), and one has to define a vector of capacities. To fully exploit the information provided by the different capacities and their spectra, it is convenient to adopt a Bayesian approach: a given point (x, y) in the image is characterized by a couple (t, λ), where t denotes the type of the singularity, and λ its height.

Let A be the vector of Hölder exponents at point (x, y). For instance, $A = (\alpha_{sum}, \alpha_{max}, \alpha_{iso})$. As is usual, one writes:

$$\Pr((t,\lambda)/A) = \frac{\Pr(A/(t,\lambda)) \times \Pr(t,\lambda)}{\Pr(A)}$$

and one looks for the couple (t, λ) that maximizes the left hand side. Since $\Pr(A)$ is constant, this is equivalent to maximizing the product $\Pr(A/(t, \lambda)) \times \Pr(t, \lambda)$. The

Figure 17.28: Original image with histogram equalization.

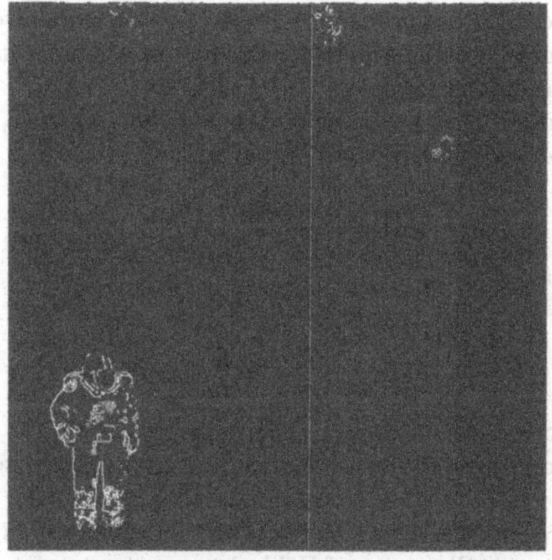

Figure 17.29: Smooth edge points obtained with *max* capacity.

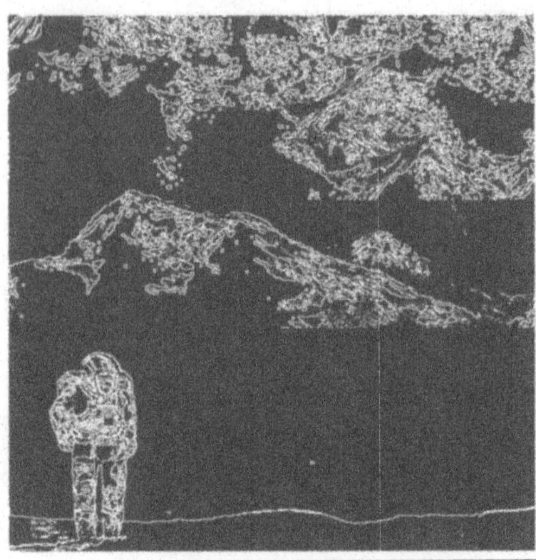

Figure 17.30: Irregular edge points obtained with *max* capacity.

computation of the conditional probability $\Pr(A/(t,\lambda))$ is quite involved in the general case. Only a few simple situations allow for an analytical computation. In other cases, numerical simulations are performed.

The computation of $\Pr(t,\lambda)$ is based on the fact that, when (x,y) does not lie in a smooth region, t and λ are independent:

$$\Pr(t,\lambda) = \Pr(t) \times \Pr(\lambda)$$

It may be shown that:

$$\begin{cases} \Pr(t \in T) & = & \Pr(\alpha_n^{iso} \in I(T)) \\ \Pr(\lambda \in \Lambda) & = & \Pr(\alpha_n^{max} \in M(\Lambda)) \end{cases}$$

with explicit relations between T and $I(T)$, and Λ and $M(\Lambda)$. It is thus possible to evaluate the prior probabilities using the f_g spectra related to the *iso* and *max* capacities.

A different approach that allows to refine the segmentation is to use multifractal correlations. In this case, the vector of Hölder exponents at point (x,y) is $A = (\alpha,\alpha')$ as defined in section 17.3.5. More complex situations call for the use of correlations of order higher than 2. A complete presentation of this topic is however beyond the scope of this paper.

Acknowledgments

Parts of the work described in sections 17.3 and 17.4 were done in collaboration with my students J.P. Berroir, J. Bestel, C. Canus, B. Guiheneuf, P. Mignot and R. Vojak.

I warmly thank M. Akian and R. Riedi for many stimulating discussions. Parts of this research were funded by Alcatel Espace, Alcatel ISR, INRETS and Dassault Aviation.

Bibliography

[1] S. O. Aase and T. A. Ramstad. On the optimality of nonunitary filter banks in subband coders. *IEEE Trans. Image Processing*, December 1995.

[2] N. Ahmed, T. Natarajan, and K. R. Rao. Discrete cosine transform. *IEEE Trans. on Computers*, 23:90–93, January 1974.

[3] M. Akian. Personal communication.

[4] M. Akian. Densities of idempotent measures and large deviations. Technical Report 2534, INRIA, April 1995.

[5] V. A. Allen and J. Belina. ECG data compression using the discrete cosine transform (DCT). *Computers in Cardiology Proceedings*, 1992.

[6] L. Alvarez, F. Guichard, P. L. Lions, and J. M. Morel. Axioms and fundamental equations of image processing. *Archi. for Rat. Mech.*, 3(123):199–257, 1993.

[7] M. Antonini, M. Barlaud, and P. Mathieu. Image Coding Using Wavelet Transform. *IEEE Trans. Image Proc.*, 1(2):205–220, April 1992.

[8] M. Arbeiter and N. Patzschke. Random self-similar multifractals. *Adv. Math.*, to appear.

[9] A. Arnéodo, G. Grasseau, and M. Holschneider. Wavelet transform analysis of invariant measures of some dynamical systems. In J.M. Combes and A. Grossmann, editors, *Proc. Int. Conf. Marseille, 1987*, pages 182–196. Springer-Verlag, Berlin, 1987.

[10] S. Arya, D. M. Mount, N. S. Netanyahu, R. Silverman, and A. Wu. An optimal algorithm for approximate nearest neighbor searching. In *Proc. 5th Annual ACM-SIAM Symposium on Discrete Algorithms*, pages 573–582, 1994.

[11] M. C. Aydin, A. E. Çetin, and H. Köymen. ECG data compression by sub–band coding. *Electronics Letter*, 27:359–360, February 1991.

[12] J. Babaud et al. Uniqueness of the Gaussian kernel for scale-space filtering. *IEEE Trans. Patt. Anal. Mach. Intell.*, Vol. PAMI-8(1):26–33, 1986.

[13] E. Bacry, J.F. Muzy, and A. Arnéodo. Singularity spectrum of fractal signals from wavelet analysis: Exact results. *J. Stat. Phys.*, 70(3/4):635–674, 1993.

[14] Z. Baharav, D. Malah, and E. Karnin. Hierarchical interpretation of fractal image coding and its application to fast decoding. In *Proc. Digital Signal Processing Conference*, Cyprus, July 1993.

[15] Z. Baharav, D. Malah, and E. Karnin. Hierarchical interpretation of fractal image coding and its applications. In [87], pages 91–117.

[16] B. Bani-Eqbal. Combining tree and feature classification in fractal encoding of images. submitted for publication.

[17] B. Bani-Eqbal. Speeding up fractal image compression. In *Proceedings from IS&T/SPIE 1995 Symposium on Electronic Imaging: Science & Technology*, volume Vol. 2418: Still-Image Compression, 1995.

[18] M. F. Barnsley and A. Jacquin. Application of Recurrent Iterated Function Systems to Images. *Proc. SPIE*, 1001:122–131, 1988.

[19] M.F. Barnsley. *Fractals Everywhere*. Academic Press, New York, 1988.

[20] M.F. Barnsley and S. Demko. Iterated function systems and the global construction of fractals. *Proc. Roy. Soc. London*, A399:243–275, 1985.

[21] M.F. Barnsley, S.G. Demko, J. Elton, and J.S. Geronimo. Invariant measures for markov processes arising from iterated function systems with place-dependent probabilities. *Ann. Inst. H. Poincaré*, 24:367–394, 1988.

[22] M.F. Barnsley and L.P. Hurd. *Fractal Image Compression*. A.K. Peters, Wellesley, Mass., 1993.

[23] K. U. Barthel, J. Schüttemeyer, T. Voyé, and P. Noll. A new image coding technique unifying fractal and transform coding. In *IEEE Conf. on Image Processing*, pages 112–116, Texas, 1994.

[24] K.U. Barthel. Private communication, 1995.

[25] G. Baudoin and M. Chaouche. A portable system for digital recording of electrocardiograms. In *Signal Processing IV: Theories and Application*, pages 1275–1278, 1988. North-Holland.

[26] T. Bedford. PhD thesis, University of Warwick, 1984.

[27] T. Bedford, F.M. Dekking, M. Breeuwer, M.S. Keane, and D. v. Schooneveld. Fractal coding of monochrome images. *Signal Process.*, 6:405–419, 1994.

[28] T. J. Bedford, F. M. Dekking, and M. S. Keane. Fractal image coding techniques and contraction operators. *Nieuw Arch. Wisk.*, (4), 10(3):185–218, 1992.

[29] T. Bell, J. G. Cleary, and I. H. Witten. *Text Compression*. Prentice Hall, Englewood Cliffs, NJ, 1990.

[30] M.J. Best and K. Ritter. A quadratic programming algorithm. *Zeitschrift für Oper. Res.*, 32:271–297, 1988.

[31] J. Bestel and J. Lévy Véhel. A *Max − −plus* approach to large deviation multifractal spectrum computation. Technical report, INRIA, September 1996.

[32] B. Bielefeld and Y. Fisher. Personal communication.

[33] B. Bielefeld and Y. Fisher. A convergence model. In [87], pages 215–228.

[34] P. Billingsley. *Ergodic Theory and Information*. John Wiley, 1965.

[35] A. Bogdan and H. E. Meadows. Kohonen neural network for image coding based on iteration transformation theory. *SPIE*, Vol. 1766-39, 1992.

[36] R. D. Boss and E. W. Jacobs. Archetype classification in an iterated transformation image compression algorithm. In [87], pages 79–90.

[37] G. Brager. Fractal coding of grey tone images, 1990. Master's Thesis (in Norwegian), The Norwegian Institute of Technology, Department of Telecommunications.

[38] G. Brown, G. Michon, and J. Peyrière. On the multifractal analysis of measures. *Journal of Stat. Phys.*, t. 66:775–790, 1992.

[39] C.A. Cabrelli, B. Forte, U.M. Molter, and E.R. Vrscay. Iterated fuzzy set systems: a new approach to the inverse problem for fractals and other sets. *J. Math. Anal. Appl.*, 171:79–100, 1992.

[40] C.A. Cabrelli, U.M. Molter, and E.R. Vrscay. Moment matching for the approximation of measures using iterated function systems. preprint, 1990.

[41] J. F. Canny. Finding edges and lines in images. Technical Report T.R. 720, MIT, June 1983.

[42] C. Canus and J Lévy Véhel. Change detection in sequences of images by multifractal analysis. In *ICASSP*, May 1996.

[43] C. Canus and J. Lévy Véhel. Sub-optimal choice a sequence of capacities for multifractal image analysis. Technical report, INRIA, September 1996.

[44] G. Caso, P. Obrador, and C.-C. J. Kuo. Fast methods for fractal image encoding. In *Proc. SPIE Visual Communication and Image Processing '95*, Vol. 2501, pages 583–594, 1995.

[45] M.E. Cates and J.M. Deutsch. Spatial correlations in multifractals. *Phys. Rev. A.*, 35(11):4907–4910, 1987.

[46] R. Cawley and R.D. Mauldin. Multifractal decomposition of Moran fractals. *Adv. Math.*, 92:196–236, 1992.

[47] Wen-Hsiung Chen and William Pratt. Scene adaptive coder. *IEEE Trans. Commun.*, COM-32(3):225–232, March 1984.

[48] C.K. Chui. *An Introduction to Wavelets.* Academic Press, Boston, MA, 1992.

[49] R.J. Clarke. *Transform Coding of Images.* Academic Press, London, 1990.

[50] J.M. Combes and A. Grossmann. Wavelets time-frequency methods and phase space. In *Proc. Int. Conf. Marseille, 1987.* Springer-Verlag, Berlin, 1987.

[51] J. Crowe, N. Gibson, M. Woolfson, and M. Somekh. Wavelet transform: A potential tool for ECG analysis and compression. *Journal of Biomedical Engineering*, pages 268 – 272, 1992.

[52] C.D. Cutler. The Hausdorff dimension distribution of finite measures in Euclidean space. *Can. J. Math.*, 38:1459–1484, 1986.

[53] I. Daubechies. *Ten Lectures on Wavelets.* SIAM, Philadelphia, PA, 1992.

[54] G. Davis. Self-Quantization of Wavelet Subtrees. In Harold H. Szu, editor, *Proc. SPIE Wavelet Applications II, Orlando, April 1995*, volume 2491, pages 141–152, 1995.

[55] G. Davis. Self-Quantization of Wavelet Subtrees. In Andrew F. Laine and Michael A. Unser, editors, *Proc. SPIE Wavelet Applications in Signal and Image Proc. III, San Diego, July 1995*, volume 2569, pages 294–307, 1995.

[56] G. Davis. Self-Quantization of Wavelet Subtrees: a Wavelet-Based Theory of Fractal Image Compression. In James A. Storer and Martin Cohn, editors, *Proc. Data Compression Conference, Snowbird, Utah, March 28–30, 1995*, pages 232–241. IEEE Computer Society, 1995.

[57] G. Davis. A wavelet-based analysis of fractal image compression. *Preprint*, see http://www.cs.dartmouth.edu/~gdavis, 1996.

[58] F. Davoine and J.-M. Chassery. Adaptive Delaunay triangulation for attractor image coding. In *12th International Conference on Pattern Recognition (ICPR)*, 1994.

[59] R. Degani, G. Bortolan, and M. Rossana. Karhunen-Loève coding of ECG signals. *IEEE*, pages 395–398, 1991.

[60] M. Dekking. An inequality for pairs of martingales and its application to fractal image coding, technical report 95-10 of the faculty of technical mathematics and informatics, Delft University of Technology, 1995. Also to appear in the Journal of Applied Probability, 1996.

[61] M. Dekking. Fractal image coding: some mathematical remarks on its limits and its prospects. In Y. Fisher, editor, *Fractal Image Encoding and Analysis. NATO ASI Series F*, volume 159. Springer-Verlag, Berlin, 1998 (this volume).

[62] R. A. DeVore, B. Jawerth, and B. J. Lucier. Image Compression through Wavelet Transform Coding. *IEEE Trans. Info. Theory*, 38(2):719–746, March 1992.

[63] L. Devroye. The double kernel method in density estimation. *Ann. Inst. Henri Poincaré*, 4(25):533–580, 1989.

[64] P. Diamond and P. Kloeden. Metric spaces of fuzzy sets. *Fuzzy Sets and Systems*, 35:241–249, 1990.

[65] D.L. Donoho. De-noising by soft thresholding. *IEEE Trans. on Info. Theory*, (41):613–627, 1995.

[66] B. Dubuc, C. Roques-Carmes, S.W. Zucker, and C. Tricot. Evaluating the fractal dimension of profiles. *Phys. Rev. A*, 39:1500–1512, 1989.

[67] F. Dudbridge. Least-squares block coding by fractal functions. In [87], pages 229-241.

[68] F. Dudbridge. Linear time fractal coding schemes. In Y. Fisher, editor, *Fractal Image Encoding and Analysis. NATO ASI Series F*, volume 159. Springer-Verlag, Berlin, 1998 (this volume).

[69] G.A. Edgar and R.D. Mauldin. Multifractal decompositions of digraph recursive fractals. *Proc. London Math. Soc.*, 3(65):604–628, 1992.

[70] J. Elton. An ergodic theorem for iterated maps. *Erg. Th. Dyn. Sys.*, 7:481–488, 1987.

[71] C.J.G. Evertsz and B.B. Mandelbrot. Multifractal measures. In H.-O. Peitgen, H. Jürgens, and D. Saupe, editors, *Chaos and Fractals: New Frontiers in Science*, pages 199–257. Springer-Verlag, New York, 1992.

[72] K.J. Falconer. Random fractals. *Math. Proc. Cambridge Philos. Soc.*, 100:559–582, 1986.

[73] K.J. Falconer. The Hausdorff dimension of self-affine fractals. *Math Proc. Cambridge Philos. Soc.*, 103:339–350, 1988.

[74] K.J. Falconer. *Fractal Geometry – Mathematical Foundations and Applications.* John Wiley, 1990.

[75] K.J. Falconer. Bounded distortion and dimension for non-conformal repellers. *Math. Proc. Cambridge Philos. Soc.*, 115:315–334, 1994.

[76] K.J. Falconer. The multifractal spectrum of statistically self-similar measures. *Journal of Theo. Prob.*, 7:681-702, 1994.

[77] K.J. Falconer and T.C. O'Neil. Vector-valued multifractal measures. *Proc. Roy. Soc. Lond. A.*, 452:1433–1457, 1996.

[78] F. Family. *Phys. Rev. E*, 47:2281, 1993.

[79] D. J. Field. Scale-invariance and self-similar 'wavelet' transforms: an analysis of natural scenes and mammalian visual systems. In M. Farge, J. C. R. Hunt, and J. C. Vassilicos, editors, *Wavelets, Fractals, and Fourier Transforms.* Oxford University Press, Oxford, 1993.

[80] D.J. Field. Relations between the statistics of natural images and the response properties of cortical cells. *J. Opt. Soc. Am.*, A4:2379–2394, 1987.

[81] Y. Fisher. A discussion of fractal image compression. In H.-O. Peitgen, H. Jürgens, and D. Saupe, editors, *Chaos and Fractals.* Springer-Verlag, New York, 1992.

[82] Y. Fisher. Fractal image compression with quadtrees. In [87], pages 55–77. 1995.

[83] Y. Fisher. Mathematical background. In [87], pages 25–53.

[84] Y. Fisher, B. Jacobs, and R. Boss. Fractal Image Compression Using Iterated Transforms. In J. Storer, editor, *Image and Text Compression*, pages 35–61. Kluwer Academic, 1992.

[85] Y. Fisher and S. Menlove. Fractal encoding with HV partitions. In [87], pages 119–136.

[86] Y. Fisher, T. P. Shen, and D. Rogovin. A comparison of fractal methods with DCT (JPEG) and wavelets (EPIC). In *SPIE Procedings, Neural and Stochastic Methods in Image and Signal Processing III*, volume 2304-16, San Diego, CA, July 28-29 1994.

[87] Y. Fisher(Ed.). *Fractal Image Compression: Theory and Application.* Springer-Verlag, New York, 1995.

[88] R. Fjørtoft. Complexity reduction in the decoder part of an attractor image coder, 1992. Student's Project Report (in Norwegian), The Norwegian Institute of Technology, Department of Telecommunications.

[89] B. Forte and E.R. Vrscay. Solving the inverse problem for functions and image approximation using iterated function systems i. theoretical basis. *Fractals*, 2:325–334, 1994.

[90] B. Forte and E.R. Vrscay. Solving the inverse problem for functions and image approximation using iterated function systems ii. algorithm and computations. *Fractals*, 2:335–346, 1994.

[91] B. Forte and E.R. Vrscay. Solving the inverse problem for functions and image approximation using iterated function systems. *Dyn. Cont. Impul. Sys.*, 1:177–231, 1995.

[92] B. Forte and E.R. Vrscay. Solving the inverse problem for measures using iterated function systems: A new approach. *Adv. Appl. Prob.*, 27:800–820, 1995.

[93] J. M. Franklin. *Matrix Theory.* Prentice-Hall Series in Applied Mathematics, Englewood Cliffs, 1968.

[94] J. H. Friedman, J. L. Bentley, and R. A. Finkel. An algorithm for finding best matches in logarithmic expected time. *ACM Trans. Math. Software*, 3,3:209–226, 1977.

[95] C. Frigaard. Fast fractal 2d/3d image compression. Technical report, Institute for Electronic Systems, Aalborg University, 1995.

[96] U. Frisch and G.Parisi. *Turbulence and predictability in geophysical fluid dynamics and climate dynamics*, page 84. M. Ghil, R. Benzi and G. Parisi, Amsterdam (Holland), 1895.

[97] A. Gersho and R.M Gray. *Vector Quantization and Signal Compression*. Kluwer Academic Publishers, 1992.

[98] D. Götting, A. Ibenthal, and R. Grigat. Fractal image coding and magnification using invariant features. *Fractals*, 5, April 1997. NATO ASI Conf. Fractal Image Encoding and Analysis, Trondheim, July 1995.

[99] S. Graf. Statistically self-similar fractals. *Probab. Th. Rel. Fields*, 74:357–394, 1987.

[100] S. Graf, R.D. Mauldin, and S.C. Williams. The exact Hausdorff dimension in random recursive constructions. *Mem. Am. Math. Soc.*, 71(381), 1988.

[101] P. Grassberger, R. Badii, and A. Politi. Scaling Laws for Invariant Measures on Hyperbolic and Nonhyperbolic Atractors. *Journal of Statistical Physics*, 51(1/2):135–178, July 1988.

[102] R.M. Gray. Vector quantization. *IEEE ASSP Magazine*, 1(2):4–29, April 1984.

[103] B. Guiheneuf and J. Lévy Véhel. Image enhancement through multifractal analysis. Technical report, INRIA, July 1996.

[104] I. Habboush, G. Moody, and R. Mark. Neural networks for ECG compression and classification. *IEEE Computer Society Press*, pages 185 - 188, 1992.

[105] T.C. Halsey, L.P. Jensen, M.H.and Kadanoff, I. Procaccia, and B.I. Shraiman. Fractal measures and their singularities: The characterization of strange sets. *Phys. Rev. A*, 33(1141), 1986.

[106] R. Hamzaoui. Codebook clustering by self-organizing maps for fractal image compression. *Fractals*, 5, April 1995. NATO ASI Conf. Fractal Image Encoding and Analysis, Trondheim, July 1995.

[107] J.C. Hart and T.A. De Fanti. Efficient antialiased rendering of 3-d linear fractals. *Computer Graphics*, 25, 1991.

[108] T. Haugan. Compression of ECG signals by means of optimized fir filterbanks and entropy allocation. Master's thesis, Rogaland University Center, 1995 (in Norwegian).

[109] J. G. Heber, S. O. Aase, and J. H. Husøy. ECG data compression by modelling of subbands. In *Proc. Computers in Cardiology*, pages 621–624, Bethesda, Maryland, USA, September 1994.

[110] H.G.E. Hentschel and I. Procaccia. The infinite number of generalized dimensions of fractals and strange attractors. *Physica 8D*, 1983.

[111] I. Heuter and S. Lalley. Falconer's formula for the Hausdorff dimension of a self-affine set in R^2. *Ergod. Th. Dynam. Sys.*, 15:77–97, 1995.

[112] R. Holley and E. Waymire. Multifractal dimensions and scaling exponents for strongly bounded random cascades. *Ann. App. Prob.*, 2(4):819–845, 1992.

[113] M. Holschneider. On the wavelet transform of fractal objects. *J. Stat.Phys.*, 50(5/6):963–993, 1988.

[114] M. Holschneider and P. Tchamitchian. Régularité locale de la fonction non-differentiable de Riemann. In P.G. Lemarie, editor, *Lecture Notes in Mathematics: Les ondeletes en 1989*, volume 1438. Springer-Verlag, Berlin, 1990.

[115] T. Holter. Fractal modelling of speech, 1990. Master's Thesis (in Norwegian), The Norwegian Institute of Technology, Department of Telecommunications.

[116] R. A. Horn and C. R. Johnson. *Matrix Analysis*. Cambridge University Press, New York, 1990.

[117] C. M. Huang, Q. Bi, G. S. Stiles, and R. W. Harris. Fast full search equivalent encoding algorithms for image compression using vector quantization. *IEEE Trans. Image Processing*, 1,3:413–416, 1992.

[118] B. Hürtgen and C. Stiller. Fast hierarchical codebook search for fractal coding of still images. In *EOS/SPIE Visual Communications and PACS for Medical Applications '93*, pages 397–408, Berlin, 1993.

[119] J. H. Husøy, M. Bøe, and J. G. Heber. An image coding approach to ECG compression. In *Proc. BIOSIGNAL*, pages 98–100, Brno, Czech Republic, June 1994.

[120] J.H Husøy and T. Gjerde. Computationally efficient subband coding of ECG signals. *Medical Engineering and Physics*, 1995. Accepted for publication.

[121] J. Hutchinson. Fractals and self-similarity. *Indiana Univ. J. Math.*, 30:713–747, 1981.

[122] E. W. Jacobs, Y. Fisher, and R. D. Boss. Image Compression: a Study of the Iterated Transform Method. *Signal Processing*, 29(3):251–263, December 1992.

[123] A. Jacquin. *A Fractal Theory of Iterated Markov Operators with Applications to Digital Image Coding.* PhD thesis, Georgia Institute of Technology, 1989.

[124] A. Jacquin. Image coding based on a fractal theory of iterated contractive markov operators, part ii: Construction of fractal codes for digital images. Technical Report Math. 91389-017, Georgia Institute of Technology, 1989.

[125] A. Jacquin. Image coding based on a theory of iterated contractive image transformations. *IEEE Transactions on Image Processing*, 1(1):18–30, January 1992.

[126] A. Jacquin. Fractal image coding: A review. *Proceedings of the IEEE*, 81(10):1451–1465, October 1993.

[127] S. Jaffard. Multifractal formalism for functions part 1: Results valid for all functions. Preprint.

[128] S. Jaffard. Multifractal formalism for functions part 2: Self-similar functions. Preprint.

[129] S. Jaffard. Exposants de Hölder en des points donnés et coéfficients d'ondolettes. *C. R. Acad. Sci. Paris*, t.308(Série 1):79–81, 1989.

[130] S. Jaffard. Construction de fonctions multifractales ayant un spectre de singularités prescrit. *C.R. Acad. Sci. Paris*, pages 19–24, 1992. T. 315, Série I.

[131] S. Jaffard. Pointwise regularity of functions and wavelet coefficients. In *Wavelets and Applications, Research Notes in Applied Mathematics.* Springer-Verlag, 1992.

[132] S. Jaffard. Multifractal formalism for functions, part 1 and part 2. *SIAM Journal Math. Anal.*, to appear.

[133] A. K. Jain. *Fundamentals of Digital Image Processing.* Prentice-Hall, Englewood Cliffs, 1989.

[134] S. Jalaleddine and C. Hutchens. Saies - a new ECG data compression algorithm. *Journal of Clinical Engineering*, 15(1):45–51, January/Feburary, 1990.

[135] S. M. S. Jalaleddine, C. G. Hutchens, R. D. Strattan, and W. A. Coberly. ECG data compression techniques – a unified approach. *IEEE Transactions on Biomedical Engineering*, 37(4):329–343, April 1990.

[136] B. Jawerth and W. Sweldens. An overview of wavelet based multiresolution analyses. *SIAM Review*, 36(3):377–412, 1994.

[137] N. S. Jayant and Peter Noll. *Digital Coding of Waveforms.* Prentice-Hall, Englewood Cliffs, 1984.

[138] J.-P. Kahane and J. Peyrière. Sur certaines martingales de B. Mandelbrot. *Adv. Math.*, 22:131–145, 1979.

[139] M.D. Kelly. Edge detection in pictures by computer using planning. *Machine Intelligence*, 6:397–409, 1971.

[140] J. King. The singularity spectrum for general Sierpinski carpets. *Adv. Math.*, 116:1–11, 1995.

[141] J. J. Koenderink. The structure of images. *Biol. Cybern*, (50):363–370, 1984.

[142] J. Kominek. Advances in fractal compression in multimedia applications. Submitted to Multimedia Systems Journal.

[143] J. Kominek. Algorithm for fast fractal image compression. In *Proceedings from IS&T/SPIE 1995 Symposium on Electronic Imaging: Science & Technology Vol. 2419 Digital Video Compression: Algorithms and Technologies*, pages 296–305, 1995.

[144] H. P. Kramer and M. V. Matthews. A linear coding for transmitting a set of correlated signals. *IRE Trans. on Information Theory*, 23:41–46, September 1956.

[145] E. Kreyszig. *Introduction to Functional Analysis with Applications*. Robert E. Krieger Publishing Co., Malabar, Florida, 1989.

[146] H. Krupnik, D. Malah, and E. Karnin. Fractal representation of images via the discrete wavelet transform. In *Proc. IEEE Conv. EE, Tel-Aviv, 7-8 March 1995*, 1995.

[147] H. Lebesgue. La mesure des grandeurs. *Monographie de l'enseignement Mathématique*, 1956.

[148] C.-H. Lee and L. H Chen. Fast closest codeword search algorithm for vector quantization. In *IEE Proc.-Vis. Image Signal Process.*, volume 141, 3, pages 143–148, 1994.

[149] S. Lepsøy. *Attractor Image Compression – Fast Algorithms and Comparisons to Related Techniques*. PhD thesis, The Norwegian Institute of Technology, April 1993.

[150] S. Lepsøy, P. Carlini, and G. E. Øien. On a weakness of fractal compression. In *Fractal Coding and Analysis*. Springer-Verlag, July 1995.

[151] S. Lepsøy and G. E. Øien. Fast attractor image encoding by adaptive codebook clustering. Chapter 9 in [87], pages 181–201.

[152] S. Lepsøy, G. E. Øien, and T. A. Ramstad. Attractor image compression with a fast noniterative decoding algorithm. In *Proc. Int. Conf. Acoust. Speech, Signal Proc.*, pages V–337 – V–340, April 1993.

[153] J Lévy Véhel. Numerical computation of the large deviation multifractal spectrum. In *CFIC*, September 1996.

[154] J. Lévy Véhel and C. Canus. Hausdorff dimension estimation and application to multifractal spectrum computation. Technical report, INRIA, June 1996.

[155] J. Lévy Véhel and R Vojak. Multifractal analysis of Choquet capacities: Preliminary results. *Adv. Appl. Math.*, to appear.

[156] Jae S. Lim and Alan V. Oppenheim. *Advanced Topics in Signal Processing*. Prentice-Hall, Englewood Cliffs,

[157] H. Lin and A. N. Venetsanopoulos. A pyramid algorithm for fast fractal image compression. In *Proceedings of 1995 IEEE Intern. Conf. on Image Processing (ICIP)*, Washington, 1995.

[158] Y. Linde, A. Buzo, and R.M. Gray. An algorithm for vector quantizer design. *IEEE Transactions on Communications*, COM-28(1):84–95, January 1980.

[159] C.-H. Ling. Representation of associative functions. *Publ. Math. Debrecen*, 12:189–212, 1965.

[160] A.O. Lopes. The dimension spectrum of the maximal measure. *SIAM J. Math. Anal.*, 20:1243–1254, 1989.

[161] A.K. Louis, R. Maaß, and A. Rieder. *Wavelets*. Teubner, Stuttgart, 1994.

[162] L. Lundheim. A discrete framework for fractal signal modeling. In [87], pages 137–151.

[163] L. M. Lundheim. *Fractal Signal Modelling for Source Coding*. PhD thesis, The Norwegian Institute of Technology, October 1992.

[164] J. Makhoul, S. Roucos, and H. Gish. Vector quantization in speech coding. *Proceedings of the IEEE*, 73(11):1551–1587, November 1985.

[165] S. Mallat. A theory for multiresolution signal decomposition: the wavelet representation. *IEEE Trans. Pattern Analysis and Machine Intelligence*, 11(7):674–693, July 1989.

[166] S. Mallat and W.L. Hwang. Singularity detection and processing with wavelets. *IEEE Trans. Inform. Theor.*, 38(2):617–643, 1992.

[167] S. Mallat and S. Zhong. Characterization of signals from multiscale edges. *IEEE Trans. Patt. Anal. Mach. Intell.*, 14(7):710–732, 1992.

[168] C. P. Mammen and B. Ramamurthi. Vector quantization for compression of multichannel ECG. *IEEE Transactions on Biomedical Engineering*, 37(9):821–825, September 1990.

[169] B. B. Mandelbrot. Multiplications aléatories itérées et distributions invariants par moyenne pondérée. *C. R. Acad. Sci. Paris*, 278:289–292 and 355–358, 1974.

[170] B.B. Mandelbrot. Intermittent turbulence in self similar cascades; divergence of high moments and dimension of the carrier. *J. Fluid Mech.*, 62(331), 1974.

[171] B.B. Mandelbrot. *Les objects fractals: forme, hasard et dimension.* Flammarion, Paris, 1975.

[172] B.B. Mandelbrot. *Fractals: Form, Chance, and Dimension.* W.H.Freeman and Co., San Francisco, 1977.

[173] B.B. Mandelbrot. *The Fractal Geometry of Nature.* W.H. Freeman and Co., New York, 1983.

[174] B.B. Mandelbrot. Self-affine fractals and fractal dimension. *Physica Scripta*, 32:257–260, 1985.

[175] B.B. Mandelbrot. An introduction to multifractal distribution functions. In H.E. Stanley and N. Ostrowsky, editors, *Fluctuations and Pattern Formation (Cargese 1988)*, pages 345–360. Kluwer, Dordrecht, 1988.

[176] B.B. Mandelbrot. A class of multinomial multifractal measures with negative (latent) values for the dimension $f(\alpha)$. In *Fractals (Proceedings of the Erice meeting).* L.Pietronero, New York, 1989.

[177] B.B. Mandelbrot. Fractal measures (their infinite moment sequences and dimensions) and multiplicative chaos: early works and open problems. Technical report, Physics Department, IBM Research Center, Mathematics Department, Harvard University, Cambridge, MA 02138, USA, 1989.

[178] B.B. Mandelbrot. Multifractal measures for the geophysicist. *Pure and Applied Geophysics*, 131:5–42, 1989.

[179] B.B. Mandelbrot and J.W. van Ness. Fractional brownian motion, fractional noises and applications. *SIAM Review*, 10:422–437, 1968.

[180] D. Marr. *Vision.* W. H. Freeman and Co., New York, 1982.

[181] D. Marr and E. Hildreth. Theory of edge detection. *Prc. R. Soc. Lond.*, B:187–217, 1980.

[182] P.R. Massopust. *Fractal functions, fractal surfaces and wavelets.* Academic Press, San Diego, 1995.

[183] R.D. Mauldin and S.C. Williams. Hausdorff dimension in graph directed constructions. *Trans. Amer. Math. Soc.*, 309:811–829, 1988.

[184] C. McMullen. The Hausdorff dimension of general Sierpinski carpets. *Nagoya Math. J.*, 96:1–9, 1984.

[185] M.J.Sabin and R.M. Gray. Product code vector quantizers for waveform and voice coding. *IEEE Trans. on Acoustics, Speech and Signal Processing*, ASSP-32:474–488, June 1984.

[186] D. M. Monro and F. Dudbridge. Fractal approximation of image blocks. In *Proc. ICASSP*, volume 3, pages 485–488, 1992.

[187] D.M. Monro. A hybrid fractal transform. In *Proc. ICASSP 5*, pages 162–172, 1993.

[188] D.M. Monro and F. Dudbridge. Fractal block coding of images. *Electron. Lett.*, 28:1053–1054, 1992.

[189] D.M. Monro and S.J. Woolley. Fractal image compression without searching. University of Bath preprint, 1994.

[190] T. Murakami, K. Asai, and E. Yamazaki. Vector quantizer of video signals. *Electronic Letters*, 18(23):1005–1006, November 1982.

[191] J.F. Muzy, E. Bacry, and A. Arnéodo. The multifractal formalism revisited with wavelets. *Intern. J. Bifurc. Chaos*, 4(2):245–302, 1994.

[192] G. Nårstad. Fractal-based compression of ECG signals. Master's thesis, Rogaland University Center, June 1995.

[193] F. Normant. *Analyse fractale de courbes et convexité.* PhD thesis, École Polytechnique de Montréal, 1994. Ph.D. Thesis.

[194] F. Normant and C. Tricot. Simplifications of curves using convex hull. *Geographical Analysis*, 25:118–129, 1992.

[195] M. Novak. *Attractor coding of images.* PhD thesis, Linköping University, May 1993. Dept. of Electrical Engineering.

[196] G. Øien and S. Lepsøy. On the benefits of basis orthogonalization in fractal compression. In Y. Fisher, editor, *Fractal Image Encoding and Analysis. NATO ASI Series F*, volume 159. Springer-Verlag, Berlin, 1998 (this volume).

[197] G. Øien, S. Lepsøy, and T. A. Ramstad. An inner product space approach to image coding by contractive transformations. In *Proc. Int. Conf. Acoust. Speech, Signal Proc.*, pages 2773–2776, May 1991.

[198] G. E. Øien. *L_2 Optimal Attractor Image Coding with Fast Decoder Convergence.* PhD thesis, The Norwegian Institute of Technology, Trondheim, Norway, April 1993.

[199] G. E. Øien. Parameter quantization in fractal image coding. In *Proc. ICIP-94 (1st IEEE International Conference on Image Processing)*, November 1994.

[200] G. E. Øien, Z. Baharav, S. Lepsøy, E. Karnin, and D. Malah. A new improved collage theorem with applications to multiresolution fractal image coding. In *Proc. ICASSP-94*, volume 5, pages 565–568, May 1994.

[201] G. E. Øien and S. Lepsøy. Analysis of a complexity reduction algorithm in fractal image coding. In *Proc. ICSPAT-94 (5th International Conference on Signal Processing Applications and Technology)*, pages 910–915, October 1994.

[202] G. E. Øien and S. Lepsøy. Fractal-based image coding with fast decoder convergence. *Signal Processing*, 40:105–117, 1994.

[203] G. E. Øien and S. Lepsøy. A class of fractal image coders with fast decoder convergence. Chapter 8 in [87], pages 157–179.

[204] L. Olsen. Self-affine multifractal Sierpinski sponges. To appear.

[205] L. Olsen. Random geometrically graph directed self-similar multifractals. *Longman*, 1994.

[206] L. Olsen. *Random Geometrically Graph Directed Self-Similar Multifractals.* Pitman Research Notes in Mathematics, 1994.

[207] L. Olsen. A multifractal formalism. *Adv. Math.*, 116:82–196, 1995.

[208] K. Parthasarathy. *Probability Measures on Metric Spaces.* Academic Press, New York, 1967.

[209] P. Perona and J. Malik. Scale-space and edge detection using anisotropic diffusion. *IEEE Trans. PAMI*, 7(12):629–639, 1990.

[210] D. C. Popescu and H. Yan. MR image compression using iterated function systems. *Magnetic Resonance Imaging*, 11:727–732, 1993.

[211] M. Rabbani and P.W. Jones (Eds.). *Digital Image Compression Techniques.* SPIE Press, Tutorial Texts in Optical Engineering, Vol. TT7, Bellingham, Washington, 1991.

[212] T. A. Ramstad, S. O. Aase, and J. H. Husøy. *Subband Compression of Images – Principles and Examples.* Elsevier Science Publishers (North Holland), 1995.

[213] D. Rand. The singularity spectrum $f(\alpha)$ for cookie-cutters. *Ergod. Th. Dynam. Sys.*, 9:527–541, 1989.

[214] A.H. Read. The solution of a functional equation. *Proc. Roy. Soc. Edinburgh Sect. A*, 63:336–345, 1952.

[215] F. Reif. Irreversible processes and fluctuations, chapter 15. In *Fundamentals of Statistical and Thermal Physics.* Mc-Graw Hill, New York, 1965. F.N.H. Robinson (Ed.), Noise and Fluctuations, Clarendon Press, Oxford, 1974.

[216] R. Riedi. An improved multifractal formalism and self-similar measures. *Journal Math. Anal. Appl*, (189):462–490, 1995.

[217] R.H. Riedi. Multifractal formalism for infinite multinominal measures.

[218] C.A. Rogers and Taylor S.J. Additive set functions in Euclidean space. *Acta Math.*, 101:273–302, 1959.

[219] B.E. Rogowitz and R.F. Voss. Shape perception and low dimension fractal boundary contours. In B. E. Rogowitz and J. Allenbach, editors, *Proc. Conf. on Human Vision: Methods, Models and Applications*, volume 1249, Santa Clara, 1990. SPIE/SPSE Symposium on Electronic Imaging.

[220] W. Rudin. *Functional Analysis.* McGraw-Hill, second edition, 1994.

[221] D. Saupe. Breaking the time-complexity of fractal image compression. Technical Report 53, Institut für Informatik, Universität Freiburg, May 1994.

[222] D. Saupe. From classification to multi-dimensional keys. In [87], pages 302–305.

[223] D. Saupe. Accelerating fractal image compression by multi-dimensional nearest neighbor search. In J. A. Storer and M. Cohn, editors, *Proceedings Data Compression Conference, March 28–30, 1995, Snowbird, Utah.* IEEE Computer Society Press, 1995.

[224] D. Saupe. Lean domain pools for fractal image compression. In *Proceedings IS&T/SPIE 1996 Symposium on Electronic Imaging: Science & Technology – Still Image Compression II*, volume 2669, Jan. 1996.

[225] D. Saupe and R. Hamzaoui. Complexity reduction methods for fractal image compression. In J. M. Blackledge (ed.), editor, *I.M.A. Conf. Proc. on Image Processing; Mathematical Methods and Applications*, Sept. 1994. To appear with Oxford University Press.

[226] D. Saupe and R. Hamzaoui. A review of the fractal image compression literature. *ACM Computer Graphics*, 28,4:268–276, Nov. 1994.

[227] J. Schmeling and R. Siegmund-Schultze. The singularity spectrum of self-affine fractals with a Bernoulli measure. To appear.

[228] M. Schmitt and J. Mattioli. *Morphologie mathématique.* Masson, 1994.

[229] M.R. Schroeder. *Number Theory in Science and Communication.* Springer-Verlag, Berlin, 1990.

[230] L. Schwartz. *Théorie des distributions.* Hermann, Paris, second edition, 1950.

[231] J. Serra. Image analysis and mathematical morphology. *Academic Press*, 1982.

[232] J. Signes. Geometrical interpretation of fractal image coding. *Fractals*, 5, April 1997. NATO ASI Conf. Fractal Image Encoding and Analysis, Trondheim, July 1995.

[233] B. Simon. Explicit link between local fractal transform and multiresolution transform. In *Proc. IEEE Conf. Image Processing*, Oct. 1995.

[234] D. Stauffer. *Introduction to Percolation Theory.* Taylor and Francis, London, 1985.

[235] G. Strang. *Linear Algebra and Its Applications.* Harcourt Brace Jovanovich Publishers, San Diego, 2nd edition, 1980.

[236] R. Strichartz. *A Guide to Distribution Theory and Fourier Transforms.* CRC Press, Boca Raton, 1994.

[237] S. C. Tai. AZTDIS – a two phase real-time ECG data compressor. *Journal of Biomedical Engineering*, 15:510–515, November 1993.

[238] S. C. Tai. Slope – a real-time ECG data compressor. *Medical & Biological Engineering & Computing*, 29:175–179, March, 1991.

[239] N. Thakor, Y. Sun, and P. Caminal. A multiresolution wavelet-based ECG data compression algorithm. *IEEE Computer Society Press*, pages 393 – 396, 1993.

[240] L. Thomas and F. Deravi. Pruning of the transform space in block-based fractal image compression. In *Proceedings of ICASSP*, volume 5, pages 341–344, 1993.

[241] C. Tricot. *Sur la classification des ensembles boréliens de mesure de Lebesgue nulle.* PhD thesis, Université de Genève, 1975. Thèse de doctorat.

[242] C. Tricot. Two definitions of fractional dimension. *Math. Proc. Cambridge Philos. Soc.*, 91:57–74, 1982.

[243] C. Tricot. *Curves and Fractal Dimension.* Springer-Verlag, 1995.

[244] C. Tricot, J.F. Quiniou, D. Wehbi, C. Roques-Carmes, and B. Dubuc. Evaluation de la dimension fractale d'un graphe. *Rev. Physi. Appl.*, 23:111–124, 1988.

[245] A. van de Walle. Relating fractal compression to transform methods. Master of Mathematics Thesis, Department of Applied Mathematics, University of Waterloo, 1996.

[246] A. van de Walle. Merging fractal image compression and wavelet transform methods. *Fractals*, 5, April 1997. NATO ASI Conf. Fractal Image Encoding and Analysis, Trondheim, July 1995.

[247] A. van de Walle. Reduction of tiling effects in fractal image compression using wavelet decomposition. NATO ASI Conf. Fractal Image Encoding and Analysis, Trondheim, July 1995.

[248] M. Vetterli and J. Kovačević. *Wavelets and Subband Coding.* Prentice Hall, Englewood Cliffs, NJ, 1995.

[249] G. Vines. Orthogonal basis ifs. In [87], pages 199–214.

[250] R. Vojak and J. Lévy Véhel. Higher order multifractal analysis. *submitted to SIAM Journal on Math. Anal.*

[251] R. F. Voss, R. B. Laibowitz, and E. I. Alessandrini. Fractal (scaling) clusters in thin gold films near the percolation transition. *Phys. Rev. Lett.*, 49:1441–1444, 1982.

[252] R.F. Voss. Random fractals: characterization and measurement. In R. Pynn and A. Sjeltorp, editors, *Scaling Phenomena in Disordered Systems.* Plenum New York, 1985.

[253] R.F. Voss. Fractals in nature: from characterization to simulation. In H.-O. Peitgen and D. Saupe, editors, *The Science of Fractal Images.* Springer-Verlag, New York, 1988.

[254] R.F. Voss. Multifractals and the local connected fractal dimension: classification of early Chinese landscape paintings. In A. J. Crilly, R. A. Earnshaw, and H. Jones, editors, *Applications of Fractals and Chaos: the shape of things.* Springer-Verlag, New York, 1993.

[255] P. Waldemar. Complexity reduction in the encoder part of an attractor image coder, 1992. Student's Project Report (in Norwegian), The Norwegian Institute of Technology, Department of Telecommunications.

[256] M. Wickerhauser. *Adapted Wavelet Analysis from Theory to Software.* A.K. Peters, Wellesley, MA, 1995.

[257] A. Witkin. Scale-space filtering. In *Proc. Int. Joint Conf. Artif. Intell.*, pages 1019–1021, Karlsruhe, Germany, 1983.

[258] B. E. Wohlberg and G. de Jager. Fast image domain fractal compression by DCT domain block matching. *Electronic Letters*, 31:869–870, 1995.

[259] S.J. Woolley and D.M. Monro. Rate/distortion performance of fractal transforms for image compression. *Fractals 2*, pages 395–398, 1994.

[260] A.L. Yuille and T.A. Poggio. Scaling theorems for zero crossings. *IEEE Trans. Patt. Anal. Mach. Intell.*, PAMI-8(1):15–25, 1986.

[261] M. Zaoui. *Dimension fractale d'une réunion d'arcs trinomiaux.* PhD thesis, Université de Montréal, 1994.

List of Participants

INVITED SPEAKERS

Izhak Baharav
Department of Electrical Engineering
Technion-Israel Institute of Technology
Haifa 32000, Israel

Michael Barnsley
Iterated Systems, Inc.
5550A Peachtree Parkway, Suite 650
Norcross, GA 30092
USA

F. M. Dekking
Dept. of Prob., Stat. & O.R.
Faculty of Math. & Computer Science
Delft University of Technology
Mekelweg 4, 2628 CD Delft
The Netherlands

Frank Dudbridge
Institute for Nonlinear Science
University of California, San Diego
La Jolla, CA 92093-0402
USA

Kathrin Berkner
Center for Complex Systems
and Visualisation
University of Bremen
Bibliothekstr. 1
D-28359 Bremen, Germany

Kenneth J. Falconer
University of St Andrews
North Haugh
St Andrews
Fife KY16 9SS
United Kingdom

Yuval Fisher
Institute for Nonlinear Science
University of California, San Diego
La Jolla, CA 92093-0402
USA

Bruno Forte
University of Verona
Faculty of Science
Cavignal
Strada le Grazie
I-37134 Verona, Italy

Arnaud Jacquin
AT&T Bell Laboratories
Room 2D-337
60 Mountain Avenue
Murray Hill, NJ 07974-0636
USA

Don Monro
School of Electronic
and Electrical Engineering
University of Bath
Claverton Down
Bath BA2 7AY
United Kingdom

Jacques Lévy-Véhel
INRIA
F-78153 Le Chesnay Cedex
France

Geir E. Øien
Rogaland University Center
Department of Electrical and Computer Engineering
Signal Processing Group
P. B. 2557 Ullandhaug
N-4004 Stavanger
Norway

Dietmar Saupe
Institut für Informatik
Universität Freiburg
Am Flughafen 17
D-79110 Freiburg, Germany

Claude Tricot
Département de Mathématiques
Ecole Polytechnique de Montréal
Case Postale 6079, succ. Centre-ville
Montréal, Québec, Canada H3C 3A7

Edward R. Vrscay
Department of Applied Mathematics
University of Waterloo
Waterloo, Ontario, Canada N2L 3G1

DISTINGUISHED SPEAKERS

Skjalg Lepsøy
Norges Forskningsråd
Postboks 2700 St. Hanshaugen
N-0131 Oslo
Norway

Lars Lundheim
Sø-Trøndelag College
Department of Electrical and Electronic Engineering
N-7005 Trondheim
Norway

Richard Voss
Dept. of Applied Physics
Yale University, Becton Center
P.O. Box 2157
New Haven, CT 06520
USA

INVITED PARTICIPANTS

Volkan Atalay
Dept. of Computer Engineering
Middle East Tech. Univ. (METU)
TR-06531 Ankara
Turkey

Stéphane Baldo
Département de Mathématiques
Ecole Polytechnique de Montréal
Case Postale 6079, succ. Centre-ville
Montréal, Québec, Canada H3C 3A7

Kai Uwe Barthel
Institut für Fernmeldetechnik
Technische Universität Berlin
Einsteinufer 25
D-10587 Berlin
Germany

Dmitry Berg
Applied Biophysics Laboratory
Ural State Technical University -UPI
Yekaterinburg 620002
Russia

Don Bone
Division of Information Technology
Commonwealth Scientific and Industrial
Research Organisation
GPO Box 664
Canberra
A.C.T. 2601
Australia

Helmut Buley
TZ141a
Deutsche Telekom AG, FTZ
Postfach 100003
D-64276 Darmstadt
Germany

Paolo Carlini
Information Engineering
University of Pisa
Via Diotisalvi, 2
I-56126 Pisa
Italy

Gregory Caso
Department of Electrical Engineering Systems
University of Southern California, MC 2564
Los Angeles, CA 90089
USA

Jean-Marc Chassery
Laboratoire TIMC-IMAG.
URA C.N.R.S 1618
Institut Albert Bonniot
Faculté de Médecine
Université Joseph Fourier
F-38706 La Tronche Cedex
France

Bing Cheng
Mathematics and Statistics
University of Kent at Canterbury
Kent CT2 7N
United Kingdom

Leszek Cieplinski
The Franco-Polish School of New
Information & Communication Technologies
ul Mansfelda 4, PO Box 31
PL-60-854 Poznan 6
Poland

Guillaume Cretin
I.N.R.I.A.
Bat 24
Domaine de Voluceau
BP105
F-78153 Le Chesnay Cedex
France

Stephan Dahl
Computer Science
University of Copenhagen
Statholdervej 15 4V
DK-2400 Copenhagen NV
Denmark

Geoffrey Davis
Department of Mathematics
Dartmouth College
6211 Sudikoff Laboratory
Hanover, NH 03755-3510
USA

Franck Davoine
Laboratoire TIMC-IMAG (INFODIS)
Institut Albert Bonniot
Domaine de la Merci
F-38706 La Tronche Cedex
France

Gary Dickson
School of Electronic
and Electrical Engineering
University of Bath
Claverton Down, Bath, BA2 7AY
United Kingdom

Attilio Fanelli
Scienza Ed Ing Dello Spazio "Luigi G.
Napolitano"
P.L.E. Vincenzio Techio 80
I-80125 Napoli
Italy

Fredric M. Gilbert
INRIA Rocquencourt
Domaine de Voluceau
B.P. 105
F-78153 Le Chesnay Cedex
France

Daniele Giusto
Department of Electrical & Electronic
Engineering
University of Cagliari
Piazza D'Armi
I-09123 Cagliari
Italy

Raouf Hamzaoui
Institut für Informatik
Universität Freiburg
Am Flughafen 17
D-79110 Freiburg
Germany

John C. Hart
School of EECS
Washington State University
Pullman, WA 99164-2752
USA

Felix Henry
Theoretical Research
CANON Research Centre
Rue de la Touche Lambert
F-35517 Cesson Sevigne Cedex
France

Erwin Hocevar
Pattern Recognition and Image Processing
Technical University of Vienna
Treitlstr.3
A-1040 Vienna
Austria

Frank Horowitz
Exploration and Mining
C.S.I.R.O.
P.O. Box 437
Nedlands, W.A. 6020
Australia

Lyman P. Hurd
Iterated Systems Inc
3525 Piedmont Road
Suite 600 Bldg 7
Atlanta, GA 30305
USA

Achim Ibenthal
Philips Semiconductors GmbH
PCALH/VS
Stresemannallee 101
D-22529 Hamburg
Germany

Thomas Kaijser
Information Systems Technology
National Defence Research Establishment
Box 1165
S-581 11 Linkøping
Sweden

C. Ozgen Karacan
Middle East Technical University
Petroleum Engineering Department
TR-06531 Ankara
Turkey

John Kominek
Department of Computer Science
University of Waterloo
Waterloo, Ontario Canada N2L 3G1

Sukhamay Kundu
Computer Science Department
Louisiana State University
Baton Rouge, LA 70803
USA

Ning Lu
Iterated Systems, Inc.
3525 Piedmont Road
Seven Piedmont Center, Suite 600
Atlanta, GA 30305-1530
USA

Maria Isabel Martins
Electrical and Computers Engineering
Faculty of Engineering
University of Porto
Largo Mompilher 22, Apartado 4433
P-4007 Porto Codex
Portugal

Erika Müller
University of Rostock
Department of Electrical Engineering
Institute of Communications
Richard-Wagner-Strasse 31
D-18109 Rostock
Germany

Harald Nautsch
Department of Electrical Engineering
Linkøping University
S-581 83 Linkøping
Sweden

François Normant
Département de Mathématiques
Ecole Polytechnique de Montréal
Case Postale 6079, succ. Centre-ville
Montréal, Québec, Canada H3C 3A7

Hagai Krupnik
Department of Electrical Engineering
Technion - Israel Institute of Technology
Haifa 3200
Israel

Haibo Li
The Image Coding Group
Department of Electrical Engineering
Linkøping University
S-581 83 Linkøping
Sweden

Evelyne Lutton
Projet Fractales
INRIA
B.P. 105
F-78153 Le Chesnay Cedex
France

Charles Moreman
Iterated Systems Inc
3525 Piedmont Road
Seven Piedmont Center, Suite 600
Atlanta, GA 30305-1530
USA

Jagan Narayanan
TMM Inc.
32400 Seaside Drive
Union City, CA 94587
USA

Jeremy Nichols
Video Coding Group
School of Electronic & Electrical Engineering
University of Bath
Claverton Down
Bath BA2 7Y
United Kingdom

Mark Orzechowski
Mathematics Institute
North Haugh
St Andrews
Fife KY 16 9SS
United Kingdom

Eric Polidori
Institut EURECOM
Institut Non Linéaire de Nice
2229 route des Cretes
B.P. 193
F-06904 Sophia Antipolis Cedex
France

Anthony Popov
Faculty of Mathematics & Informatics
Sofia University
5, James Bouchier Blvd.
1126 Sofia
Bulgaria

Julien Signes
4 rue du Clos Courtel
F-35517 Cesson Sevigne Cedex
France

Benoit Simon
Telecommunications and Remote Sensing
Université Catholique de Louvain
2 Place Du Levant
F-1348 Louvain - La Neuve
Belgium

Kevin Smith
Division of Information Technology
CSIRO
PO Box 664
Canberra ACT 2601
Australia

Axel van de Walle
Department of Applied Mathematics
Faculty of Mathematics
University of Waterloo
Waterloo, Ontario, Canada N2L 3G1

Gert van de Wouwer
Vision Lab
Ruca (University of Antwerp)
Groenenborgerlaan 171
B-2020 Antwerp
Belgium

Horst Walter
Network and Signal Theory
Darmstadt University of Technology
Merckstrasse 25
D-64282 Darmstadt
Germany

Niclas Wadstromer
Image Coding Group
Department of Electrical Engineering
Linkøping University
S-581 83 Linkøping
Sweden

Paul Wakefield
Video Coding Group
School of Electronic & Electrical Engineering
University of Bath
Claverton Down
Bath BA2 7Y
United Kingdom

Cornelius Willers
Electro-Optics
Kentron
Esther Street 682
PO Box 38196
Garsfontein
0042
South Africa

Index

NATO ASI Series F

Including Special Programmes on Sensory Systems for Robotic Control (ROB) and on Advanced Educational Technology (AET)

Vol. 120: Reliability and Safety Assessment of Dynamic Process Systems. Edited by T. Aldemir, N. O. Siu, A. Mosleh, P. C. Cacciabue and B. G. Göktepe. X, 242 pages. 1994.

Vol. 121: Learning from Computers: Mathematics Education and Technology. Edited by C. Keitel and K. Ruthven. XIII, 332 pages. 1993. *(AET)*

Vol. 122: Simulation-Based Experiential Learning. Edited by D. M. Towne, T. de Jong and H. Spada. XIV, 274 pages. 1993. *(AET)*

Vol. 123: User-Centred Requirements for Software Engineering Environments. Edited by D. J. Gilmore, R. L. Winder and F. Détienne. VII, 377 pages. 1994.

Vol. 124: Fundamentals in Handwriting Recognition. Edited by S. Impedovo. IX, 496 pages. 1994.

Vol. 125: Student Modelling: The Key to Individualized Knowledge-Based Instruction. Edited by J. E. Greer and G. I. McCalla. X, 383 pages. 1994. *(AET)*

Vol. 126: Shape in Picture. Mathematical Description of Shape in Grey-level Images. Edited by Y.-L. O, A. Toet, D. Foster, H. J. A. M. Heijmans and P. Meer. XI, 676 pages. 1994.

Vol. 127: Real Time Computing. Edited by W. A. Halang and A. D. Stoyenko. XXII, 762 pages. 1994.

Vol. 128: Computer Supported Collaborative Learning. Edited by C. O'Malley. X, 303 pages. 1994. *(AET)*

Vol. 129: Human-Machine Communication for Educational Systems Design. Edited by M. D. Brouwer-Janse and T. L. Harrington. X, 342 pages. 1994. *(AET)*

Vol. 130: Advances in Object-Oriented Database Systems. Edited by A. Dogac, M. T. Özsu, A. Biliris and T. Sellis. XI, 515 pages. 1994.

Vol. 131: Constraint Programming. Edited by B. Mayoh, E. Tyugu and J. Penjam. VII, 452 pages. 1994.

Vol. 132: Mathematical Modelling Courses for Engineering Education. Edited by Y. Ersoy and A. O. Moscardini. X, 246 pages. 1994. *(AET)*

Vol. 133: Collaborative Dialogue Technologies in Distance Learning. Edited by M. F. Verdejo and S. A. Cerri. XIV, 296 pages. 1994. *(AET)*

Vol. 134: Computer Integrated Production Systems and Organizations. The Human-Centred Approach. Edited by F. Schmid, S. Evans, A. W. S. Ainger and R. J. Grieve. X, 347 pages. 1994.

Vol. 135: Technology Education in School and Industry. Emerging Didactics for Human Resource Development. Edited by D. Blandow and M. J. Dyrenfurth. XI, 367 pages. 1994. *(AET)*

Vol. 136: From Statistics to Neural Networks. Theory and Pattern Recognition Applications. Edited by V. Cherkassky, J. H. Friedman and H. Wechsler. XII, 394 pages. 1994.

Vol. 137: Technology-Based Learning Environments. Psychological and Educational Foundations. Edited by S. Vosniadou, E. De Corte and H. Mandl. X, 302 pages. 1994. *(AET)*

Vol. 138: Exploiting Mental Imagery with Computers in Mathematics Education. Edited by R. Sutherland and J. Mason. VIII, 326 pages. 1995. *(AET)*

Vol. 139: Proof and Computation. Edited by H. Schwichtenberg. VII, 470 pages. 1995.

Vol. 140: Automating Instructional Design: Computer-Based Development and Delivery Tools. Edited by R. D. Tennyson and A. E. Barron. IX, 618 pages. 1995. *(AET)*

Vol. 141: Organizational Learning and Technological Change. Edited by C. Zucchermaglio, S. Bagnara and S. U. Stucky. X, 368 pages. 1995. *(AET)*

NATO ASI Series F

Including Special Programmes on Sensory Systems for Robotic Control (ROB) and on Advanced Educational Technology (AET)

Vol. 142: Dialogue and Instruction. Modeling Interaction in Intelligent Tutoring Systems. Edited by R.-J. Beun, M. Baker and M. Reiner. IX, 368 pages. 1995. *(AET)*

Vol. 143: Batch Processing Systems Engineering. Fundamentals of Chemical Engineering. Edited by G. V. Reklaitis, A. K. Sunol, D. W. T. Rippin, and Ö. Hortaçsu. XIV, 868 pages. 1996.

Vol. 144: The Biology and Technology of Intelligent Autonomous Agents. Edited by Luc Steels. VIII, 517 pages. 1995.

Vol. 145: Advanced Educational Technology: Research Issues and Future Potential. Edited by T. T. Liao. VIII, 219 pages. 1996. *(AET)*

Vol. 146: Computers and Exploratory Learning. Edited by A. A. diSessa, C. Hoyles and R. Noss. VIII, 482 pages. 1995. *(AET)*

Vol. 147: Speech Recognition and Coding. New Advances and Trends. Edited by A. J. Rubio Ayuso and J. M. López Soler. XI, 505 pages. 1995.

Vol. 148: Knowledge Acquisition, Organization, and Use in Biology. Edited by K. M. Fisher and M. R. Kibby. X, 246 pages. 1996. *(AET)*

Vol. 149: Emergent Computing Methods in Engineering Design. Applications of Genetic Algorithms and Neural Networks. Edited by D.E. Grierson and P. Hajela. VIII, 350 pages. 1996.

Vol. 150: Speechreading by Humans and Machines. Edited by D. G. Stork and M. E. Hennecke. XV, 686 pages. 1996.

Vol. 151: Computational and Conversational Discourse. Burning Issues – An Interdisciplinary Account. Edited by E. H. Hovy and D. R. Scott. XII, 202 pages. 1996.

Vol. 152: Deductive Program Design. Edited by M. Broy. IX, 467 pages. 1996.

Vol. 153: Identification, Adaptation, Learning. Edited by S. Bittanti and G. Picci. XIV, 553 pages. 1996.

Vol. 154: Reliability and Maintenance of Complex Systems. Edited by S. Özekici. XI, 589 pages. 1996.

Vol. 155: Cooperation: Game-Theoretic Approaches. Edited by S. Hart and A. Mas-Colell. VIII, 328 pages. 1997.

Vol. 156: Microcomputer-Based Labs: Educational Research and Standards. Edited by R.F. Tinker. XIV, 398 pages. 1996. *(AET)*

Vol. 157: Logic of Computation. Edited by H. Schwichtenberg. VII, 396 pages. 1997.

Vol. 158: Mathematical Methods in Program Development. Edited by M. Broy and B. Schieder. VIII, 528 pages. 1997.

Vol. 159: Fractal Image Encoding and Analysis. Edited by Y. Fisher. XIV, 368 pages. 1998.

Vol. 160: Discourse, Tools, and Reasoning: Essays on Situated Cognition. Edited by L.B. Resnick, R. Säljö, C. Pontecorvo and B. Bunge. XII, 474 pages. 1997. *(AET)*

Vol. 161: Computational Methods in Mechanical Systems. Edited by J. Angeles and E. Zakhariev. X, 425 pages. 1998.

Vol 162: Computational Intelligence: Soft Computing and Fuzzy-Neuro Integration with Applications. Edited by O. Kaynak, L.A. Zadeh, B. Türkşen, I.J. Rudas. IX, 538 pages. 1998.

Vol. 163: Face Recognition: From Theory to Applications. Edited by H. Wechsler, P.J. Phillips, V. Bruce, F. Fogelman Soulié, T.S. Huang. IX, 626 pages. 1998.

Vol. 164: Workflow Management Systems and Interoperability. Edited by A. Doğaç, L. Kalinichenko, M.T. Özsu, A. Sheth. XVII, 481 pages. 1998.